SUBSURFACE CONDITIONS

SUBSURFACE CONDITIONS

Risk Management for Design and Construction Management Professionals

EDITED BY

David J. Hatem, Esq.

Burns & Levinson LLP

A WILEY-INTERSCIENCE PUBLICATION

JOHN WILEY & SONS, INC.

New York / Chichester / Weinheim / Brisbane / Singapore / Toronto

This text is printed on acid-free paper.

This publication is designed to provide accurate and authoritative information in regard to the subject matter covered. It is sold with the understanding that the publisher is not engaged in rendering legal, accounting, or other professional services. If legal advice or other expert assistance is required, the services of a competent professional person should be sought.

The views expressed in each chapter of this book are those of the respective authors, and not necessarily of any firm affiliated with the authors; nor do the views embodied in this book necessarily express the practices of the firms with which the authors are affiliated.

Library of Congress Cataloging-in-Publication Data

Hatem, David J.
Subsurface conditions : risk management for design and construction management professionals / David Hatem.
p. cm.
"A Wiley Interscience publication."
Includes index.
ISBN 0-471-15607-8 (cloth : alk. paper)
1. Civil engineering contracts–United States. 2. Risk management–United States. 3. Civil engineering–Management. 4. Earthworks–Management. 5. Insurance, Engineers' liability--United States. I. Title.
KF2928.3.H38 1998
624′.0681–DC21 97-12584

Printed in the United States of America

10 9 8 7 6 5 4 3 2

CONTENTS

CONTRIBUTORS

David J. Hatem, Esq., is a partner with Burns & Levinson LLP in Boston, Massachusetts. Mr. Hatem is chairman of the firm's Professional Practices Department. He concentrates his practice on the liabilities of professionals, particularly architects, engineers, construction management professionals, and environmental professionals, as well as lawyers and accountants. He frequently lectures on issues of professional liability and has authored numerous articles. Mr. Hatem is an affiliated member of the American Society of Civil Engineers. He is legal counsel to the American Consulting Engineers Council of New England and a corporate advisor to the Boston Society of Architects. Mr. Hatem teaches at Tufts University, School of Engineeering, on the legal aspects of engineering.

Eldon L. Abbott, P.E., is a vice president and a principal project manager with Parsons Brinckerhoff Quade & Douglas, Inc. He has extensive experience in the management and design of marine structures, piers, bulkheads, underground structures, parking garages, and emergency ventilation systems for subway tunnels. Mr. Abbott currently serves as project manager for the firm's engineering services for the Boston Harbor 9.5-mile, 24-foot-diameter Ocean Outfall in Boston, Massachusetts. He has also served as project manager for the design and construction of twenty emergency ventilation shafts on the MBTA's Green Line subway and for two "top down" multi-level below-grade parking garages, including the award-winning Post Office Square Garage and Park in Boston.

Gary S. Brierley, Ph.D., P.E., is director of tunnel design for Haley & Aldrich, Inc. He has served as president of the American Underground Construction Association, chairman of the Underground Construction Research Council, and as a member of the U.S. National Committee on Tunneling Technology. He has participated in the design and construction management of more than 100 tunneling projects. Dr. Brierley was president of the Denver-based consulting firm of Brierley & Lyman prior to merging with Haley & Aldrich, Inc. in 1991.

Hugh P. Caspe, P.E., is vice president and northeastern manager of infrastructure services with Woodward-Clyde Consultants. Mr. Caspe has been involved in some of the largest subsurface projects in the Northeast and has been responsible for the design of over 1.5 billion dollars of construction value work. He is chairman of the American Consulting Engineers Council of New England's Ethics Committee, a member of the board of directors of the American Underground Construction Association, and a member of the MOLES. Mr. Caspe has also received the Ralph W. Horne Award for Public Service from the Boston Society of Civil Engineers Section of the American

Society of Civil Engineers and has authored numerous articles on tunneling. Mr. Caspe is registered as a professional engineer in numerous states.

Timothy F. Gilmartin, Esq., is the insurance manager for Stone & Webster Engineering Corporation, a worldwide engineering, design, and construction company with annual revenue in excess of one billion dollars. Mr. Gilmartin has over twenty years' experience in risk management and construction insurance. He is responsible for the company's liability and property insurance, including any project-related insurance on a worldwide basis. He is an attorney and civil engineer.

Joseph D. Guertin, P.E., is a senior principal at GZA GeoEnvironmental, Inc. His geotechnical engineering experience ranges from high-rise building foundations to major subsurface projects for transportation, utilities, and nuclear waste. He is a member of the MOLES, a former president of the American Underground Construction Association, and a former chairman of the Underground Technology Research Council.

James E. Monsees, Ph.D., P.E., is a senior vice president and technical director for underground structures at Parsons Brinckerhoff Quade & Douglas, Inc. Dr. Monsees is a member of the National Academy of Engineering and chair of its Awards Committee. He has served as a board member for the American Underground Construction Association, as a founding board member of the American Rock Mechanics Association, as a member of the U.S. National Committee for Rock Mechanics, as a board member for the Rapid Excavation and Tunneling Conference, and as a member of numerous professional committees related to tunneling and underground construction. His career has included work on over 200 underground projects.

Thom L. Neff, Ph.D., P.E., is a senior professional associate and assistant vice president with Parsons Brinckerhoff Quade & Douglas, Inc. He has 30 years' experience in the planning, design, and construction of large civil works projects in the United States and overseas. He has degrees in civil, structural, and geotechnical engineering and has authored numerous publications in these areas. Dr. Neff is currently the manager of geotechnical services for the $10 billion Central Artery/Tunnel Project in Boston and a member of the team of Bechtel/Parsons Brinckerhoff, Management Consultant for the Massachusetts Highway Department.

Steve J. Poulos, Ph.D., P.E., was an assistant professor of soil mechanics at Harvard from 1965 through 1970. He then started GEI Consultants, Inc., with three partners in 1970. The firm now has eight offices throughout the United States. As a principal of the firm, Dr. Poulos has carried out hundreds of projects relating to all aspects of foundation and earth dam design and construction, including numerous evaluations of claims for both owners and contractors.

Richard M. Simon, Ph.D., P.E., has been employed by GZA GeoEnvironmental, Inc. (formerly Goldberg-Zoino & Associates, Inc.) since 1977 and currently serves as risk manager and executive vice president for professional practice. Dr. Simon is registered as a professional engineer in nine states and has experience in projects with both a geotechnical and a hazardous waste emphasis. He is a fellow of the American Society of Civil Engineers, a member of the Risk and Insurance Management Society, and a member of the Practice and Emerging Trends Committee of ASFE/Professional Firms Practicing in the Geosciences. Dr. Simon serves on the Geotechnical Advisory Committee for the Massachusetts State Board of Building Regulations and Standards and has been appointed to the Panel of Arbitrators of the American Arbitration Association.

Timothy P. Smirnoff, Ph.D., P.E., is a senior supervision engineer with Parsons Brinckerhoff Quade & Douglas, Inc. He has over 25 years' experience on tunneling and underground projects. He has been responsible for the design and analysis of structures, tunnels, and ground supports in soft ground and rock, as well as directing geotechnical investigations and conceptual studies and preparing designs and contract plans and specifications for over 100 miles of tunnels and appurtenant structures. He is an active member of the American Arbitration Association's Panel of Arbitrators, American Society of Civil Engineers, the American Underground Space Association, the International Society of Rock Mechanics, and the Underground Technology Research Council. Dr. Smirnoff is registered as a professional engineer in 12 states.

David E. Thompson, P.E., is chairman and CEO of Haley & Aldrich, Inc., a national engineering firm specializing in the design and construction management of subsurface structures. Mr. Thompson has over 35 years' experience in underground engineeering and is frequently involved in risk management and dispute resolution for subsurface projects. He is a past president of ASFE/Professional Firms Practicing in the Geosciences. Mr. Thompson is currently the president of the National Council for Geo-Engineering and Construction, a director of the Institute for Professional Practice, a director of the Terra Insurance Company, and a member of the MOLES.

ACKNOWLEDGMENTS

This book was the result of a collaborative effort by a team of professionals. I wish to acknowledge and thank the contributing authors for sharing experiences and insights and making practical recommendations. These thoughtful, dedicated professionals worked very hard and skillfully in writing their respective portions of the book, and the many discussions and exchanges among us strengthened the text. I owe a tremendous amount of gratitude to my associate Constantia T. Papanikolaou, who continually and thoroughly reviewed all the authors' drafts and provided support for us all. As editor, I enjoyed and was enriched by working with all of these individuals, whose collective contributions made the book possible.

DAVID J. HATEM, ESQ.

Boston, Massachusetts
November 1997

INTRODUCTION

David J. Hatem
Burns & Levinson LLP

SUBSURFACE CONDITIONS: RISK MANAGEMENT FOR DESIGN AND CONSTRUCTION MANAGEMENT PROFESSIONALS

Subsurface projects—that is, projects involving the design and construction of substantial and/or complex underground work—present an enormous risk potential for the primary project "stakeholders," i.e., the owner and contractor. The risk inherent in subsurface projects arises because the nature, quality, or quantity of subsurface conditions—that is, the physical conditions located underground at the site, or the behavior of those conditions during the construction process—typically cannot be precisely predicted prior to the start of excavation or related construction activities. In addition, the behavior of subsurface conditions may be dependent on the actual construction means and methods selected by the contractor, which typically will not be known until sometime after the design is completed and the construction contract is awarded.

In most instances, reasonable efforts are undertaken by the project owner during the planning phase to retain qualified professionals to investigate subsurface conditions at the site, and to prepare data, interpretive, or other reports, the purpose of which is to convey to the contractor information and/or opinions about the *subsurface conditions expected to be encountered* during construction, and in some instances to communicate assumptions regarding the anticipated construction means and methods of the contractor. However, despite efforts to prepare and to disclose to the contractor such information, *the actual subsurface conditions encountered* during construction may differ (to a greater or lesser degree) from those indicated in the reports.

In addition, the construction means and methods selected by the contractor may differ from those assumed in the reports, and any variation may significantly impact the anticipated behavior of the subsurface conditions during the construction process. When these circumstances occur, there is a substantial risk that the economic and/or temporal expectations of one or more of the project participants (especially the owner and contractor) will be seriously disappointed, disrupted, and unbalanced.

Although this risk for both the owner and contractor is inherent in subsurface projects, it can be effectively managed and fairly allocated between them. It is equally accurate to observe, however, that the risk for both owners and contractors may be significantly heightened due to the absence of effective management and fair allocation of risk. In addition, subsurface projects provide considerable opportunities for risk "gambling" by the owner and the contractor, in circumstances in which the stakes can be high and have a significant potential for seriously disappointed cost, time, and performance expectations.

There are several alternative approaches to risk allocation for subsurface conditions. The specific approach selected by the owner will have a material impact on various aspects of the project, including

- whether an adequate number of contractors will be interested in bidding the project;
- the amount of the bids received;
- whether the bids will contain any contingencies (disclosed or not) for subsurface conditions and the amount of any such contingencies;
- the ability to manage and effectively anticipate and resolve conflicts, disputes, and claims arising out of subsurface conditions; and
- and the availability of project funding, with respect to both the amount of the initial contract award, as well as the payment of additional costs (due to subsurface conditions) during the construction process.

In some instances, risks are consciously allocated and assumed, as when an owner does not authorize an adequate preconstruction site investigation, does not disclose to the contractor pertinent information regarding actual or anticipated site conditions, or chooses not to carry any contingency for differing site conditions that might be encountered during construction; or when a contractor bids a project without conducting any site inspection, acts in conscious disregard of disclosed or available information regarding site conditions, or enters into a contract without a differing site conditions clause and carries no contingency for the risk of differing site conditions. In other instances, risk is allocated and assumed not through conscious decision making by the owner or contractor, but rather through inadvertence, inattention, or pure ignorance. Whatever the precise circumstances, the allocation and assumption (conscious or not) of heightened or intolerable risk for either the owner or the contractor on a subsurface project bears a direct relationship to the increased probability of disputes, claims, "pathological" project relationships, and eventual litigation or arbitration.

In addition, the unfair or disproportionate allocation of risk for subsurface conditions between the owner and contractor often results in a significant increase in the potential professional liability exposure for design and construction management professionals involved in subsurface projects. Actual and documented experience demonstrates that when owners and/or contractors do not timely and adequately identify and fairly allocate risk for subsurface conditions, a substantial probability exists that, in some manner or respect, the design and/or construction management professional will be subjected to a significantly increased risk of professional liability claims by either its client or third parties. While these professionals are not in a position to control the decision making of owners and contractors concerning risk identification, allocation, and management, they often do have the opportunity, if not the obligation, to consult with and advise their clients about those important subjects and thereby influence the decision making.

Realistically, not *all* risk for subsurface conditions can be entirely avoided or eliminated. The emphasis in this book is on the timely and adequate identification, management, and fair allocation of risk for subsurface conditions. Virtually all of the generic types of risks associated with subsurface conditions can be identified, even if not always specifically or precisely quantified or characterized. Risks are manageable and controllable, provided that they are identified in a timely fashion during the planning phase by means of adequate site investigation, and then appropriately addressed in the design, and provided that the pertinent results of the investigation are disclosed to bidders. In addition, the financial and temporal risks associated with the encountering of actual conditions materially different from those reasonably anticipated should be addressed and fairly allocated in the agreement between the owner and contractor through, for example, the inclusion of differing site conditions clauses. Finally, the construction contract should include a fair and impartial procedure for the timely submission and resolution of disputes and claims that may arise concerning site conditions.

The authors candidly acknowledge that many of these recommendations reflect "additional" cost at the "front end" of a subsurface project; however, we are equally cognizant that these recommendations, if properly implemented, will reduce the ultimate project cost and enhance the overall objectives and relationships of all project participants. In addition, the goal of fair allocation of risk for subsurface conditions involves much more than timely and adequate risk identification, and contractual risk allocation at the front end of the project. Achieving that goal requires a continuous, disciplined, and diligent effort by all of the project participants, to deal and interact with each other in good faith and equitably, throughout the entire planning, design, and construction process.

The design and construction management professional plays a critical role in the successful delivery of subsurface projects, whether success is measured by the schedule, the budget, or positive project participant relationships. That role includes the education of owners with respect to the timely and adequate identification and fair allocation of risk for subsurface conditions; the need for the preparation of appropriate reports and data, and adequate contract documents for the execution of the design intent; the importance of the professional's observational and administrative services during the construction phase; and the inclusion in the contract documents of fair and impartial procedures for the timely submission and resolution of disputes and claims involving differing site conditions.

The authors of this book collectively have substantial experience in the performance of design and construction management professional services in connection with subsurface projects, and in associated risk identification, management, and allocation issues. Although most of the authors are engineers, it is the intention of the book to address these risk issues in a balanced manner, since it has been our experience that this attitude and approach are in the best overall interest of all project participants. It is especially our expectation that this book will be read and recognized by project owners (and those who represent their interests) as a balanced treatment of the subject matter and as a set of guidelines for risk identification, allocation, and management in the planning, design, and construction of subsurface projects.

Although it is expected that legal counsel for design and construction management professionals and owners will certainly have an interest in the subject matter of this book, it should be emphasized that it has been written primarily for non-lawyers.[1] The reader is advised to consult with knowledgeable and experienced legal and insurance advisors in his or her particular state. This book is not intended as a substitute for such consultation.

A number of the specific subjects addressed in this book—such as how much information should be disclosed in the bid and contract documents; whether the accuracy and/or completeness of any such disclosed information should be disclaimed; what types of geotechnical reports should be prepared; and whether dispute review boards should be included as part of the construction contract documents—are subjects on which considerable debate currently exists in the design and construction industry involved in subsurface projects. The various authors have expressed their experienced and individual views on these subjects. We acknowledge and anticipate that there will be developments, new ideas, and additional information on the subjects addressed in this book and on related topics, which will require periodic updating of the book. In this regard, we invite your comments, criticisms, and/or suggestions as to how it may be improved to better serve your needs.

NOTE

1. Chapter 10, which deals with the legal aspects of professional liability and risk allocation and management considerations for design and construction management professionals, contains substantial reference material that may be of use to legal counsel.

1

CLIENT-CONSULTANT SELECTION

David E. Thompson, P.E.
Haley & Aldrich, Inc.

1.0 INTRODUCTION

When Ford Motor Company implemented Dr. W. Edwards Deming's "14 Points for Quality" in 1982, it focused on quality and found that when quality went up, costs went down. The Ford Taurus was developed under the Deming principles and became the best-selling American-made car in the United States after the Honda Accord, also manufactured under the Deming "14 Points for Quality" approach. According to the Deming principles, "price has no meaning except in terms of quality of the product."

1.0.1 Subsurface Engineering—What's Different?

In approaching the issues surrounding the procurement of professional services for subsurface projects, both the client and the consultant should recognize that design of subsurface projects differs significantly from design of aboveground structures, particularly with regard to the risk inherent in the work and the allocation of that risk between the client and the consultant.

Services for subsurface projects can generally be broken down into five phases, as follows:

Subsurface Conditions: Risk Management for Design and Construction Management Professionals, edited by David J. Hatem
ISBN 0471-15607-8

1. Define, to a reasonable level of confidence, the subsurface stratigraphy, including information on soil types and stratum depth, bedrock depth and quality, groundwater levels, contamination of soils, rock, and groundwater, if any; type and limits of previous construction; adjacent structures; and other factors. This is usually done by performing a limited number of randomly situated test borings that sample a very small portion of the site and by making assumptions with respect to conditions that exist between boreholes.
2. Define, to a reasonable level of confidence, the engineering and chemical properties of the subsurface soils, bedrock, groundwater, and so forth, including their strength, stress-strain properties, permeability, unit weight, chemical properties, and other parameters. This is usually done by laboratory testing performed on selected samples of materials on the assumption that they are representative of all other similar materials.
3. Assess, generally based on previous experience and empirical relationships, how the various subsurface materials will behave when subjected to a range of reasonably anticipated construction means and methods that may vary significantly, depending on the successful contractor's approach to construction. For example, some construction methodologies may require the soil to stand unsupported for minutes or days in order to place the permanent structure. The consultant must assess the risk of allowing construction techniques that require long "stand-up" times and, if acceptable, allocate that risk to the party benefiting financially from the successful implementation of the process.
4. Design a structure that will accommodate not only the final loads, but also the construction methodologies that are compatible with the subsurface conditions and will be sufficiently robust to accommodate reasonable variations in subsurface conditions and normal variations in construction workmanship. The design of most subsurface facilities is determined by the existing ground and groundwater conditions and the means and methods anticipated to build the structure.
5. Provide an appropriate scope of services through construction, since the design of a subsurface structure is not complete until construction is complete. Since variations in subsurface conditions are inevitable, and may affect the design, and since construction methods and workmanship are a large component of the successful completion and quality of the design, it is likely that design modifications will be required to adapt to subsurface conditions actually encountered. Therefore, it is imperative that the consultant's services include involvement through the construction phase. The failure to recognize the need for continuity from design through construction is one of the primary causes of problem projects and cost escalation on subsurface projects.

Contrast the above process with, for example, the design of a building above ground. For the building, the environment is known and totally visible; the loadings on the structure are given by code (versus estimated from test borings); the strength, weight, and stress-strain properties of the materials are guaranteed to meet or exceed specified minimums (versus estimated from laboratory tests on a few samples); the performance of the final structure is entirely unrelated to the construction methods used; and there are few or no latent site conditions likely to become known during construction or variations in construction means and methods that would require design modification.

1.0.2 Relationship of Client-Consultant Selection to Risk Management

There are many more unknowns in the design of subsurface facilities than for most other types of structures. Unknowns represent risk in the construction industry and must be assessed and managed primarily through appropriate risk allocations between the owner and the construction contractor. The ability to identify risks and develop appropriate strategies to manage and allocate these risks is one of the most important services a consultant brings to a client on a subsurface project. The consultant's ability to provide this service is a function of its qualifications, experience, and the adequate definition of an agreed-upon scope of services.

A mutual understanding by the consultant and the client of the risks, risk management strategies, risk tolerance of the client, contract approach, and other factors is essential to a successful client-consultant relationship. As a general rule, the more risk a client (owner) is willing to take, the lower the initial (bid) construction cost will be. Conversely, the less risk assumed by the owner, the higher the initial (bid) construction cost as a result of defensive designs, contractor contingency costs, and so on. (See Figure 1.1.) The optimum position for the client (or owner) in this relationship varies significantly from one proj-

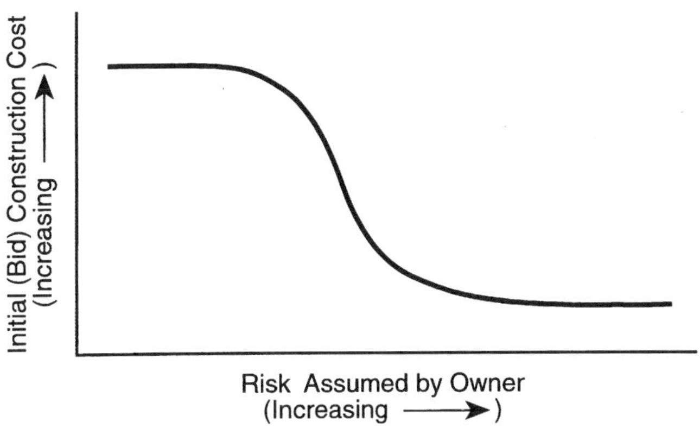

Figure 1.1 Risk versus initial (bid) cost.

ect to another depending on the risk tolerance and attitude of the client (or owner).

The client-consultant selection process for subsurface projects must recognize the importance of managing risk in subsurface projects and the attributes of the relationship between the client and consultant that are necessary to facilitate a constructive and cost-effective solution. In order to achieve a mutual understanding of the primary risks on a project, and the alternatives and costs to manage these risks, it is essential that the client and the consultant discuss options and arrive at a mutually agreeable position with respect to risk management. For private-sector owners, where reasonable risks can be accepted and funded (if required), a more aggressive (i.e., accepting more risk in exchange for lower initial costs) position may be appropriate. In contrast, for public owners, where changes are not easily accommodated, a more conservative (i.e., accepting less risk and paying higher initial costs) approach may be appropriate.

The consultant selection process used by the client and consultant should facilitate a mutual understanding of risks and risk management options, and the development of scope and design approaches that are responsive to the client's needs and acceptable to the consultant. Risks result from disappointed clients. Disappointed clients result from unmet expectations. The client-consultant selection process should establish realistic mutual expectations of both parties, thereby maximizing the probability of a successful project and a satisfied client.

1.0.3 ASFE Experience

No group of design professionals has demonstrated more success in minimizing risk exposures than geotechnical engineers. In 1970, so many of these engineers were being faced with meritless claims that professional liability insurance was all but unavailable to them as a profession. By 1980, 10 years after the formation of the Association of Soil and Foundation Engineers (ASFE), members had the best professional liability record of all design professionals. ASFE's primary mission is risk management for subsurface construction. Extensive educational and training materials are provided on the subject. Although the focus of ASFE is from the geotechnical perspective, the strategies and methods are equally applicable to all members of the design profession. ASFE (now ASFE/Professional Firms Practicing in the Geosciences) offers extensive literature on subsurface risk management and client-consultant selection procedures.

1.0.4 The First Step in Risk Management

The process of client selection and procurement of professional services involves a mutual process whereby a client and a consultant agree to work together on a project, ideally with mutual goals and realistic expectations. It is a mutual and consensual process, regardless of the selection procedure employed.

The consultant always has the option not to pursue the project, and the client can always select another consultant.

From the client's and consultant's perspective, client selection and procurement are the first—and possibly the most important—steps in project risk management. For a consultant, the decision to pursue a particular project or client involves balancing the need for new business with an appropriate level of risk tolerance and acceptance. At one extreme, if no projects are undertaken, there is no risk; however, this is not a viable option for a business. At the other extreme, consultants must balance the need for new work with judgments relative to projects that may be inherently risky, at the limits of a firm's technical expertise, or sponsored by a client who may have a risk management goal that is unreasonable.

There are times when the consultant should recognize that a certain client or project may not be compatible with the consultant's goals, and that the project or client should not be pursued further. This is not an easy decision for most consultants; however, in a few situations it is, in fact, the only effective approach to risk management.

1.0.5 A Fundamental Problem

Consultants are trained in a service industry and, as a result, they are service-oriented. When a consultant is requested by a client to do something, the response is usually affirmative, for fear of upsetting the client or appearing in an unfavorable light compared to other competing professionals. In addition, all new work is perceived as good work, to be pursued vigorously, without a lot of consideration for the character of the work or the client. Price is usually set "to get the job." As a result, staff interpret company policy to mean aggressive pricing, and pursuing all possible projects. This dilutes resources and affects the ability of a firm to deliver quality work for more desirable clients.

1.0.6 The Cost of Design Professional Services

The cost of the full range of engineering services typically amounts to less than 1 percent or 2 percent of the life-cycle cost of most construction projects. It is, therefore, in the client's best interest to engage the most qualified and reputable consultant available to perform an adequate scope of services for the specific project, since doing so will result in far superior quality. And even assuming that the increased scope adds 25 percent to the engineering budget, that increased cost will add no more than $\frac{1}{4}$ to $\frac{1}{2}$ percent to the total cost of the project.

The client should select a consultant that can and will support the project from conception through design, construction, and project startup, if appropriate. Continuity of service aids in developing a relationship between the consultant and the client, which adds to the quality of the project and reduces the risk of misunderstandings.

Qualifications, experience, reputation, and quality of client service are of critical importance in the selection of a consultant. Selection based primarily on cost of services, with limited consideration of the competence or expertise of the consultant, can result in seriously disappointed client expectations, unsatisfactory service to the client, and higher overall project costs. The consultant's competence in specialty fields, performance on other projects of a similar nature, interrelationship with the design team members, personnel assignments, provision for independent reviews, and quality assurance will all influence short- and long-term quality, including the construction process, operation, and maintenance costs, insurance and other annual charges, and the resulting life-cycle costs of the project. It is important that the client recognize that selecting a consultant based on quality and expertise is somewhat subjective. It is imperative that clients assign those individuals who are best suited within their organizations to make that selection.

1.1 WHAT CLIENTS SHOULD LOOK FOR IN A CONSULTANT

During the process of selecting a consulting firm, the client can—and should—require information from potential consultants that will facilitate the selection of the most qualified firm. The client's first step is to define the proposed goals and objectives of the project. In some cases, this may be a general statement of performance and, in other cases, a detailed definition of goals and objectives may be necessary. By defining the goals and objectives of the engagement, the consultant can propose an approach and an appropriate scope of services that best illustrate its ability to respond to the client's needs. Some facts that a client should look for and request information on from a consultant are discussed below.

1.1.1 Previous Experience with Similar Projects

Subsurface engineering involves the application of experience and judgment to the solution of design problems to a significantly greater degree than in any other engineering discipline. Previous experience with similar projects in similar ground conditions is essential to building the basis for sound engineering judgments and good designs. Clients and owners should request consultants to document and describe previous projects of a similar nature and their participation in those projects.

1.1.2 Assignment of Key Personnel

The experience of a consulting firm on previous similar projects resides in the key personnel who performed on the previous projects more than it does with the firm itself. During the proposal phase of the selection process, prospective consultants should be required to identify key professionals to be assigned to the project and their individual experience. These key individuals should be present during the interview process, if not leading the presentation, and should have significant involvement in the project execution. The client should include an approval clause in the contract for any proposed change in key personnel assigned to the project.

1.1.3 References

Prospective consultants should be required to submit the names of clients for whom they have previously completed similar work. During the selection process, references should be contacted and an evaluation of the consultant's previous performance solicited.

1.1.4 Scope of Services

A clear and unambiguous scope of services, which is considered adequate to achieve the project objectives, should be agreed to between the prospective consultant and the client. The scope should provide for continuity from ground characterization through design and construction-phase services. Subsurface engineering frequently involves significant construction phase elements, including design modifications required to address variables such as differing site conditions or the unanticipated behavior of site conditions arising out of the construction means and methods selected by the contractor. In many cases, measurement instruments will be installed to monitor critical strains, stresses, water levels, vibrations, and the like, and construction-phase field engineering will be required to adapt the design and construction to actual ground performance. Construction-phase participation by the consultant is an essential part of subsurface engineering and should be a defined portion of the scope.

1.1.5 Schedule and Deliverables

The prospective consultant's proposal should include a schedule of planned activities, including interim milestone dates and a list of deliverables. It is essential to establish a clear mutual understanding of the scope and content of each deliverable before a contract is executed.

1.1.6 Peer Review

For most significant projects, the client, owner, and consultant will benefit from an independent external review of the design concept, criteria, and methodology. Project peer review program guidelines have been developed and published by the American Society of Civil Engineering (ASCE) and the American Consulting Engineers Council (ACEC) to facilitate the use of peer reviews. Clients should encourage or require the use of project peer review programs particularly for complex projects. Peer reviews are most beneficial when implemented early in the design, while fundamental concepts are evolving.

1.1.7 Financial Data

At some point during the selection process, the client is entitled to certain financial data, bank references, and insurance information. The client should request this information if the size of the project or engineering assignment justifies the concern for financial resources. Fee expectations should not be solicited or discussed until consultants have been ranked in order of qualifications (see discussion of QBS procedures in Section 1.3).

1.2 WHAT CONSULTANTS SHOULD LOOK FOR IN A CLIENT

The client-consultant selection process should be viewed as a *mutual* selection procedure. Obviously, the client will use a methodology to screen prospective consultants and, ultimately, to select one. At the same time, consultants should consider a number of questions in evaluating a particular project opportunity. While a negative response on some issues may be acceptable, multiple negative responses should raise questions relative to the desirability of the client and/or project.

1.2.1 Is the Project Consistent with Our Business Plan?

There are times when projects in one particular market area are more abundant than in other areas. The temptation to pursue projects that are outside the limits of the expertise and/or personnel qualifications and availability of a consulting firm can be great, and can lead to overextending the technical capabilities of the firm. Every consulting firm should have a business plan that identifies its project market areas. Except in the situation where a planned diversification effort is being implemented, projects should:

- Be consistent with the type of projects successfully completed by the firm previously;
- Be consistent with the skills and experience of available staff;
- Require internal support systems similar to other projects completed; and
- Require specialized equipment owned by or available to the consultant.

Projects that do not fall within these criteria should be outside the target markets in a good business plan and should not be pursued, since the project requirements may be inconsistent with the abilities and resources of the firm.

1.2.2 What Is the Client's Consultant Selection Process?

If the client's perception of the appropriate selection process is to request fee bids on a specified scope of work, this is a good indication that the appropriate client-consultant relationship is not understood by the client. Client education is the first priority in this situation. In the end, the selection process should:

- Facilitate mutual and realistic development of the work scope, with a full understanding of risk management goals;
- Facilitate a selection based primarily on qualifications; and
- Recognize that professional services are not a commodity, which can be compared on price based on the assumption of identical quality.

1.2.3 Is the Scope of Work Adequate?

Subsurface engineering services, unlike the design services provided by other engineering disciplines, are ongoing from the initial ground characterization phase through construction. In addition, cutting corners in the ground characterization phase can lead to increased cost resulting from a conservative design and can increase the risk of a significant differing site conditions claim, or adverse performance of the structure. Design-phase costs can almost always be reduced by restricting the scope of services undertaken. Similarly, eliminating the consultant's role during construction also will reduce cost. Within normal limits, however, the costs "saved" by such scope reductions are more than offset by the costs of the added construction risk and claims, and other costs associated with the resultant risk of diminished quality of the design and construction of the project.

Projects that are run by client managers who do not recognize the value of continuity from design into construction, and focus on minimizing design-phase cost elements without considering the cost of the resulting construction phase risk, are high-risk projects. Consultants should be particularly wary of clients who change consultants continually, and particularly those who change consultants at different phases of a project.

1.2.4 Have We Worked with This Client Previously?

As in any other relationship, the client-consultant relationship develops with time. Key personnel representing both parties become familiar with each other's goals and objectives, mutual risk allocation preferences, and policy; in addition, individual personalities, authority, and reporting relationships become known and communications are much more effective and efficient.

Clients and consultants who work together frequently on similar projects can develop a close "partnership," which promotes efficiency and can reduce disputes and risk. Conversely, new clients present a challenge in communications, which can increase the likelihood of disputes and risks in a new assignment.

1.2.5 Has the Client Done a Project Like This Before?

For a client who is not familiar with subsurface work, the myriad of unknowns and risks may be difficult to comprehend and accept, and may be inconsistent with available funding mechanisms. Therefore, clients new to the world of subsurface construction can become dissatisfied and frustrated with the inherent uncertainties in the work.

As discussed earlier in this chapter, subsurface engineering is not like most other types of engineering; the subsurface project is different, typically involving much more judgment related to unknowns, risk, and risk management. Clients and their project managers who have completed several subsurface projects can appreciate the unknowns and participate intelligently in the realistic and fair assessment and allocation of risk.

1.2.6 What Are the Client's Performance Expectations?

Some of the worst words a consultant can hear from a client is a statement to the effect that he or she expects "no movement" of the structure during or after construction. This, of course, is practically impossible to achieve without creating extraordinary costs that are unnecessary. Such a statement by a client indicates that the client simply doesn't understand the fundamental issues. Unreasonable and/or unrealistic expectations are generally not met, and unmet expectations create unhappy clients, and, consequently, risk to the consultant. A client with unreasonable and/or unrealistic performance expectations will require a heavy investment in education, and represents a major risk management concern for the entire project duration, and thereafter.

1.2.7 What Is the Client's Attitude Toward Risk?

Subsurface construction entails more risk than most other types of construction, primarily because it is impossible to predict, with total certainty, the subsurface conditions that may affect the construction methods. The inevitable risks must be acknowledged and allocated in some manner. In general, the risk of a legitimate (unforeseen) differing subsurface condition should reside with the owner. If the owner wishes to transfer this risk entirely to the contractor, the owner should expect to pay a premium for the initial construction (bid) cost, but should still expect disputes to arise if the contractor did not carry enough of a contingency in its bid to cover the costs. Similarly, if the owner or client believes that a differing site conditions claim is automatically a defect in the design for which the consultant should pay, there is a major problem to be confronted. Likewise, if the owner has unfairly allocated risk to the contractor, this may increase the likelihood of professional liability risk for the consultant (see Chapter 10, Section 10.2.4). This is more often the case when an owner's or client's attorneys become involved in the development of contract terms that seek to transfer all risk away from those best able to control and manage it.

Unfair or unrealistic risk allocation attitudes should warrant a serious review of the desirability of a particular client.

1.2.8 What Schedule Does the Client Expect?

More often than not, a client's schedule expectations are an indication of the client's understanding of the scope and complexity of a project. Unreasonable schedule expectations, in most cases, are a preamble to totally different concepts of the work scope and costs; occasionally, however, educated clients simply desire an accelerated schedule and are willing to accept the enhanced risk associated with that acceleration. Adequate time is essential for the completion of any preconstruction investigation and/or design. An adequate preconstruction investigation and design are essential prerequisites to proper risk identification and fair allocation of risk. If a reasonable performance period cannot be agreed to, it may indicate a client who cares little about fair allocation of risk (among all project participants) or even the quality of the completed project.

1.2.9 What Is the Financial Track Record of the Client?

In providing services to a client, a consultant is essentially extending credit to that client. Frequently, services having a value of up to $100,000 are extended prior to receipt of the first payment. How many financial institutions would extend this amount of credit without checking the financial record of the recipient? Financial checks on clients are not customary in the engineering business and therefore, are considered awkward by some consultants. In recent years, however, on-line sources and services have made extensive financial information on many organizations readily available at no cost or for a small fee.

For publicly traded companies, annual reports and 10K forms are available on-line. Hoover's Corporate Profiles provides reasonably extensive information on many companies at no charge. Dun & Bradstreet Companies On-line allows for searches of individual companies for a small fee. In addition, more and more firms are establishing home pages on the World Wide Web, which can provide extensive insight into the firms.

On some projects, an owner may create a special legal entity solely for the purpose of developing the project and may seek to limit its liability to the assets of that specially and newly formed entity. These types of project development arrangements need to be carefully evaluated by the consultant, particularly as they relate to the specially formed entity's financial resources and the availability of financial guarantees from principals or affiliates of the owner, if considered necessary.

For clients without a financial track record in the construction sector, it may be necessary to obtain a retainer in advance of providing services.

1.2.10 How Much Access Will There Be to the Client's Decision Makers?

Risk management requires direct and candid discussions of various risk elements, how they may be managed, and an agreement—or, at least, an understanding—of their existence and ultimate disposition. If communications with the client's key decision makers relative to these issues are filtered through a third party or are nonexistent, there is a high probability that important concepts will not be properly communicated and, therefore, the client or owner may have unreasonable expectations. Direct access to key decision makers on significant risk management issues is important.

In addition, it is critically important that the consultant have direct access to the representatives of the owner who have pertinent information required for the adequate and timely performance of the consultant's services; deprivation and/or unreasonable restrictions on that access (such as where the client requires that all information or communications be channeled through a lawyer) significantly increases the probability of risk for the consultant.

1.2.11 How Stable Are Key Members of the Client's Staff?

As discussed above, risk management involves communications with key decision makers on the client's staff. If there is a large turnover rate in the client's organization, what is the likelihood that the key contact at the client organization will still be there when there is a problem? A lack of continuity in the client organization can lead to a lack of institutional memory as to why certain decisions were made, and by whom. Unless the project history is extraordinarily well documented, previous communications and decisions may be forgotten, questioned, and could be disavowed when problems arise.

1.2.12 Are There Any Indications of Business Ethics Issues?

Intentional ethical breaches are relatively infrequent; if they occur, however, they can signal a serious issue with the client. The promise of project award if an influential political figure is retained, or if a firm or supplier is engaged when services are not required, is not only unethical and illegal, but also can dilute resources that should be allocated to the design effort and can seriously damage a consultant's reputation.

In a similar vein, the client must acknowledge that the consultant, as a registered professional, owes obligations to the public that, in certain circumstances, may transcend and potentially conflict with the individual interests of the client.

1.2.13 Is the Client Type a Risk Factor?

Some professional liability insurance companies and professional associations have studied the frequency and severity of professional liability claims and have determined that some client types are more likely to litigate against a consultant than others. Client types representing a high risk factor generally fall into the categories of persons inexperienced in the design and construction process and/or consisting of a group of laypersons or attorneys. For example, particularly risky clients include residential homeowners, condominium associations, church organizations or other charitable organizations, municipal clients with a building committee such as a school building committee, and the like. Not all clients of these types are a problem, but reliable statistics demonstrate that they represent a higher than average risk to design professionals.

1.3 THE CONSULTANT SELECTION PROCESS

1.3.1 Goals of the Selection Process

To yield satisfactory results, the selection process must facilitate:

- A rigorous evaluation of the consultant's qualifications, previous experience with similar projects, dedicated staff, and so on;
- A basic understanding, by the consultant, of the client's experience, expectations and risk preferences, and decision-making process; and
- The mutual development of an adequate service scope with an appropriate fee. The process of service scope development and associated fee negotiations will identify critical issues, avoid confusion, minimize delays, and reduce unreasonable expectations by both parties, as well as provide the consultant with an opportunity to educate the client concerning the advisability of fairly and realistically allocating risk for subsurface conditions.

1.3.2 Qualifications-Based Selection (QBS)

Qualifications-based selection (QBS) is a process for selection and retention of design professionals that accomplishes the above goals. As set forth in greater detail below, pursuant to the QBS process, the client is encouraged to conduct discussions with a number of prospective consultants regarding the client's expectations and needs, and each consultant's qualifications and experience. The goal of the QBS process is not to minimize consultant fees, but rather to attempt to procure the most highly qualified consultant to provide the services requested at a compensation rate that is fair and reasonable. The QBS process also allows the consultant the opportunity to discuss and understand the client's experience, expectations, and risk preferences, and to point out options and project risks that the client may not have considered.

The QBS procedure is endorsed by many professional organizations, including the American Consulting Engineers Council (ACEC), the American Society of Civil Engineering (ASCE), the Association of Firms Practicing in the Geosciences (ASFE), the National Society of Professional Engineers (NSPE), the American Institute of Architects (AIA), and many others.

1.3.3 The Brooks Law for Federal Procurements

The Brooks Law, 40 U.S.C. §§543, 544 (1997), embodies QBS into federal law and requires the QBS process for the selection of design professionals on federal design contracts. The law, named after its principal sponsor, Rep. Jack Brooks (D-Tex.), is relatively simple and reads as follows:

> **§543. Requests for data on architectural and engineering services.** In the procurement of architectural and engineering services, the agency head shall encourage firms engaged in the lawful practice of their profession to submit annually a statement of qualifications and performance data. The agency head, for each proposed project, shall evaluate current statements of qualifications and performance data on file with the agency, together with those that may be submitted by other firms regarding the proposed project, and shall conduct discussions with no less than three firms regarding anticipated concepts and the relative utility of alternative methods of approach for furnishing the required services and then shall select therefrom, in order of preference, based upon criteria established and published by him, no less than three of the firms deemed to be the most highly qualified to provide the services required.
>
> **§544. Negotiation of contracts for architectural and engineering services.** *(a) Negotiation with the highest qualified firm.* The agency head shall negotiate a contract with the highest qualified firm for architectural and engineering services at compensation which the agency head determines is fair and reasonable to the Government. In making such determination, the agency head shall take into account the estimated value of the services to be rendered, the scope, complexity, and professional nature thereof.
>
> *(b) Negotiation with second and third, etc., most qualified firms.* Should the agency head be unable to negotiate a satisfactory contract with the firm considered to be the most qualified, at a price he determines to be fair and reasonable to the Government, negotiations with that firm should be formally terminated. The agency head should then undertake negotiations with the second most qualified firm. Failing accord with the second most qualified firm, the agency head should terminate negotiations. The agency head should then undertake negotiations with the third most qualified firm.
>
> *(c) Selection of additional firms in the event of failure of negotiation with selected firms.* Should the agency head be unable to negotiate a satisfactory contract with any of the selected firms, he shall select additional firms in order of their competence and qualification and continue negotiations in accordance with this section until an agreement is reached.

1.3.4 American Bar Association Model Procurement Code

The American Bar Association Model Procurement Code for State and Local Governments, Section 5-501, specifically outlines the QBS method as the selection process to be used. Unfortunately, uninformed lawyers still advise client school boards, municipalities, and state agencies that they must award contracts to consultants based on competitive bids. How is it possible to procure the highest level of professional competence at the lowest price for a service that cannot be precisely defined?

1.3.5 The Massachusetts Inspector General's Findings

In a 1989 report undertaken for the Commonwealth of Massachusetts' inspector general's office by the Ward Commission to identify sources of corruption in state construction, the Commission wrote:

> The goals of the designer selection process should not include minimizing designer's fees. Making fees part of the selection process will not save the taxpayer's money. Rather, the practice is likely to increase the ultimate cost of the public building system due to reduction in design quality, while at the same time offering no actual benefit insofar as reducing the opportunities or occurrence of corruption. (Office of the Inspector General, 1989)

1.3.6 The QBS Procedure

The QBS procedure is described in documents published by many design professional societies. The American Society of Civil Engineers Manual No. 45, "Consulting Engineering: A Guide for the Engagement of Engineering Services," contains a detailed description, which is abbreviated here.

The selection procedure is considerably enhanced when the client is fully familiar with the purpose and nature of the proposed project, can describe the project in detail, and can prepare a project scope and outline of services expected from the consultant. In some cases, the client may not have professional staff available to define the project scope and describe the required services. The client should still be familiar enough with the project requirements to understand what is expected of the consultant. The selection procedure, however, can be modified to suit the circumstances.

The client's usual steps in the selection procedure are presented below. If the client has had a satisfactory experience with one or more consulting engineers, it may not be necessary to follow all of the outlined steps.

1. By invitation or by public notice, state the general nature of the project, the services required, and request statements of qualifications and experience from consultants who appear to be capable of meeting the project requirements. The client may issue a request for qualifications (RFQ)

or request for proposals (RFP). RFQs are normally used to ascertain the general qualifications in a selected area of expertise, while RFPs are used to seek consultants for creating "short lists" for selecting a consultant for a specific project.

2. Prepare a budget for the staff time and costs that can be expected from potential consultants prior to receipt of the RFQs or RFPs.
3. Evaluate the statements of qualifications received. Select at least three consultants that appear to be best qualified for the specific project. It should be noted that, often, more than three consultants may appear to be equally qualified—in which case more consultants may be considered. In fairness to those not selected, however, it is usually best to make a conscientious effort to keep the number selected for further consideration to a realistic minimum, in view of the cost and time required to prepare competent proposals.
4. Write a letter to each consultant selected for further consideration, describing the proposed project in detail, including a project scope and outline of services required, and asking for a proposal describing in detail the consultant's plan for managing and performing the required services, the personnel to be assigned, the proposed schedule, experience with similar projects, office location in which services are to be performed, financial standing, present workload, and references. Each consultant should have an opportunity to visit the site, review all pertinent data, and obtain clarification of any items as required. For major or complex projects, a pre-proposal conference may be desirable to explain details of the proposed scope of services and to answer questions.
5. On receipt of proposals, invite the consultants to meet individually with the selection committee for interviews and discussions of the desired end results of the project and the professional services required. These interviews may be held at the consultant's office. The client may consider supplementing the selection committee with personnel who have specialized expertise to advise the committee, when appropriate. During each interview, the selection committee should review the qualifications and experience of each consultant, the capability to provide the services within the time allotted, and the key personnel to be assigned to the project.
6. Check with recent clients of each consultant to determine the quality of its performance. This check need not be limited to references listed by the consultant.
7. List the consultants in the order of preference, taking into account their approach and understanding of the project, reputation, experience, financial standing, size, personnel available, quality of references, workload, location, and other factors pertinent to the project being considered.
8. Invite the consultant considered to be best qualified to develop a detailed

scope, list of deliverables, and schedule, and to negotiate fair compensation for the services.

9. The compensation proposed by the consultant should be evaluated on the basis of the client's experience and budget estimate, taking into account the range of charges reported by other users of engineering services and the experience data reported in the literature. Consideration should be given to the project's special characteristics and the agreed upon scope of services. Fair and reasonable compensation to the consultant is vital to the success of the project, since it will enable the consultant's expertise to be fully utilized.
10. If a satisfactory agreement is not reached with the first consultant selected, the negotiations should be terminated and that consultant notified in writing to that effect. Similar negotiations should then be held with the second consultant and, if necessary, with the third. If no accord is reached, the client should seek outside assistance before continuing with the selection process. Such a procedure will usually result in development of a satisfactory contract. All such negotiations should be on a strictly confidential basis, and in no case should the compensation discussed with one consultant be disclosed to another.
11. When agreement has been reached on scope, schedule, and compensation, the client and selected consultant should formalize their agreement in a written contract.

1.3.7 QBS Facilitators

Engineering organizations in several states have organized QBS facilitator programs, which provide trained professionals to guide clients and owners through the process of selecting a consultant. In states without facilitator programs, assistance and materials are generally available through the local sections of the American Consulting Engineers Council (ACEC) or the American Society of Civil Engineers (ASCE).

The Connecticut QBS Council presents the four-step summary of the QBS process on page 22.

The development of a mutually acceptable and adequate service scope is the key element in the QBS process. The process fosters a positive relationship between the client and the consultant, based on trust and mutual respect. It promotes a free exchange of information through which the consultant can point out options and project risks that the client may not have considered; conversely, the client can discuss any constraints that limit its funding abilities, and can consider tradeoffs or reduction in scope. It also provides an opportunity for the client to conduct business with key professionals of the selected consultant's firm before entering into a contractual relationship, as well as an opportunity for the consultant to lay the foundation for educating the client about the principles of fair and realistic risk allocation on subsurface projects

Step 1: Request for Qualifications

Prepare a formal request for qualifications (RFQ) for a proposed project and distribute to a select list of firms. Ask respondents to select list of firms. Ask respondents to indicate their interest in the project and illustrate their relevant project experience and overall capabilities to perform the required services. The initial RFQ should include the following:

- ☐ A description of the proposed project
- ☐ Any special expertise or unusual services that might be required
- ☐ A time schedule for the overall project, including the selection process to retain a consultant
- ☐ A list of selection criteria (e.g., geographic location of firm, experience on similar projects, or current workload of staff)

Step 2: Evaluation of Firms

Evaluate the firms' qualifications and select the top three to five firms to be interviewed. Base the evaluation on the following factors:

- ☐ Qualifications of personnel to be assigned to the project team
- ☐ Overall experience of personnel as a team
- ☐ Ability to perform within time contraints
- ☐ Firm's awareness of project's issues, opportunities, and contraints
- ☐ Project team's experience on similar projects
- ☐ Quality of past work
- ☐ Performance evaluation from past clients

Step 3: Interview

Arrange for interviews with the top firms. An evaluation form with predetermined selection criteria should be used to allow interviewers to uniformly evaluate the consultants. The evaluation form should include the following:

- ☐ Project team management plant
- ☐ Experience of project team
- ☐ Location and quality of office facilities
- ☐ Project control measures to monitor schedule and budget
- ☐ Availability of staff based on firm workload
- ☐ Project approach
- ☐ Quality of work on similar projects
- ☐ Awards and past client recommendations
- ☐ Financial stability of firm
- ☐ Overall team rating

Step 4: Selection and Negotiation

Select the top-ranked firm and initiate the negotiations for the scope of work, the terms and conditions of a contract, and compensation. The negotiation process includes the following:

- ☐ Preparation of comprehensive proposal by the top-ranked firm
- ☐ Presentation of scope of work, schedule, and level of effort by the top-ranked team
- ☐ Agreement on scope of work, schedule, fee, and areas of responsibility and liability
- ☐ Preparation of written contract binding both parties to agreement

and how those principles relate to the scope of services required of the consultant.

1.4 THE PROBLEM WITH COMPETITIVE FEE BIDDING

More and more frequently in recent years, clients have attempted to capitalize on the responsive nature of most consultants by pitting them against each other by competitive fee bidding.

Clients are frequently naive, uneducated, and/or intentionally misdirected in their selection process, frequently using methods that attempt to translate the low-bid process for procuring materials or construction to the procurement of professional services. These procedures inherently assume that engineering services are a commodity that can be specified precisely when, by their nature, the definition of professional services requires professional judgment and flexibility.

As an example of the absurd outcome frequently resulting from the competitive fee bidding approach, the author's firm was requested to bid on the foundation engineering services required for the design of a 10-story building to be constructed over 60 feet of moderately overconsolidated clay underlain by glacial till and rock. The fee proposed was about $12,000 and included the cost of several 60-foot-deep borings, laboratory consolidation and strength testing of the clay, and settlement analysis for a mat foundation system with one basement level in the building. Extensive previous experience in the area indicated that a shallow mat foundation would be feasible and highly economical.

The contract was awarded to another firm for a fee of $1,500. One 10-foot-deep boring was conducted in the center of the building, and the structure was built on a pile foundation driven to rock. The owner paid about $500,000 more for the building than necessary, but saved $10,500 in engineering fees. The contractor had a claim pending against the owner for longer piles than he had anticipated as a result of a "differing site condition." Most frustrating of all, the benefits of years of technological development, engineering design evolution, and experience were never utilized on the project—and the owner never knew it.

Fee bidding is any selection process that considers price in the selection of one firm over others. This includes the full range of procedures, from outright bidding for a defined scope of professional services to a selection process in which price is only one of several evaluation factors. In all cases, however, the client creates a "bidding environment" within which the consultant must respond, and this is the fatal flaw in all fee bidding systems. When a consultant perceives a "bidding environment," it strongly tends to price the minimum service to accomplish the project scope. Any other strategy would be contrary to the firm's immediate financial interest, since it must ensure that its price is "competitive" or the investment in the proposal is completely wasted.

Price-based selection has appeal because it appears to provide a common denominator for comparison of proposals. In the case of a pure bid, it is the sole factor in determining the selection. When price is one of several evaluation factors, it acts as a strong magnet, tending to draw a disproportionate share of attention away from other considerations. Therefore, even though the price selection procedure may not require the selection of the lowest price offered, it creates a price competition that can have effects that do not serve the client's long-term interests.

1.4.1 Price Generally Controls the Decision

Fee bidding creates a situation in which it is difficult not to select the low price. In many instances, the procurement committee must prepare a justification for selecting on a basis other than low price. This creates a strong incentive to select the lowest price proposed. In addition, for public procurements, public pressures make it difficult to make selection on any basis other than price, and public officials become subject to criticism for acting otherwise.

1.4.2 No Basis for Comparison of Bids

No clear basis exists for comparing bids on engineering services. No scope of work can ever be written that allows the complete quantification of price. A certain amount of creativity and interpretation based on the firm's experience must factor into the development of the proposal and approach. There is a significant difference between selection for engineering services and the procurement of a construction contractor (who will build based on a set of specific obligations defined in the contract documents) and the purchase of products (which is based on specifications defining the characteristics or quality of the product).

1.4.3 Failure to Recognize Professional Judgment

Bidding does not recognize professional judgment, which is the key difference between providing professional services and the furnishing of products. Judgment is an essential ingredient in quality engineering services. Equally competent consultants may have very different design approaches to do the same project based on the exercise of different professional judgments.

1.4.4 Minimal Scope is Certain

It is virtually impossible to completely detail, in advance, the scope of services required for an engineering project, especially for the study and preliminary phases, without lengthy discussions and negotiations with the selected consultant. Lacking specifics, the bidding consultants must, in order to be competitive, submit a price for the least effort envisioned. The resulting service performed is likely to be tailored to fit the minimal requirements of the bid documents and will not necessarily suit the client's needs or expectations.

1.4.5 Short-Cut Analysis and Conservative Design

In-depth studies and analyses by the consultant are not likely to be performed. The consultant selected by lowest bid will often provide only the minimum services necessary to satisfy the client's scope of services. The design phase of a project represents the best opportunity to refine the project to achieve a combination of cost-effective construction and long-term operating cost efficiency. Saving a few dollars in design can result in substantial increase in construction and life-cycle costs.

1.4.6 Limited Flexibility and Creativity

The consultant's ability to be flexible and creative in meeting the client's requirements is severely limited. Conversely, a QBS selection process allows consultants to include, in the proposal, innovative ideas and approaches for accomplishing the work—ideas that may reduce the cost of construction, prolong the life of the facility, conserve resources, create a more aesthetic structure, minimize the life-cycle costs, and achieve a higher level of energy effectiveness.

1.4.7 Lack of Detailed Design

The engineering designs are likely to be minimal in completeness, with the details left to the contractor. This produces a lower first cost for the design, but tends to add to the cost of the completed project. The lack of design details also can—and frequently does—lead to a greater number of change orders during construction and to contractor claims at a later date.

1.4.8 Adversarial Relationship Created

Price selection tends to create an adversarial relationship. If selection of a consultant is predicated disproportionately on price, the relationship tends to be a "forced marriage." The consultant will carefully guard what it perceives as the responsibilities on which it based its fee, the opportunity for cooperation is diminished, and the utterance of "It's not in our scope" becomes more prevalent.

1.4.9 Recent Trends

In recent years, fee bidding has become more prevalent in the industry, as the economy has softened and work volume has decreased. Clients have sought to capitalize on the situation by pitting consultants against each other in fee bidding. In some cases, projects are broken up into two or three phases and separate fee bidding is undertaken for each phase, frequently resulting in two or three different firms providing sequential services on the same project. The result has been poor designs, conflicting or confused design logic, inconsistency between design and construction phases, failure to modify designs when differing site conditions are recognized, and significantly more disputes. It is impossible to achieve quality in a constructed facility when the design professional is different for each phase of work. The commitment to quality must be consistently built in from the beginning through completion of construction of the project.

1.5 AN ANALYSIS AND COMPARISON OF THE MARYLAND AND FLORIDA PROCUREMENT SYSTEMS

1.5.1 The Maryland and Florida Systems

In 1985, the American Institute of Architects (AIA) prepared a report that compared the experience of Maryland and Florida in procurement of engineering services. The state of Maryland selected architects and engineers (A/Es) on the basis of price and other factors. Florida utilized a QBS system, in which A/Es were selected on the basis of qualifications, then price was negotiated with the selected firm.

In the Maryland process, engineering firms were required to submit SF254/255s in response to a request for qualifications. The firms judged most qualified were placed on a short list and asked to submit a technical proposal and a price. The firms were then ranked, and an award made with "neither the price proposal nor the technical proposal becoming the sole criterion for selection."

1.5.2 The Conclusions of the Study

The study reached several conclusions, based on an analysis of data supplied by the Maryland Department of General Services, Florida's Department of General Services, and the State University System of Florida. The conclusions included:

- “Maryland’s A/E selection process is significantly more time-consuming and expensive than Florida’s. In Maryland, the necessity of preparing detailed programs on which A/Es can base price proposals results in added expense to the state in the form of administrative staff, time delays and consultant costs.”
- “Total costs of the A/E portion of Maryland’s capital construction process average 13 percent of estimated construction costs. Total costs in Florida average 6.7 percent of the estimated construction costs. A/E fees are lower in Maryland than in Florida; however, the added costs of the Maryland process far outweigh the savings in A/E fees that result from a process in which the state develops detailed programs ...”
- “The Maryland A/E selection system ... also results in extraordinary costs to the A/E firms that compete but are not awarded a contract.”
- “Maryland’s A/E selection process takes considerably longer to complete than Florida’s. This delay occurs while the detailed program descriptions are being prepared, during the selection process and during the design and approval phase.”
- “By law Maryland considers both price and technical competence in the award of state contracts, but it appears that price is becoming the dominant factor. Of the last 40 projects awarded by the Maryland Department of General Services prior to June 1983, 33 (83 percent) went to the firm with the lowest price proposal.” This corroborates the perception by engineering firms that a bidding environment exists when price is a factor.” (AIA, 1985)

1.5.3 The Result

Maryland changed its A/E procurement law in April 1986. It no longer asks for price, but requires the use of QBS for all Department of Transportation projects and for all other projects over $100,000.

1.6 CONCLUSION

The design and construction of subsurface projects entails risks that are unique to working in the underground and are largely unavoidable due to the environment within which the construction must take place. Engineering and science cannot fully anticipate what exists below the ground. At best, subsurface conditions are estimated based on a limited number of randomly placed explorations and the results extrapolated for hundreds of feet.

Unlike other types of construction, the design of a subsurface project is not complete until construction is complete. Unanticipated conditions are inevitable and require modifications or adaptations of the design in the field to accommodate the new condition. Furthermore, the performance of subsurface construction is closely related to construction means and methods, timing, and other factors that may not become known until construction is under way. Therefore, it is imperative that the consultant's services include continuous involvement from the initial ground characterization phase through construction. The failure by owners and/or clients to recognize the need for continuity from design through construction is one of the primary causes of claims, cost escalation, and poor-quality subsurface projects.

The ability to identify risks and to develop appropriate strategies to manage and allocate them is one of the most important services a consultant brings to a client for a subsurface project. The consultant-client selection process is generally the first, and probably one of the most important, steps in the risk management process.

A consultant-client selection process that facilitates the communication of mutual goals, an understanding of the owner's risk tolerances, the empowerment of the consultant to propose innovative solutions, and a general spirit of teamwork is the ultimate goal.

The mutual development of a work scope is a key element in project success. The process fosters a positive relationship between the owner and the consultant, based on trust and mutual respect. It promotes a free exchange of information in which the consultant can point out options and risks that the owners may not have considered; conversely, the owners can discuss any constraints that limit its risk tolerances and its preferences in scope of services. The qualifications-based selection (QBS) process, mandated for federal procurement of professional services, facilitates this interaction.

REFERENCES

Office of the Inspector General, Commonwealth of Massachusetts, *Designing and Constructing Municipal Facilities: Legal Requirements, Recommended Practices, Sources of Assistance* (October 1989).

American Institute of Architects, *Selecting Architects and Engineers for Public Bidding Projects: An Analysis and Comparison of the Maryland and Florida Systems* (Washington D.C.: American Institute of Architects, 1985).

2

THE PROFESSIONAL SERVICES AGREEMENT

James E. Monsees, Ph.D., P.E., Timothy P. Smirnoff, Ph.D., P.E.
Parsons Brinckerhoff Quade & Douglas, Inc.

2.0 INTRODUCTION

The size, complexity, and cost of subsurface construction and the litigious climate of today's world have all but ended the days of handshakes and verbal contracts. This chapter provides an overall perspective of the general nature of contract law and necessary constituents of the professional services agreement, with particular attention to subsurface projects. This chapter is not a substitute for assistance of competent legal counsel, nor is it meant to be totally inclusive of contract law. Knowledge of contract provisions and the ability to understand and quickly assess risks contemplated at the time of contracting or arising out of situations that reasonably may be anticipated to arise during the course of the work are essential. This knowledge promotes delivery of services in accordance with clearly defined standards and expectations, preserves the professional's rights, and defines, allocates, and mitigates risk exposure.

Subsurface Conditions: Risk Management for Design and Construction Management Professionals, edited by David J. Hatem
ISBN 0471-15607-8

2.1 THE CONTRACT (THE WHO, WHAT, WHEN, AND HOW MUCH)

A contract is a legally enforceable agreement between two or more consenting and competent parties to do or not to do a particular thing. A contract creates an obligation for performance and payment. Ideally, it provides a clear, written understanding of the obligations and responsibilities of each party, standards for performing the service, and for reporting, accepting, and paying for the services. The contract should also include provisions that address the procedures for changing the terms of the contract and its requirements, resolving disputes, indemnification (and/or other risk allocation provisions, such as limitation of liability), insurance requirements, and termination and default.

Contracts are legal documents and not engineering documents. As such, they are founded on long-standing legal precedent, and are ultimately interpreted by lawyers, juries, and judges. To be binding and enforceable, the terms and conditions of contracts must not necessarily be fair, nor must they always be equitable or founded on logic. As long as all the parties to the contract agree on the terms, they become binding. Either or both parties may waive, forfeit, or otherwise surrender rights, immunities, or obligations normally considered as inalienable rights. Many large clients (e.g., local, state, and federal government entities) have standard form contracts that contain provisions that are not advantageous for the engineer.[1] Changes to these provisions are often not subject to negotiation; rather, such provisions often must be accepted to obtain the engagement. This does not mean that they cannot be changed if both parties agree. However, it must be assumed that, once an agreement is made, all terms and conditions will be strictly and fully enforced. The almost certain and inevitable differences between any two subsurface projects make it especially essential to review, revise, and appropriately modify and tailor terms and conditions of standard contracts or to write new contracts to meet each project's specific needs.

2.1.1 Essential Considerations

For a contract to exist, the parties must agree on the essential terms and conditions of that contract prior to execution. The signing or execution of a contract typically is proof positive of this mutual acceptance. Verbal contracts should never be relied on—usually they truly are "worth the paper they are written on." Some of the essential terms in a contract are:

- A description of the parties included in the contract and the specific duties and responsibilities of each party (*who*);
- The quantity and nature of the services to be performed, i.e., a scope of services (*what*);
- The time in which these services will be performed (*when*);

- The compensation for the services (*how much*) (a gratuitous promise lacks consideration and is generally not enforceable against the party who made the promise);
- Allocations of the various uncertainties that may be associated with the project. Absent such terms, allocation will be inadvertent, following the rules and interpretations of complex statutory and common law;
- Other matters that are essential, depending on the purpose or specific requirements of the project or because of the requirements imposed by any of the contracting parties, e.g., a requirement for specific expert technical personnel to be assigned to the project or a requirement that the owner furnish pre-existing site information that may be relied on by the professional in the performance of his or her services.

These are the terms of a contract that must be understood and that form the basis of the contract. In its most general form, the contract will consist of several essential parts:

- The project: A general description of the project
- The parties: A detailed and clear description of the legal entities constituting the client and other parties
- Scope: A detailed description of the services to be provided and the deliverables
- Schedule: A detailed timetable for delivery of work products by the engineer and return of work product to be reviewed by the client
- Compensation: Provision for the payment for services provided
- Indemnity and insurance
- Procedures for changes, termination, and resolution of disputes

The project description covers location and general magnitude of the work and is fairly self-explanatory; the other portions of the contract require some elaboration.

2.1.2 The Parties (Who)

Before the parties can agree to the work, they must understand their prospective roles and interrelationships. Any contract must explicitly express the relationship among the parties. This requires a clear and concise statement as to for whom, and for what purpose, the services are being performed and who may be entitled to rely on or receive the work product of the engineer.

The engineer may function in a variety of roles depending on the project and its circumstances. In the position of prime professional or subconsultant, a number of relevant questions must be answered when considering the contract and its scope to ensure that the outcome and deliverables are commensurate with the project needs, the intent and expectations of the parties, required standards, and level of effort required. These questions have to do with (1) the client's expectations, (2) the experience and sophistication of the client and his or her normal range of practice, (3) the client's ability to fund the project, and (4) the client's ability to regularly pay invoices and other legitimate project costs.

In addition to identifying the parties, the contract must specify the duties and responsibilities of each party. Typically, the client provides basic site survey information, pre-existing site information, planning and concept reports, environmental impact and assessment reports, rights of entry and access to sites and required permits, scheduling of meetings and coordination with other agencies, review of submittals and work products in a timely manner, office space, furniture, and equipment. For any owner-furnished information, the engineer must quickly and thoroughly evaluate the quality, applicability, and completeness of that information. In general, the engineer should be entitled to rely on owner-furnished information. Any and all shortcomings must be discussed with the client and the evaluation and resolution of the shortcomings documented in writing. Duties of the engineer include providing services pursuant to the professional standard of care, using up-to-date design methods, meeting codes, keeping information confidential, providing efficient and cost-effective design, and producing the final package of documents making up the construction contract.

2.1.3 The Scope (What)

The key business of any engineer is the project; the scope of services is the principal vehicle that defines that project. The budget and schedule are derived mainly from the scope. The financial success and continued profitability of an engineering firm is dependent on the adherence to the budget and schedule. Typically, the scope defines the project in terms of a number of areas:

- Measurable tasks
- Task resources
- Deliverables and production methods
- Standards and quality control
- Definitions and assumptions
- Constraints
- Permits and approvals
- Project resources
- Project management
- Client liaison
- Contract changes
- What is excluded (not to be done)
- Client scope and interface definition

For the professional to fulfill contractual obligations, the scope of services must be clearly and completely defined. A clearly and concisely drafted scope provision often provides the basis for an effective defense to professional liability claims asserted against the engineer by either clients or third parties. Ambiguous and poorly defined tasks or obligations result in the addition of work that was not intended, add cost and time to the project, and increase the risk of professional liability claims (see Chapter 10, Section 10.3.1). It is important that the description of services to be performed be consistent with the compensation to be paid. It should not create the basis for an unintended liability due to an ambiguous or unclear expression of the scope of services to be performed. Beyond those services explicitly described, the scope of services will be interpreted to include work that is plainly and obviously necessary for the complete delivery of the required professional service effort.

The interpretation of the contract after execution is often a focal point of disputes and disagreements. In a disagreement or dispute over scope of services, the client may try to apply what he or she "meant" to say if he or she cannot find anything in the documents that precisely indicates the item in dispute. To prevent time-consuming and nonproductive discussions of the actual scope, the client may invoke general contract clauses such as, but not limited to, the following: "It is the intent ... to include all items necessary for proper execution and completion of the work ..." Allocation of responsibility for the cost to provide these items is a frequent source of dispute between owners and design and construction management professionals.

Preventing such problems requires knowledge of a few basic rules of contract interpretation that generally apply to all contracts and are used to determine how conflicts will be resolved. Even if these ideas are not expressly provided as strict rules, the principles are useful to apply in determining the practical resolution of various situations and options.

A clear and definitive scope of services is the front line of defense against a claim. The first general principle of contract law is to protect the reasonable expectations of the parties to the contract. To do this, the courts have sought to develop an objective standard of reasonable expectations, which would normally be contemplated by intelligent and informed people competent in their profession, with adequate knowledge of all related facts. The application of the principle generally provides that:

- The ordinary meaning of language is given to the words unless (as in the case of technical terms) their technical context indicates differently.
- The writing is interpreted as a whole, including the entire text of all other contract documents.
- Specific terms are given greater weight than general language, and added or separately negotiated terms are given greater weight than standardized terms.

In the case of ambiguities, the law applies the principle that they are resolved against the drafter. In so doing, the law assumes that drafters will provide for their own interests and that they have reason to be aware of uncertainties. This places the burden on the drafter of the contract. The drafter should have a clear idea of the intent of the contract and take every opportunity to be sure that those intentions are clearly defined to allow reasonable, competent people to understand and accommodate them. This principle is particularly applicable to public procurements and contracts produced by one party and presented generally as a take-it-or-leave-it form, with little or no opportunity for negotiation. Since the design and construction management professional has the most intimacy with the details of the scope of services he or she should take the initiative in precisely defining both the services included and those excluded.

Further, the law generally holds that if there is more than one reasonable interpretation available, there is no duty to continue to search for other reasonable interpretations. The interpretation need only be reasonable; it does not have to be the "most reasonable." Interpretation of the contract language may be limited by one or more special exceptions. This includes the so-called parol evidence or integration rule, which prohibits later introduction of evidence to modify or contradict the otherwise definable and clear terms of a written contract that was intended to be the final, total expression of the agreement. Under that rule, items such as preliminary memoranda, correspondence, proposals, and drafts generally are irrelevant and not controlling unless specifically incorporated as part of the final written agreement.

2.1.3.1 Scope of Geotechnical Investigations The proper design of a subsurface project requires adequate knowledge of subsurface conditions at the site of the project. More than any other factors, subsurface conditions determine the degree of difficulty, the cost, and the schedule for designing and constructing subsurface projects. The geotechnical investigation provides the basis for estimates of ground mass and intact rock and soil characteristics to evaluate loads, ground interaction, bearing capacity, and other pertinent design parameters. The geotechnical investigation also provides information for selecting and judging construction means, methods, and performance; short- and long-term response of groundwater; and the influence of the project on adjacent structures and the environment. In addition, the geotechnical investigation provides the foundation for the preparation of data in interpretive reports, and establishes baselines or other contractual mechanisms to define and allocate risk for site conditions between the owner and contractor. As discussed in Chapter 3, it is the height of false economy to scrimp on either the scope of or the quality control associated with the subsurface investigation needed for the proper design and construction of a major subsurface project.

The designer and the geotechnical engineer should understand the areas of importance and special concern for each structure designed. This will aid in the development of a site investigation program that provides data appropriate for the combination of subsurface conditions and specialized design and construction methods employed. Chapter 3 discusses geotechnical investigations and reports that should be used in developing the details of the scope of work for a subsurface project.

The types of subsurface information required for design include:

- Geologic profile (stratigraphy, structure, principal rock and soil types, and their characteristics) to define the extent, depth, and thickness of each identifiable stratum
- Depth to rock and the characteristics of the rock, including lithology, extent, depth and thickness, strike, dip, spacing of joints and bedding planes, faults and shear zones, and the state of weathering and decomposition
- Location of groundwater and the presence and magnitude of artesian pressures, aquifers, and reservoirs
- Engineering properties of the soil and rock in situ

Geotechnical data acquisition is generally performed in a phased sequence:

Reconnaissance investigations provide data (1) for site selection, (2) to define the concept, feasibility, and approximate cost, and (3) to plan the explorations for succeeding phases. Reconnaissance may include performing office research and library studies of available geotechnical data, including soil survey maps, aerial photographs, geologic and topographic maps, field reconnaissance and mapping, remote sensing, geophysical explorations, and a limited number of borings or soundings.

Detailed explorations for preliminary and final design must be carried out during the engineering design phases to acquire data necessary and sufficient for the design of the subsurface project, and for a contractor to fairly evaluate the project work and to select the construction means and methods, and to estimate the cost of those means and methods. The contractor may be fairly called on to perform additional explorations necessary for his or her design and implementation of his or her selected means and methods. However, sufficient information must be obtained in the preliminary phase to permit the selection of the type, location, and principal dimensions of all major structures. Planning of the investigation requires coordination with engineering geologists, geotechnical engineers, the designer, and engineers knowledgeable of construction means and methods and other related constructability considerations. Most geotechnical data for design are usually obtained during preconstruction planning and engineering. Supplemental explorations, as well as explorations and testing for purposes of construction, will be carried out in later stages.

Because of the complexities of geology and the variety of functional demands, no two projects are alike. Therefore, it is not possible to give hard and fast rules about the required intensity of exploration, or the most appropriate types of exploration. However, common sense would dictate:

- Plan explorations that attempt to define the full range of conditions to be encountered—the best, the worst, and the average—for the construction of the work. Locate and define the conditions that pose a threat to safety, represent special or very different conditions, or pose great difficulty during construction.
- Use qualified personnel to produce geologic interpretations. This allows the development of a geologic model that can be used as a framework to organize the data and extrapolate conditions to the locations of the pertinent parts of the subsurface structures. Geologic interpretation is clearly a professional service and should be in accordance with the professional standard of care (see Chapter 10, Section 10.1.1.2). Reasonable individuals may disagree on interpretations and none may be totally accurate.
- Use cost-effective methods to derive the required information.
- Carry the explorations to a depth sufficient to define the stratum and nature of materials that could either be affected by the loads imposed by the structure or could affect the integrity of the structure.
- Anticipate the methods of construction and obtain data required for the contractor to select those methods and estimate costs and production.
- Anticipate potential failure modes, limiting design cases, and requisite analyses for the completed structures. Obtain the data necessary for those analyses.
- Apportion borings and exploration with the greatest intensity where special circumstances or problems are apparent, or special design evaluations are required.

2.1.3.2 Contract Modifications Depending on the contract type, the scope, definitions, and contingencies may be altered as the work progresses. Consideration of the various types of contracts and scope variations are outside the range of this chapter. As terms and conditions, the parties may specify certain items that postpone performance, e.g., starting work only after notice to proceed or establishing overhead rates after a final audit at the end of the project. Contracts are "living" documents and will probably require revision before the completion of the project, so the contract must contain a provision for changes requested by either the owner or the engineer. Every adjustment or revision to scope, schedule, payment, or any other section requires a change or amendment to the contract. It is therefore incumbent on the contract drafter to ensure a sufficiently detailed scope such that extras (out-of-scope work) can be easily identified and the corresponding additional compensation justified.

The contract should provide for changes that are required and/or requested by either party. Changes should all be in writing and be signed by the respective representatives of both parties with the legal authority to bind those parties. Changes must provide for adjustments to contract scope, price, and completion time. Once executed, changes have the same weight as the original contract and are binding on both parties.

2.1.4 The Schedule (When)

The schedule defines the time line for the project and its completion and is a condition precedent used in most contracts. The schedule should be spelled out in detail, either by dates or, preferably, by durations of specific work items; such undefined phrases as "time is of the essence" should be avoided. These phrases may be interpreted by the client (owner) as implying a condition or warranty in a contract that, if not met and regardless of other terms or events, will relieve the client from the obligation to pay, and potentially may allow the owner to collect damages directly related to untimely (even if excusable) performance.

It is important that the schedule be both reasonable and integrated with the entire effort and within the control of the engineer—e.g., not depending on the contractor's performance, since the contractor's performance is not within the control of the engineer. Similarly, the ability to complete the design on schedule may be dependent on receipt of owner-provided information or of owner or public regulatory approval, neither of which is in the control of the engineer. In reviewing schedules, care must be taken to avoid, prevent, or eliminate any actual or implied conditions to the schedule that cannot be met or are not within the exclusive control of the engineer. Often, time for quality assurance reviews, or other owner/client review of submittals, is not included in schedules, or the time taken for reviews is not controlled. The impact of such delays and omissions is often not appreciated by client owners and may lead to overruns in schedule and in budget.

Any events that are likely to cause a delay, including permitting, obtaining rights of access or entrance, obtaining special jurisdiction or governmental approvals, or regulatory compliance, should be clearly delineated and an appropriate duration assigned. If events are delayed or require longer durations, careful attention to the initial schedule detail will pay off later should these delays become contentious or result in overruns. Some items—e.g., force majeure, weather, strikes, and the like—usually cannot be given a duration up front. In these cases, the contract must clearly state how the resultant delays will be identified and evaluated, and how appropriate time extensions are to be incorporated into the contract.

2.1.5 The Budget (How Much)

Professionals are in the business of providing their services in such a way as to fulfill their clients' needs and to make a profit. Thus, it is often assumed that the budget for a project is the most important item in a professional services contract. In reality, the project contract should be approached considering the triad of *scope*, *schedule*, and *budget*. These three are not independent variables; rather, they are totally dependent on each other such that a change in any one of the three must result in a change to at least one of the others. Thus, for example, an enlarged scope held to the same schedule *must* result in an enlarged budget—i.e., a changed project.

The appropriate order to follow in reaching an agreement on a budget for a contract is simplified as follows:

- Agree on scope rather than price
    - — Measurable tasks
    - — Defined deliverables
    - — Defined constraints
    - — Client-provided items
- Agree on schedule rather than price
    - — Defined milestones
    - — Defined resource loading
- Agree on project staffing, rather than price
- Agree on risks/rewards

When these agreements are reached, the client should then be presented with the pricing and costs for the project.

The detail of this presentation should consist only of the minimum cost and pricing information required to support the proposed budget and produce a fair profit for the professional. It is vitally important that each cost element relate, one by one, to the elements of the scope of work. In fact, developing the budget details will serve as a check on the completeness and adequacy of the scope of work. Correspondence between items in the budget and those in the scope of work should be both clear and obvious. Significant sums under headings such as "general expenses" or "miscellaneous" may signal problems in this one-to-one correspondence and may also lead to disagreements during negotiations and/or producing the work. Any item of work that can be identified (e.g., a design report) should be listed and priced separately rather than "included" under some other item(s), such as design drawings, which also can be priced individually.

All negotiations should be conducted in a win-win atmosphere, so that the client is convinced the project will be done professionally and at a fair price, and the engineer is convinced that the scope, schedule, and budget realistically allow for doing the required work on time and at a fair profit.

2.2 TERMS AND CONDITIONS

The terms and conditions for any engineering service undertaking on a subsurface project must contain the same standard terms and conditions as any professional services contract. Generally, the scope will be grouped into the following components: geotechnical and technical components and contingent and reserve provisions. Frequently, the technical personnel think only of the technical and project-specific components and overlook the standard terms and conditions. The technical scope and its components are those specific activities that must be completed to meet the client's project-specific objectives. Additional items that expand the technical scope into a comprehensive scope of services include project responsibilities and roles, project administration (invoicing, progress payments), quality control, and quality assurance plan.

Geotechnical investigations and studies are the basis for overall project design decisions and are invariably based on a limited sampling of the total geologic mass at the project site. As a result, there is an unavoidable risk in performance of geotechnical investigations and design, and the fee or compensation should be commensurate with the risk. The exact extent of the investigation program will vary from project to project. The number and spacing of the explorations should be consistent with achieving the objectives of the program at a minimum cost.

In many cases, the process for the development of a scope and the precursor to a contract exists in the request for proposal (RFP) issued by the client or owner. This sets forth a description of the project (scope and schedule), and contains a proposal form that provides a standardized format and content for the proposal to assist in evaluating and selecting the winner. The RFP commonly contains a sample contract form as well. The subsequent proposal is an offer to perform, and may or may not become part of the contract. It is incumbent on the proposer that the proposal conforms to the scope represented by the RFP and does not promise more than can be delivered consistent with the agreed-upon budget and schedule. In negotiations, any or all parts of the proposal and scope are open to discussion. Removal of an item of scope from the proposal must be documented and agreed upon by both parties, as should any understandings that arise during the negotiations. The scope of the investigation should be detailed in the proposal and negotiated. If the engineer thinks that the investigation required for the project is more detailed than the owner is willing to fund, this should be clearly documented in the record of negotiations and in the reduced scope of the final contract.

Agreements with subconsultants (e.g., other engineering firms, drillers, or testing laboratories), as they relate to the scope of services, indemnification, insurance, and the like, must be given the same thought as the agreement with the owner. For the most part, applicable portions of requirements for scope items, indemnification, insurance, and so on should be passed directly to these subconsultants (see Chapter 11, Section 11.1.4). In doing so, the "part/whole rule" must be considered—i.e., incorporating a proposal as opposed to attaching it as an exhibit. Both options should be carefully considered and evaluated. For each given situation, the preferred choice is the one that results in the "cleanest" subcontract, that is, the subcontract that best defines the work to be done and eliminates conflict or confusion between details contained in the contract with the owner and those in the subconsultant's proposal.

2.3 RISK AND REWARD

With the assumption of any project comes a risk to the engineer that cannot be escaped. The key is to identify and effectively manage that risk, and to be adequately compensated for the risk assumed. A measure of that risk is the fee and the profit that is returned to the engineer. The smallest of mistakes may lead to expensive design changes or claims for differing site conditions that result in expensive and time-consuming litigation, for which there may be no compensation. In many instances, the time and costs to be extricated from litigation may be several times the fee received for the actual work. The current trend in subsurface projects is for the contractor to contest every geotechnical report and to provoke the owners to assume the risk and direction of the work. This trend has created an almost assured climate of conflict between owners and design, geotechnical, and construction management professionals. Many subsurface projects require extensive investigations and state-of-the-art solutions for difficult and varying geotechnical problems and the risk of conflict and/or claim may outweigh any fee that may reasonably be collected. Almost invariably, the value of a firm's personnel time in defending a claim is not covered by insurance, which covers (after satisfaction of deductible) only legal defense costs and settlement or judgments. An owner may or may not choose to compensate a design or construction management professional for his or her time in defending a claim from a contractor if the owner feels that, based on the allegations regarding the professional's actions, the professional may be even partly responsible for the claim.

When an RFP or request for quotation is received, one of the first actions that must be taken is a careful and thoughtful review of the job risks and history of the client: What is the experience of the owner's engineer responsible for the project? What is his or her track record for not only the commercial issues, such as payment and treatment of contractors, but also on how claims and litigation have been settled; do they result in the engineers/geotechnical engineers being sued and having to defend themselves? Is the owner looking for solutions that are "cheap," "conservative," or "state-of-the-art"? Does the owner understand the risk associated with each design philosophy and the effect that it will have on the scope, cost, and risk associated with the geotechnical investigation? (For further information on risks and risk allocation, see Chapter 5, Sections 5.4 and 5.7.)

Consideration of risk is necessary, essential, and proper when determining fee and profit. Innovative and/or state-of-the-art designs require a higher level of effort and risk for the design professional and, therefore, a higher fee for services rendered. In any contract, it is important to provide appropriate mechanisms for the handling of contingency and reserve. Contingency generally is set aside for unknown and uncertain events, e.g., design of piles for a foundation, dewatering system design, and the like. Reserves are set aside for known or anticipated events or scope of defined magnitude, e.g., preparation of an environmental impact report or preparation of well-water quality or National Pollutant Discharge Elimination System (NPDES) permits.

In addition to the customary and anticipated risk inherent with the prediction of nature and the errors and omissions of employees, the contract scope definition and contract type may also play a significant role in defining the amount and nature of financial risk and exposure the professional is facing. Acceptance of a lump-sum contract, while offering a reward for efficiency, makes it absolutely mandatory to have a fully detailed scope with defined deliverables so that the total effort is definable. Other contract types, such as cost plus fixed fee, may allow recovery of all costs, but if overruns or changes occur, they may drastically alter the profit margin. The language of the contract may also increase the firm's liability and characterize performance in such a way that the usual and customary standard of care is increased by the client and the public. Such terms as "assure," "insure" and "ensure," "all," and the like, should be avoided, as they often are considered to be a guarantee or warranty to the client and, thus, increase the engineer's liability. (See Chapter 10, Section 10.1.1.1.)

Many times, the issues of risk and liability are determined by the level of insurance required by the owner or agency contracting for the service. The larger the project, the larger the errors and omissions (E and O) liability insurance requirement. While such insurance is a necessity, it must be recalled that it does not cover the engineer for all occurrences and project liability exposures. (See Chapter 10, Subsection 10.1.1.2, and Chapter 11, Section 11.2. See also (Hatem, 1996).) Many larger projects may include an owner-directed insurance program or wrap-up policy that covers all the firms involved in the project: designers, construction managers, contractors, and subcontractors. Many contracts require indemnification of the owner/client by the engineer/geotechnical engineer. This indemnity is the shifting of liability risk from one party to another, usually from the client to the engineer. For a price, insurance shifts this risk from one party to another—e.g., from the engineer to an insurance company.

The current trend in litigation is to go for the deep pockets, generally the largest firms and those with the most money. A small firm may not present a large enough target for a lawsuit, but this is not a shield to hide behind. Depending on the firm structure (be it a sole proprietorship, partnership, or corporation), the level of joint and several responsibilities may vary considerably.

2.3.1 Limitation of Liability

Chapter 4 contains a detailed discussion on and recommendations for identifying, assessing, and allocating risks. In theory, at least, it is possible to assign risks between the owner and the contractor in a manner similar to that in Figure 2.1, which was recommended for the Superconducting Super Collider project. However, it must be recognized that actual allocation of risk in the field often is not as simple in practice as it is in theory.

For the design and construction management professional, the first—and, possibly, most important—tool in limiting liability is a clear, concise, and complete scope of work in the professional services agreement. Under such conditions:

Risk	Allocation: Contractor	Owner	Shared	How Risk is Allocated/Managed
A. Resources and Prerequisites				
Adequate project funding		×		Termination for convenience clause
Adequate financial resources	×			Surety bond requirements
Adequate labor force	×			
Permits/licenses				
Project-wide		×		Advance planning and identification
Contractor-specific	×			Permits clause
Site access; rights-of-way; infrastructure		×		Advance planning
Insurance		×[a]		
Bonds	×			
Suppliers and Subs	×			
B. Performance-Related				
Feasibility/sufficiency of design		×		Changes clause
Owner-furnished equipment availability/performance		×		Suspension clause; expediting
Contractor-furnished equipment availability/performance	×			
Construction means and methods	×			
Supervision	×			Superintendence clause
Delay in presenting potential claims	×			Notice requirements
Delay in addressing potential claims		×		Contract-level decision-making; disputes review board
Labor productivity	×			
Differing site conditions		×		Geotechnical design summary report; differing site conditions clause
Unforeseeable, no-fault delays	×		×	Time extensions clause
Site/worker safety[b]	×			Clear contract language
C. Outside Influences				
Government (sovereign) acts		×		
Severe weather			×	Time extension clause
Acts of God; riots; vandalism			×	Time extension clause
Inflation/escalation				
>2 yr. contract		×		Escalation clause
<2 yr. contract	×			
Union strife; work rules	×			

[a]Assumes use of wrap-up insurance.
[b]Subject to stipulations laid out in the wrap-up insurance program.

Figure 2.1 Sample allocation of risks (after U.S. National Committee on Tunneling Technology, 1989)

- The professional would be held only to a "professional standard of care" in discharging design and construction management services.
- Construction contractors and owners would share other risks in accordance with a project-specific version of Figure 2.1.

In the real world, however, the professional is frequently subjected to liability exposure beyond that resulting from the "professional standard of care" by such actions as the following:

- The professional agrees to perform the services in accordance with the "highest," "best," or other superlative level.
- The construction contractor asserts the professional is responsible for the failure of construction means and methods selected by the contractor on the theory that the professional reviewed or otherwise was simply aware of those construction means and methods.
- The owner asserts that changed conditions were brought about due to the professional's inaccurate description of the geological conditions.
- The professional acquiesces to perform the work with less than recommended information (e.g., on subsurface conditions) and does not include such qualifications in the professional services agreement and in subsequent work products.

2.4 ENGINEERING SERVICES DURING CONSTRUCTION

In some situations, the role of the geotechnical professional is often ended once the project report and contractual input, such as the Geotechnical Baseline Report, are furnished. However, it is becoming increasingly more apparent, necessary, and advisable that the engineer's role be extended into the construction phase of the contract, so that the engineer may represent the owner in observing the construction of the design, review construction submittals, and be in a position to investigate and evaluate differing site conditions claims in a prompt, fair, and impartial manner. This simply reaffirms the thoughts of two geotechnical engineering pioneers regarding the observational method in subsurface projects:

> To acquire competence in the field of earthwork engineering, one must live with the soil. (Terzaghi, 1957)

> The successful use of the observational procedure requires more than a knowledge of theory and of the physical properties of earth materials. On one hand, it requires an appreciation of geologic processes. On another, it demands a knowledge of construction procedures, because the contractor's workmanship is a major variable. To determine the appropriate kinds of field observations, to select the best observational equipment and techniques, and to interpret the results, are more of an art than a science. Even though research in the field is extremely active, it seems unlikely that science will supersede the art. (Peck, 1984)

Engineers are increasingly being called on to provide a wide variety of services:

- Verification of qualifications of contractor's proposed specialty and subcontractors
- Verification of conformance for selected elements of the design, e.g., footing inspections, slope verifications, rock quality verification
- Construction management support:
    - Requests for information
    - Contract deliverables review and approval
    - Mapping of exposed conditions
    - Definition of conditions for estimates of support or payment
    - In-situ testing and sampling
    - Problem resolution
    - Review of contractor-proposed value engineering and design
    - Review of construction means and methods
- Resolution of changes and claims
- Post-construction litigation

As with any other part of the work, acceptance of such assignments requires a contract or contract modification with the appropriate development of scope, budget, and schedule and allocation of risk and responsibility. The effort for many post-award services (such as those bulleted above) is often ill-defined and highly dependent on the actions of others. The scope and budget may, therefore, be highly indeterminate. The contractor's submission of an alternative design or radically different excavation support or final support design may require intense effort for very short periods of time, which cannot be detailed in advance. Providing for such contingencies may be difficult, but contingencies are better anticipated and identified before the fact than after the services are performed. Changes, and a mechanism for change for such additional work, should be negotiated prior to the occurrence.

Historically, contract documents (i.e., the drawings and specifications), despite their volume and detail, are not intended to address and do not specify the construction means and methods to be utilized by the construction contractor to achieve the finished and completed project work. Paragraph 9.13.2 of the Standard General Conditions of the Construction Contract No. 1910-8 (EJCDC, 1990) specifically states:

> Engineer will not supervise, direct, control or have authority over or be responsible for CONTRACTOR'S means, methods, techniques, sequences or procedures of construction, or the safety precautions and problems incident thereto, or for any failure of CONTRACTOR to comply with laws and regulations applicable to the furnishing or performance of the work.

All major public works projects and agencies in the United States have historically followed this philosophy—i.e., they have used performance specifications rather than prescriptive ones in addressing construction means and methods. The differences between performance and prescriptive specifications are dramatic: In the former, the contractor understands and takes sole responsibility for the design and implementation of the means and methods to accomplish the construction; in the latter, the contractor follows the means and methods design detailed in the contract documents and takes little responsibility should they fail (he or she was "just following the specifications"). In addition, contractors are becoming more adept and successful at converting performance specifications into prescriptive ones. This is accomplished by asserting that:

- The design professional's approval of required submittals of working drawings constitutes the design professional's warranty that the means and/or methods represented will work regardless of how they are applied in the field, or
- Recommendations and comments offered by the design professional (often with the owner's encouragement) to solve problems encountered during construction constitute direction of the work and absolve the contractor of responsibility for means and methods.

If the owner is known to support the effective transfer of responsibility for means and methods to the design professional by such actions, then the scope of work and the professional's insurance and fee must be adjusted accordingly. Although difficult to do, the most prudent action to take may be to refuse the work.

The current system, wherein the owner retains a designer and independently selects a construction manager, has led to a significant and real problem. In geotechnical engineering, many of the state-of-the-art designs require an observational approach during construction to implement the design. For the intent of the design to be faithfully executed, the construction manager must have the requisite qualifications, knowledge, and experience in observation and in design to satisfactorily issue the appropriate guidance to the contractor and, thereby, adequately represent the interest of the owner in obtaining a completed project that meets the intent and quality required under the contract documents. Similarly, leaving it to the construction manager to "call" the design/geotechnical engineer when a problem arises in the field is not a satisfactory resolution, since such issues should be identified well before they become problems, and this approach presumes the ability and acumen of the construction manager to recognize and adequately (and timely) communicate those problems to the design/geotechnical engineer. Ideally, this would dictate that the designer be directly and continually involved during the construction phase. This admonition has critical importance in a subsurface project, given the high degree of correlation and interaction between the construction means and methods used by the contractor and the successful and quality implementation of the designer's intention with respect to the completed project. The engineer should emphasize this point to the client during the proposal and contract negotiation process. At a minimum, if the client does not accept the design/geotechnical engineer's recommendations regarding involvement in the contract administration phase, the recommendation and the client's action should be documented.

The lack of such direct input during the contractor's execution of the design may well demand that, as protection from liability, the design professional exercise care so that the design can be built by workers from the union hall. This usually leads to conservatism in design. To derive the fullest gain from state-of-the-art designs, the design professional must have a presence on the site to assist in managing the risk and interpreting the design.

During the course of construction, contractors often propose alternatives and changes to the design and the specifications. These proposals are, in effect, changes to the work. In such cases, a change order is appropriate and necessary. The old axiom "No good turn goes unpunished" is often invoked when reviewing these proposals. Contractors and owners often attempt to assert that, by acceptance of these alternatives, the design engineer warrants that they will work and are truly alternatives, not just for one year after construction (which is the usual limit of the contractor's guarantee period), but for the useful lifetime of the structure. The goal of the contractor is often saving time or money, but this goal may not always fully benefit the owner. Careful consideration of changes must be undertaken by the design professional and the owner to ensure that they both understand the risks inherent in such changes, and that the owner receives both an appropriate and equitable adjustment to contract price and a finished project that will fulfill his or her expectations for the full lifetime of the structure.

Changes are not necessarily bad, nor are they to be avoided. The recognition of changes as they occur during construction, and the fair and impartial resolution of changes, are an important part of the owner's and the design and construction management professionals' job. Owners inherently must realize the limitations (as previously discussed) in geotechnical investigations and the inherent vagaries of nature itself. Owners must be able to accept legitimate changes to contracts because of differing site conditions or because of conditions rightfully out of control of the contractor. The assumption of owners that there has to be someone to blame is counterproductive and nonproductive. Adversarial positions have no place on civil engineering projects. All the parties must work to solve the problems and resolve them, and that comes only with a mutual understanding and professional attitude by all.

2.5 CONCLUSION

The professional services agreement is a contract between the owner and the design or construction management professional. It has been stated that "a good contract does not merely divide the responsibilities of the project; it is a unifying force, an agreement committing both parties to a single common objective. Every position in the contract must be an acknowledgment not only of the legitimate interests of the individual parties but of their common goal." (U.S. National Committee on Tunneling Technology, 1976). This chapter is not a cookbook, nor a substitute for competent legal advice; rather, it strives to provide some guidance for professionals attempting to write good contracts for professional services related to subsurface projects.

NOTE

1. Sometimes, statutes and public policy make certain contract terms unenforceable. However, even in these circumstances, it is risky to rely on courts for relief of the burden of unfairness due to disadvantageous contract terms.

BIBLIOGRAPHY

ASFE Model Clause Library, "Geotechnical, Geoenvironmental" (Maryland: ASFE 1995).

Bachner, J. P., *The ASFE Contract Reference Guide*, 3rd ed. (Maryland: ASFE 1997).

Bachner, J. P., *Practice Management for Design Professionals* (New York: John Wiley & Sons, 1991).

Dixon, S. A., and Crowell, R. D., *DPIC's Contract Guide*, 2d ed. (CA: DPIC Companies, Inc. 1997).

Hatem, D. J., "The Owner as 'Additional Insured' Under the Design Professional's Professional Liability Policy: Can It Be Done and If So, How?" *The CA/T Professional Liability Reporter*, Vol. 2, No. 1 (Burns & Levinson LLP, Boston, September 1996), pp. 1–8.

Peck, R. B., "Subsurface Engineering in the Space Age," in *Judgment in Geotechnical Engineering*, D. U. Deere and J. Dunnicliff, eds. (New York: John Wiley & Sons, 1984).

Schinnerer Modules, *Understanding and Managing Risk, A Guide and Voluntary Educational Program for Design Professionals* (Chicago: Victor O. Schinnerer & Co., 1997).

Sweet, J., *On Construction Industry Contracts* (New York: John Wiley & Sons, 1996).

Terzaghi, K., Presidential Address at Fourth International Conference, in *From Theory to Practice in Soil Mechanics* (New York: John Wiley & Sons, 1960).

U.S. National Committee on Tunneling Technology, *Executive Presentation—Recommendations on Better Contracting for Underground and Construction*, National Research Council (Washington, D.C.: National Academy Press, 1976).

U.S. National Committee on Tunneling Technology, *Contracting Practices for the Underground Construction of the Superconducting Supercollider*, National Research Council (Washington, D.C.: National Academy Press, 1989).

3

SUBSURFACE INVESTIGATIONS AND GEOTECHNICAL REPORT PREPARATION

Gary S. Brierley, Ph.D., P.E.
Haley & Aldrich, Inc.

3.0 INTRODUCTION

All constructed facilities come into contact with the ground, and the complexity of this structure/ground interaction will determine, to a large degree, how the project is planned, how it is designed, and under what conditions it will be constructed. The "ground" can be a very difficult and risky environment in which to conduct business. This chapter provides guidelines for investigating the ground and for reporting on the results of those investigations for design and construction purposes.

Since the topic of subsurface investigation is large, it is necessary to establish parameters for the focus of this chapter as compared to other subjects addressed in this book:

1. The client for the engagement is a business entity, professional design firm, or government agency with construction project experience that

Subsurface Conditions: Risk Management for Design and Construction Management Professionals, edited by David J. Hatem
ISBN 0471-15607-8

intends to utilize a conventional design-bid-build approach for project delivery. For instance, an industrial client might be proposing to develop a large site, a sewer district might be planning a small tunneling project, or a structural engineer might need help with foundation design for a high-rise building. This chapter is not intended, however, to address the investigation and reporting requirements for clients without construction experience, such as individual homeowners, for highly complex "megaprojects" (see Chapter 5), for the situation where the services are being performed for a construction contractor (see Chapter 7), or for unconventional project delivery systems, such as design-build (see Chapter 10, Section 10.2.3).

2. The geotechnical engineer's services on the project are controlled by a written agreement such that the purpose and scope of the services have been discussed with the prospective client and a written description of the geotechnical engineer's participation in the design and construction process exists. (See Chapter 2, Section 2.1.3.)
3. It is assumed that the project owner will utilize the services of a geotechnical engineer to assist with preparation of the contract documents, to review construction-related submittals, and to observe all ground-related activities during construction. Without some significant involvement during construction, the geotechnical engineer has an overwhelming desire to change the approach of his or her report from "let's get the job done" to "let's cover my liability," which could result in a more conservative (i.e., more costly) design. Clearly, it is in the best interest of the owner to foster the former, rather than the latter, approach to report preparation by allowing for the geotechnical engineer's significant involvement during the construction phase.
4. Last, and most important, it is assumed that the site in question is not "contaminated." The impact of contamination on a site can be significant in terms of cost and schedule impacts, third-party involvement, and regulatory intervention. For these reasons, every site should be specifically assessed for contamination. This generally involves a phased approach of "environmental site assessment" undertaken specifically for those types of services. A scope for a phase I environmental site assessment is presented in ASTM E1527; however, many owners and agencies have evolved their own criteria. Lending institutions, in particular, frequently have criteria for site assessments, which are required as a condition of project financing.

It should not be assumed that a "geotechnical" investigation will reveal any contamination that might exist at a site, for two primary reasons. First, the goal of a geotechnical investigation is to determine the physical properties of the ground; the chemical properties, however, require entirely different sampling and testing procedures. Second, subsurface explorations for a structure are usually located within the footprint of the proposed structure, whereas explorations for contaminant evaluation should be located where prior site usage indicates the highest possibility of contaminant source (e.g., underground storage tanks or a former processing plant). Although it is usually preferable to undertake these two types of site studies simultaneously and with the same engineering firm in order to consolidate and coordinate the entire body of subsurface knowledge, nothing in this chapter is intended to discuss the highly complex process of evaluating sites for contamination.

3.1 THE CHALLENGE OF SUBSURFACE CONSTRUCTION

It seems intuitive to state that subsurface construction is different—sometimes, vastly different—from construction that takes place above the surface of the ground, but many owners and design professionals seem to relearn this lesson on a regular basis. Construction below the surface of the ground involves many risks that simply do not exist for aboveground structures. Given below is a listing of some of the more important aspects of subsurface construction that must be taken into account during the planning and design phases of a project and during the preparation of reports associated with that work.

1. *The ground cannot be specified*—Think of a typical specification for the concrete or steel that is used for construction and imagine writing a similar specification for the ground. Such a specification would be written without even performing subsurface investigations and would simply state that all soil at the subject site shall possess a certain strength or stiffness, that it shall exist to a certain depth, and that it shall not allow groundwater to flow into the excavation during construction irrespective of the contractor's chosen means and methods for the work. It could also be *specified* that all soil shall stand vertically to a certain depth for a predetermined period of time and that no bedrock outcrops or boulders shall be present anywhere within the limits of the proposed project.

   Clearly, such a specification is not realistic; the ground cannot be specified. The ground is the ground, and it can be utilized only as it exists in situ. The whole purpose of a subsurface investigation is, therefore, to determine the nature of the underground condition well enough so that a project can be implemented in a satisfactory manner. Although no clear parallel exists for this situation as compared to the aboveground portion of a project, it is possible to use the climatic conditions as an analogy. Imagine how different it is for a contractor to erect a building in Hawaii as compared to Alaska in the middle of winter. Hence, it might be helpful to think of the subsurface investigation as representing a "forecast" of subsurface conditions at a subject site.

2. *The ground can be changed during construction*—The ground can be changed, for better or for worse, as a result of the contractor's selected construction means and methods utilized during the process of construction. Hence, competent ground can be damaged by careless or inappropriate construction operations, or relatively unstable ground might be improved by construction dewatering or grouting. A well-written geotechnical report must discuss at least some of the more obvious possibilities for changing the ground during construction, and the effects that those changes could have on project design and/or the quality of the completed project.
3. *The ground requires the construction of temporary structures*—Before any completed facility can be constructed underground, it is necessary to create the *space* in which the completed facility can be constructed. Hence, in many cases, the major risks, costs, and impacts of underground construction are associated with the temporary structures rather than the completed facility. As a consequence, major aspects of the work for which the project owner has only limited control and no long-term vested interest could determine, to a large degree, the success or failure of the completed project.
4. *Construction means and methods*—Construction means and methods also need to be considered in the context of ground conditions. Contractors can be quite innovative at taking "short cuts" or using new procedures or techniques for the purpose of facilitating underground construction. Conversely, inappropriate construction methods can result in highly adverse ground behaviors that can damage a site, cause injuries to third parties, or result in differing site conditions claims or other expensive problems for an owner. How, then, and to what degree it is necessary to prescribe and/or to control construction means and methods is a serious issue relative to the implementation of subsurface construction projects (see also Chapter 6, and Section 10.2.2 of Chapter 10).
5. *Critical path*—Little else takes place at a site until the subsurface portion of the work is complete. Hence, the subsurface work is almost always on the critical path for project completion and difficulties with this work can have serious "ripple effects" on all remaining portions of the project. In essence, many of the largest project risks occur early in the work, and the owner is faced with the challenge of making sure that this portion of the work goes smoothly, to minimize impact to all remaining phases of the project.
6. *Third-party impacts*—All projects have impacts on adjacent properties during construction (such as noise, dust, and construction traffic); most can be identified and mitigated in advance by proper planning and coordination. For subsurface construction, however, instances exist where construction dewatering results in the settling of adjacent properties, an unanticipated loss of ground results in damage to nearby structures or utili-

ties, or vibrations from pile driving or blasting result in class action lawsuits for damages. Underground construction can and does have direct and costly impacts on adjacent third parties to a degree that is generally greater than the aboveground portion of a project.

3.2 BACKGROUND

The first professional organization to critically examine how geotechnical engineers conducted their business and how they reported on the results of their investigations was the Association of Soil and Foundation Engineers (ASFE), now known as ASFE Professional Firms Practicing in the Geosciences. ASFE was founded in 1969 specifically to investigate ways that geotechnical engineering firms could reduce their exposure to professional liability claims and losses. In addition to contractual and insurance issues, ASFE has consistently addressed the issue of professional practice, and more specifically, the appropriate standards for service delivery and report preparation. With respect to report preparation, ASFE has made numerous recommendations about both the format and content of geotechnical reports, including peer review of actual reports with appropriate feedback and suggestions to the report writers. Although it is beyond the scope of this chapter to discuss all the programs and documents made available by ASFE relative to the topic of report preparation, it is the intention of this chapter to generally follow the guidelines established by ASFE. It is recommended, however, that any serious student of this topic become directly familiar with appropriate ASFE publications (ASFE, 8811 Colesville Rd., Suite G106, Silver Spring, MD 20910).

Simultaneously with the work being performed by ASFE, five U.S.-based professional societies have addressed the issue of subsurface information and geotechnical report preparation (especially as related to tunneling projects), beginning with the now disbanded U.S. National Committee on Tunneling Technology (USNC/TT), which published a 1974 report titled *Better Contracting for Underground Construction*. The stated purpose of that report was to improve contracting methods for underground construction, particularly with respect to the following objectives:

- To help the owner obtain a completed project at lower cost while, at the same time, providing a "just" profit for the contractor
- To foster a cooperative atmosphere between the owner and the contractor
- To stimulate the use of advanced technologies and innovative construction techniques
- To provide for the equitable sharing of risks, especially those risks that are not "identifiable"

The USNC/TT report presented a total of 17 contracting practice recommendations, which were further consolidated in a later publication into six specific categories:

1. To share construction risks
2. To share financial risks
3. To expedite the handling of claims
4. To stimulate innovation in construction
5. To assure award to a qualified contractor
6. To realize other cost savings

All the recommendations concerning the subsurface aspects of construction were included in the category about the sharing of construction risks, which contained the following five subheadings:

1. *Changed conditions clauses*—Contracts for underground construction should include changed conditions (or differing site conditions) clauses permitting the owner to both assume (share) the risk of unknown subsurface conditions and reduce the need for large contingencies in the contractor's bids.
2. *Disclosure of geological data and interpretations*—All relevant subsurface data should be made available to bidders, along with the associated professional interpretations.
3. *Elimination of disclaimers*—Contracts should not contain disclaimers as to the accuracy of the factual subsurface data provided to prospective bidders.
4. *Provision for water problems*—Contracts should include a provision for the cost of coping with unexpected water problems.
5. *Responsibility for support systems*—The responsibility for the design and performance of ground support systems should be clearly defined in the contract.

Of most relevance to this chapter are the first three recommendations given above: (1) changed or differing site conditions clauses, (2) full disclosure of relevant information, and (3) the elimination of disclaimers. Most contracts for large, public infrastructure projects contain a differing site conditions clause, patterned after the Federal Government Construction Contract Standard Form 23-A, which states the following:

a. The contractor shall promptly, and before such conditions are disturbed, notify the contracting Officer in writing of: (1) subsurface or latent physical conditions at the site differing materially from those indicated in this contract, or (2) unknown physical conditions at the site, of an unusual nature, differing materially from those ordinarily encountered and generally recognized as inhering in work of the character provided for in this contract. The Contracting Officer shall promptly investigate the conditions, and if he finds that such conditions do materially so differ and cause an increase or decrease in the contractor's cost of, or the time required for, performance of any part of the work under this contract, whether or not changed as a result of such conditions, an equitable adjustment shall be made and the contract modified in writing accordingly.

b. No claim of the contractor under this clause shall be allowed unless the contractor has given the notice required in (a) above; provided, however, the time prescribed therefor may be extended by the Government.

c. No claim by the contractor for an equitable adjustment hereunder shall be allowed if asserted after final payment under this contract.

The primary objectives of the differing site conditions clause are to put the contractor on notice during bidding that the owner is willing to share some of the risk of "unknown" subsurface conditions and to encourage the contractor to reduce large contingencies in the bid for that risk. The overwhelming belief by the USNC/TT committee at that time was that both the owner and the contractor would be well served over the long term by this recommendation in that the owner could expect to receive a fairer bid for the work, and the contractor could expect to be fairly compensated if it did encounter a condition differing materially from that "indicated" in the contract document (Type 1) or of an unusual nature (Type 2). This same conclusion was reached and the same recommendation made in the five professional publications listed below:

- *Guidelines for Tunnel Lining Design* (O'Rourke, ed., 1984);
- *Avoiding and Resolving Disputes in Underground Construction* (ASCE, 1989);
- *Avoiding and Resolving Disputes During Construction, Successful Practices and Guidelines* (ASCE, 1991);
- *ITA Report: Contractual Sharing of Risk*, 1994, prepared under the auspices of the Working Group on Contractual Sharing of Risks in Underground Construction, and published by the International Tunneling Association;
- *Construction Dispute Review Board Manual* (R. M. Matyas, et al., 1996), as a continuation of the UTRC documents listed above and published by McGraw-Hill.

It is intended, therefore, to adopt in this chapter the above three recommendations and to assume, that the project owner is willing to share at least some of the risk of "unknown" subsurface conditions with the successful, low-bid contractor and that all relevant subsurface information will be made available to that contractor without disclaimers.

For anyone knowledgeable about the practice of subsurface construction in the United States, it is fair to say that not everyone is willing to accept the above recommendations. Most owners recognize the need to do subsurface investigations and to use the results of those investigations for project design. Very often, however, those same owners will plot the logs of test borings on a contract drawing or make a book of subsurface information available to the contractor, but almost always that information is disclaimed in some way or the contractor is otherwise put on notice that he or she is more or less at risk when it comes to "differing site conditions." The owner is saying, in essence, that he or she does not want to be bothered by claims for additional compensation during construction and that the contractor is fully expected to carry sufficient contingency in the bid to cover the cost of all known *and* unknown subsurface conditions that might affect the cost or schedule for the project.

In the experience of this author, most projects, both public and private, being constructed in the United States are probably being done on the basis of the above approach (or some variation thereof), rather than on the approach recommended herein, and are usually completed without major problems. To be frank, most ground conditions are not that complex, most projects do not have large amounts of underground work as compared to the overall completed project components, and most contractors are sufficiently familiar with ground conditions in their chosen area of activity to avoid major disasters. Hence, most participants in construction are able to obtain what they want from the process without putting a lot of thought into how best to handle the subsurface aspect of the work. This is not, however, the type of project for which this chapter has been written.

Project owners must recognize that there are certain types of projects that simply cry out for a more sophisticated approach to doing the subsurface portion of the work. These projects can usually be described as possessing one or more of the following characteristics:

1. *Underground work represents a high percentage of the overall project completion cost.* Projects such as underground parking garages, sewage treatment plants, tunnels, dams, and highways contain large and potentially very expensive interfaces with the ground.
2. *Highly complex, difficult to describe, or very poor ground conditions.* Sometimes the ground is so complex that there is a high probability that the contractor who wins the work will be the one who has no idea how difficult or challenging the anticipated problems are and who, obviously, has made no allowance in the bid for those problems. No owner wants to become involved with this type of project more than once.
3. *A complex or highly risky interface with adjacent third parties.* Third parties adjacent to a major construction project expect the project owner to insulate them from negative impacts and are perfectly free and willing to seek judicial relief if that does not happen. In order to protect third parties during construction, the owner needs some measure of control over the

contractor, and how best to obtain that control without interfering unduly with the contractor's ability to do the work is always a major intellectual, legal, and managerial challenge.

4. *Involvement of contractors from beyond the local area.* Large or complex subsurface projects result in a high probability that contractors from outside the immediate area will become involved with the work. Generally, larger and/or complex subsurface projects involve new or different construction practices, larger or deeper holes in the ground, or persons or firms not completely familiar with local ground conditions or construction practices. Hence, the possibility, at least, of something going wrong or of the need to carry a large contingency to cover that possibility increases.

In order to be successful with the above types of projects, it is necessary for the project owner to obtain highly reliable subsurface information that is well described and well presented in the contract documents and which takes into account construction practices that are most likely to provide a cost-effective finished product and which result in minimal impact to adjacent third parties. It is for these types of projects that this chapter is primarily intended.

The next four sections of this chapter (Sections 3.3 through 3.6) contain discussions concerning subsurface investigations, Geotechnical Data Reports (GDR), Geotechnical Interpretation, and Geotechnical Interpretive Reports (GIR). These four sections are intended to introduce the reader to the process of how one goes about investigating the ground and reporting on the results of those investigations. Although the basic approach to report writing described herein is divided into two documents, a data report and an interpretive report, it is entirely possible to combine these two reports into one document that would be referred to simply as the Geotechnical Report. Even if this is done, however, it is still useful to think of the problem of geotechnical report preparation as consisting of two distinct phases involving data accumulation and data interpretation, and that is the basis on which this topic is described herein. It is also sometimes highly beneficial to have one document consisting of nothing but "factual" information that can be transmitted to various parties for various reasons independently of the interpretations derived from those data. Further discussion of these and other issues are presented in the next four sections. Following these four sections are three additional sections entitled as follows and concerning the following topics:

Section 3.7—Information Dissemination—This section provides recommendations about how best to make subsurface information available to prospective contractors. Some discussion is also provided about the effect of the differing site conditions clause on this decision.

Section 3.8—Geotechnical Design Summary Report—Some background information and comments are provided in this section about how to prepare a GDSR as compared to a GIR.

Section 3.9—Geotechnical Baseline Report—A few comments are provided here about a new type of report, called a Geotechnical Baseline Report, being proposed for use on large, subsurface projects. As of the writing of this chapter, the concept for this type of report is still in the formative stage and definitive recommendations for the preparation and use of this report will not be provided. The Technical Committee on Geotechnical Reports of the Underground Technology Research Council has prepared suggested guidelines for preparing Geotechnical Baseline Reports (see Essex, 1997).

3.3 SUBSURFACE INVESTIGATIONS

The primary purpose of a subsurface investigation is to obtain factual information about the distribution and engineering characteristics of soil and rock deposits at a site with sufficient accuracy so that a safe and economical design and a reliable construction proposal for the work are possible. Of particular importance for most sites are the strength and stiffness characteristics of the ground, the flow and pressure distribution characteristics of the groundwater, and the probability of encountering rock or rock-like materials within the limits of proposed construction. The range of strength and stiffness characteristics for soil and rock is large, and the geologic environments that result in soil deposition and/or rock formation can be extremely complex. Hence, there is always a large measure of uncertainty associated with investigation of the subsurface environment.

Without doubt, the soil and rock property with the largest degree of variation is permeability. Many references emphasize that soil or rock properties relating to groundwater flow are the hardest properties to measure, either in the field or in the laboratory, and result in some of the most difficult problems during construction. Water inflow at a construction site is expensive to control and can have significant impact on the construction schedule. In addition, water pressure simultaneously increases the driving force of the ground while reducing the ground's ability to withstand those very forces. A good working knowledge of the groundwater conditions in the vicinity of proposed construction activities is one of the primary objectives of a subsurface investigation.

Although the encountering of rock at a site may or may not result in a problem, the presence of rock adds considerably to the complexity of the subsurface investigation. The misidentification of "bedrock" as compared to boulders or rock fragments in the soil profile is a common cause of construction-related problems, such as difficulty in driving piles or drilling piers, the need for rock excavation, and/or difficulties associated with rock blasting or blasting vibrations. Probably the most complex of all geologic environments relative to variations in both soil properties and the groundwater regime occurs at the soil/rock interface. The weathering of rock also results in a highly complex mixture of soil and rock materials that can be difficult to decipher using conventional methods of subsurface exploration.

All subsurface investigations should begin with a visit to the site, with the intention of answering such questions as:

1. Is the site in its natural state, or has it been disturbed by previous construction activities?
2. Is there standing water at or near the site or other indications of possible groundwater?
3. Are rock outcrops or other indications of near-surface rock deposits observable at or near the site?
4. Do site access restrictions exist relative to conducting the subsurface explorations?

Simultaneously with the site visit, all readily available sources of additional information concerning the site, such as geologic or topographic maps, aerial photographs, soil survey reports, construction case histories, professional papers about the area, historical documents concerning prior site usage, or discussions with local building officials or contractors should be pursued. In general, it is necessary to find out as much about the site as possible before actually starting the subsurface investigation.

For most sites, the issue of subsurface investigation usually revolves around how many test borings to drill, and to what depth. Many guidelines, such as those provided by various government agencies and in various textbooks, provide suggestions about how many test borings to drill per square foot of structure or per foot of pipeline or highway. The final decision with respect to this issue, however, is always a judgment call based on the importance and complexity of proposed construction versus the nature and complexity of assumed subsurface conditions. The more difficult and costly the proposed facility and the more complex the ground conditions, the more extensive the subsurface investigation should be.

It should be noted, however, that the most difficult and costly sites to investigate are not necessarily those with the *worst* site conditions. For instance, foundation design for either dense glacial till or thick deposits of organic peat are equally straightforward, resulting in spread footings for the former and piles or piers for the latter. The most difficult and costly sites to investigate are those that are not *obvious*, such as a multistory building underlain by thick deposits of moderately stiff clay. Without an adequate subsurface investigation and laboratory testing program, the geotechnical engineer has no choice but to employ conservative design practices, probably at great cost to the project. It takes an experienced owner to recognize this situation and to realize that a truly modest investment in additional subsurface investigations can result in large savings in time and money for the project and large reductions in risk during construction.

3.3.1 Scope of Subsurface Investigation

Two of the most important decisions that must be made relative to the scope of the subsurface investigation are the sampling interval and the need for field testing. The frequency and type of samples that must be obtained are a function of project type and ground conditions with the added proviso that these samples will be available for laboratory testing. If only soil and rock classification tests are anticipated, then disturbed sampling procedures are acceptable. If, however, laboratory strength and compressibility tests are needed, then more costly undisturbed samples will be required. Field tests that might be used consist of in-situ vane testing or pressuremeter testing in soil or in-situ pumping tests or the measurement of combustible gas in rock.

In many cases, it is necessary to supplement the above test boring program with other forms of subsurface investigation, such as test pits, groundwater pumping tests, or geophysical investigations. Test pits can be an indispensable method of subsurface investigation, especially in urban areas where multiple episodes of construction can result in thick deposits of miscellaneous fill. Test pits provide for the sampling of a large volume of ground and the exposure of undisturbed soil to observe. Test pits also provide direct evidence of the type and distribution of obstructions that exist in the ground and reveal important facts about possible ground performance characteristics such as sloughing, squeezing, or groundwater inflow in the vicinity of the test pit.

Sometimes, especially for a simple subsurface profile, all that is necessary to understand the groundwater condition at a subject site is to observe the static water level in a series of completed test borings or observation wells. These observations reveal very little, however, about how the groundwater will respond to various construction activities, such as the excavation of deep cuts or the establishment of cut slopes. In order to find out more about groundwater response, it is usually necessary to "stress" the system by conducting a pumping test. By actually pumping the water, a great deal can be learned about the continuity and hydraulic conductivity of various soil strata and, possibly, about the nature of the soil/rock contact. This information can then be used to learn more about the susceptibility of soil to disturbance, the level of difficulty anticipated for various construction activities, and the possibility of negative impact on adjacent third parties. Well conducted and well interpreted groundwater pumping tests are an excellent source of information about subsurface conditions.

There are also many innovative and cost-effective methods of investigating the ground by geophysical means. These investigations can be particularly effective in helping to interpolate subsurface conditions between the test borings and for estimating important soil or rock properties. Although a complete listing of all possible geophysical exploration methods is beyond the scope of this chapter, seismic refraction or reflection surveys, electrical resistivity studies, and some downhole investigations have been shown to be effective in some types of subsurface profiles for civil engineering applications. In general, such applications should be used if they provide useful information at nominal cost relative to proposed design and construction requirements.

Other issues of critical importance to the success of a subsurface investigation include good-quality exploration and sampling services and continuous on-site monitoring of the work by a trained professional representing the engineer. Test boring contractors are typically paid for their work per foot of drilling or in-situ test completed, but this method of compensation can run counter to the need to obtain the best quality information from the ground. In many cases, the work should be performed on a daily rate basis so that the method of compensation will not interfere with the level of care needed to obtain high-quality test results. In any case, thorough observations of field activities must be obtained so that the results of fieldwork can be interpreted in a proper manner.

Probably one of the most useful concepts for conducting a subsurface investigation is to do the work in phases, whereby each phase of the work provides the information needed to plan the next phase of study for site-specific problems. Assume, for instance, that 20 test borings are needed at a site for a complete investigation. The Phase I investigation might consist of five test borings, each with an observation well and each with samples for laboratory testing purposes. Upon completion of this work, it is then possible to establish a working model for the site and to go back to the field with a considerable amount of additional information about soil properties and stratigraphy. Hence, the Phase II investigation becomes not only an opportunity to obtain new information, but also a test of the validity of the model. Depending on the outcome of this "test," the geotechnical engineer is then able to question his or her thoughts about the site and to conduct various "what if" scenarios of important geological assumptions. Many important facts about a site have been obtained by a thorough study of multiple working hypotheses associated with sets of facts about subsurface conditions. Depending on the outcome of this work and the importance of specific conclusions relative to subsurface conditions, it might be advisable to go back for a third phase of even more detailed investigations to resolve any remaining uncertainties.

The bottom line with respect to the issue of subsurface investigation is that the geotechnical engineer must be allowed to do enough such that he or she is sufficiently comfortable with his or her knowledge of subsurface soil, rock, and groundwater conditions to provide reliable guidelines for project design and a confident assessment of the important construction-related issues associated with the work. Obviously, there are no hard and fast rules for accomplishing these objectives. Rather, what is required is a thoughtful, and sometimes imaginative, application of knowledge derived from experience with similar situations. What *is* known is that a well conceived, well implemented, and well interpreted subsurface investigation creates, at least, the opportunity for the geotechnical engineer to produce a satisfactory report for a subject project. Conversely, a poorly performed subsurface investigation literally dooms a project to costly misunderstandings and highly conservative and/or expensive design options for the work. Probably no other area of civil engineering practice depends so heavily on the knowledge, experience, and judgment of the engineer working closely with the client to produce a result that is in the best, long-term interest of the project. It is the height of false economy to scrimp on either the scope of or the quality control associated with the subsurface investigation needed for the proper design and construction of a major underground project.

3.4 GEOTECHNICAL DATA REPORT

The direct outcome of a subsurface investigation is a collection of facts: facts about what was done, how it was done, and what was learned from each activity. It is extremely important to take the time to assemble this information into one coherent document, typically called the Geotechnical Data Report (GDR), for use by all actual or possible participants in a project, including planners, designers, and contractors and, possibly, adjacent third parties or members of the legal profession. Although this process may seem laborious at times, it is crucial to success to have one good source of factual information obtained as a direct result of the subsurface investigation.

The two most important guiding principles for the preparation of a good GDR are that is be (1) thorough and (2) accurate. Each field operation and each laboratory test procedure should be thoroughly described and the outcome of that operation or procedure accurately reported. It is this information from which all subsequent opinions, conclusions, and recommendations about the project are derived. Hence, if, at some time in the future, it can be shown that these data are incomplete, inaccurate, or otherwise unsatisfactory for some reason, then the conclusions and recommendations derived from the data can also be brought into question.

A third guiding principle for the data report is to avoid data *interpretation* in this document. Although there are, again, no hard and fast rules for determining where data accumulation stops and data interpretation begins, it is very important for the author of the data report to be mindful of this distinction and to at least be aware of when the process of interpretation has begun. For instance, the log of a test boring is a factual document, which should clearly and accurately describe what happened during the drilling of *that* test boring, at *that* location, and at *that* time. The production of a soil profile from a series of test borings is, however, an act of data interpretation, involving a good deal of both geological and engineering input and judgment. Although the use of subsurface profiles is entirely justified in helping to understand the data, the production of a subsurface profile is data interpretation, not data accumulation.

A similar approach is necessary with respect to the performance and interpretation of laboratory tests. For instance, the performance of a one-dimensional consolidation test on an undisturbed sample of clay, including description of the test procedure, the load increment, and the rate at which sample deformation occurs, is factual. Any analysis of the properties of that clay as a result of the test, however, such as a derivation of the coefficient of consolidation or a determination of the degree of overconsolidation, is at least the beginning of a process of interpretation leading directly to important conclusions about such things as the magnitude of settlement of a spread footing or the length of time needed for a thick deposit of clay to stabilize under the load imposed by a layer of fill.

Another important reason to provide as clear a separation as possible between the "facts" developed from a subsurface investigation and the interpretations associated with those facts is that multiple interpretations of the facts are possible. It is possible, for instance, for other professional engineers to interpret the subsurface information in a manner that is different from the geotechnical engineer's and some allowance for that possibility must be provided in the normal course of project implementation. Although it is possible to carry the process of multiple interpretations to an extreme (especially if third parties that have strong vested interests in the work that differ materially from the interests of the project owner become involved with the project), it is also true that equally experienced professionals can have equally valid, but different, ways of looking at the same data. Hence, the best way to provide for this possibility is to assemble and to present the data independently of the interpretation.

3.4.1 Structure of the GDR

Given below is a proposed table of contents for a typical GDR with the major subheadings of *Introduction*, *Background Information*, *Field Investigations*, and *Laboratory Testing Program*, together with a listing of typical tables, figures, and appendices (Figure 3.1). Following the table of contents are brief comments about what should be included in each section of the report. Note that the table of contents for the GDR does not contain headings for an executive summary or summary and conclusions. Although an executive summary can be written for any report, such a section for a data report would be no more than a perfunctory description of the scope of work. Usually, this type of information can be included in a letter of transmittal for the report. With respect to a section for a summary and conclusions, it is pointed out that the data report is itself a summary of the subsurface exploration program and that there really are not any major conclusions that can be drawn from the presentation of factual data. All major conclusions obtained as a result of this work should be included in the interpretive report, which is described later in this chapter.

The primary purpose of the *Introduction* is to set the stage for the report and to describe the project and the scope of work. Under "*General*," the major participants in the work and the dates of contracts, meetings, or understandings should be provided. Any information made available to the geotechnical engineer by the client or by client representatives specifically for the purpose of the study should also be listed. Reference is also made in this section to the "Project Locus," which shows the plan location of the site on a topographic map.

The purpose and scope of both the study and the report itself should be

- Introduction
  - General
  - Purpose and Scope
  - Survey Control
  - Report Organization
  - Report Limit
- Background Information
  - General
  - Other Investigations
  - Regional Geological Setting
- Field Investigations
  - General
  - Test Borings
  - Test Pits
  - Observation Wells
  - Geophysical Investigations
  - In-Situ Testing Results
- Laboratory Testing Program
  - General
  - Soil Testing
  - Rock Testing
- References
- Tables
  - Summary of Subsurface Explorations
  - Summary of Observation Wells
  - Summary of Laboratory Test Results
- Figures
  - Project Locus
  - Subsurface Exploration Plan
- Appendices
  - Glossary of Technical Terminology
  - Logs of Test Borings
  - Logs of Observation Wells
  - Geophysical Investigations
  - In-Situ Test Results
  - Laboratory Soil Test Results
  - Laboratory Rock Test Results

Figure 3.1 Proposed Table of Contents—Geotechnical Data Report

discussed. For the study, it is usually best to refer to the written agreement between the geotechnical engineer and the client, which should clearly describe why the geotechnical engineer became involved with the project. For the report, it is highly recommended that the scope of services portion of the agreement be reproduced in its entirety in the report so that no confusion is allowed to exist about the scope and purpose of the geotechnical engineer's services on the project. The "*Introduction*" of the data report should also clearly describe the difference between data accumulation and data interpretation and tell the reader, if possible, where he or she should look to find the follow-up interpretive documents.

The process whereby the field investigations were located, both horizontally and vertically, should be described. For instance, were the locations obtained by field personnel measuring from nearby landmarks or as the result of a carefully controlled surveying operation? Of particular concern for this section would be an identification of the survey datum used to determine ground surface elevations. Many cities have multiple survey datums and some projects go so far as to establish a site-specific datum for the work. No confusion should be allowed to exist in the data report, however, with respect to the survey datum used for the subsurface exploration program.

It may also be useful, near the end of the "*Introduction*," to provide a brief annotation of the table of contents for the purpose of helping the reader locate various items of interest. One brief statement about what is located in each major section of the report is usually all that is necessary under "*Report Organization*."

3.4.2 Limitations

GDRs are not controlled by a long list of limitations, but, even here, some realistic limitations apply. For instance, a statement should be included that the GDR was commissioned by a specific client for a specific purpose and that no other use of the data is authorized. It is also important to describe the degree to which (if at all) the investigation was intended to search for and identify sources of soil or groundwater contamination. If that type of work was not authorized or it is known that work related to site contamination was done by others, then a statement to that effect must be included. Some mention should be made of the time frame in which the work was done and the fact that soil properties and/or groundwater conditions may change seasonally or with the passage of time. Finally, and most importantly, if the geotechnical engineer's access to the site itself or to nearby sites was limited in some way during the work, or some or all of the data were provided by others without adequate field supervision by the engineer, then these facts *must* be pointed out in the report with reference to an appropriate disclaimer.

Under "*Background Information*," the process that was undertaken to locate sources of available information about the site and the results of that process is a *factual* aspect of the investigation. For example, how intensive was the investigation to find old topographic maps, previous aerial surveys, or subsurface investigations from adjacent sites? In one instance, a series of newspaper accounts provided critical information about obstructions discovered at the site but not revealed in the data report. No one knows how hard it will be or how much effort should be expended in looking for available sources of information, but the process used and the effort expended in accomplishing this task should be described.

Any "*Other Investigations*" that were conducted at the site, either for the proposed project or for some other reason, including investigations by the firm given the task of writing the data report or by other firms, should be documented. It is sometimes difficult to know how much credence should be given to investigations by other firms, especially after the passage of considerable time and under circumstances that may not have been conducive to doing a good job, but, in any event, these reports should be investigated for possible clues to important site characteristics. The purpose here is not to utilize these documents for the purpose of drawing definitive conclusions, but merely to point out that other possible "facts" about the site have been discovered that need to be factored into the process of site evaluation.

It is also possible, at this point in the report, to say something about the "*Regional Geologic Setting*" for the site, although it is very easy to say too much about this topic and go clearly into data interpretation. Generally, it is necessary to develop some idea of the subsurface conditions from available geological maps and reports to plan the subsurface investigation. Insofar as this information helps the reader of the data report understand the process of subsurface investigation and data accumulation, then the regional geologic setting should be described in a straightforward, factual manner.

It is under "*Field Investigations*" that the writing becomes somewhat repetitive and based on a plan to reduce the data to writing as quickly and as effortlessly as possible. Reference at the beginning of this section is usually made to a plan of all investigations made at the site. Reference might also be made at this time to an appendix that contains a glossary of technical terms used in the document. Since many readers of a GDR are not trained in geotechnical or geological engineering, a brief glossary of terminology can be helpful in avoiding confusion relative to technical matters presented in the document.

In the "*General*" section under "*Field Investigations*," it is critical to describe how, and to what degree, the subsurface investigation was monitored by representatives of the engineer. As stated above, it is truly false economy to do this work without knowing for certain that it was done in accordance with accepted practice. Even if some problems do occur during drilling (and they usually do), at least the interpreter of the data will be aware of these problems and able to take the precise circumstances of data acquisition into account during data interpretation. It is crucial to both realize and accept the fact that many of the most important observations about a site can only be made in the field by a trained observer during actual performance of the subsurface investigation.

As soon as the introductory material is completed, the author of the GDR begins to describe each field procedure with reference to an appendix and/or a table where the results of that procedure are presented and/or summarized. For instance, under "*Test Borings*," the author of the GDR would explain the manner in which the work was performed, with appropriate reference to any subcontractors used to help complete the investigation. It is also beneficial during each of these write-ups to be able to refer to some standard or guideline for doing the work and to be able to say that all work was, in fact, done in accordance with that standard or guideline. If this is not the case, then a clear and thorough explanation of what was done must be provided.

3.4.3 Format

In the interest of making the report more readable, it is usually best to keep the text fairly brief and to place any detailed explanation of technique at the front of the appendix containing the logs of the investigation. In this way, the reader has the option of searching out additional detail, if required. Tables summarizing the results of sets of investigations can also be very helpful in understanding the data. For instance, it is frequently beneficial to provide a table of test boring results giving the number of the boring, the ground surface elevation, the total depth of drilling, whether or not the boring was equipped with an observation well, and other useful information such as depth below ground surface to the top of rock or the observed thickness of miscellaneous fill. A good rule of thumb for this type of summary is that if the author of the GDR feels a need for such a summary to help him or her understand the data, then a similar presentation would be equally helpful for the reader of the GDR.

Once a format for data presentation is established, it is repeated for each additional body of information. For instance, for test pits and/or the installation of observation wells:

1. The report would contain a brief write-up about the field procedure with reference to some standard, an appendix, and possibly a table;
2. The appendix would contain a more detailed description of the procedure with logs of the actual activity; and
3. A table would contain some useful summary of the data, such as a series of readings made of water levels on different dates, or the depth and character of fill deposits revealed by the test pits.

Each of these and other sections are then written until all of the data have been presented.

One possible variation on this theme is that if a specialist firm is hired to conduct an investigation based on its own proven technique or procedure for doing the work, then that firm would be expected to prepare a report summarizing the results of its work; that entire report would then be included in an appendix. For the table of contents shown above, this might be the case for some geophysical investigation or for the in-situ measurement of stress in a rock mass. Hence, it would only be necessary in the text of the GDR to briefly describe the work and to make reference to the appropriate appendix.

Finally, a format similar to the above can be used to summarize the results of the "*Laboratory Testing*" program. Slight modifications for the laboratory testing portion of the report are that there is almost always a highly precise and well established procedure that can be referenced for conducting each test, and that the laboratory test results are almost always conducive to some form of tabular summary, either individually (based on each test), or as one large table showing all of the test results.

Some effort must be expended to assemble the field and laboratory data obtained from the subsurface investigation into one coherent document for use by all project participants. Many potential problems at a site can be revealed during this work, when the person assembling the data notices some fact or some trend in the data that warrants additional investigation. As an example of this process, a project is recalled where a young engineer was asked to assemble the data from a series of observation wells for a tunnel that ran parallel to open ocean. During assembly of these data, the engineer noticed that almost all of the wells showed readings within a few inches of one another, but that one set of three wells consistently showed variations of one to two feet. Upon further investigation, it was discovered that the larger variations correlated with tidal fluctuations, suggesting a previous rock structure influenced by seawater, and further suggesting the possibility that salt water might flow into the tunnel during construction. Still further investigation suggested that this was, in fact, the case, leading to a major conclusion and several significant recommendations for project design and construction. Such is the process whereby "unknown" ground conditions become "known," and claims of "differing" site conditions are avoided. The second half of the above process is, however, data interpretation, which is the topic of the next section of this chapter.

3.5 GEOTECHNICAL INTERPRETATION

Geotechnical interpretation involves four major topics: subsurface conditions, design recommendations, construction considerations, and third-party impacts. Figure 3.2 is a simple flowchart that can be used to help evaluate these important geotechnical issues for a subject site. The chart begins with subsurface conditions and project requirements, proceeds through an evaluation of ground behavior/ground control scenarios to the consideration of important design, construction, and third-party impacts, and ends with the contract document. Each aspect of the chart is important for success, and each part of the chart results in important conclusions that must be carried forward to the next part. It is also important to realize that the use of this chart is an iterative process, whereby negative implications identified at any point in the process might be reason to go back in the chart and modify certain aspects of the project to see if a more favorable design or construction outcome is possible. In this way, a variety of approaches to the work can be evaluated, possibly leading to a single best approach, but definitely resulting in a large number of topics and issues that should be discussed in the Geotechnical Interpretive Report. Given below are brief comments about how someone would go about using this chart to assist with the evaluation of important geotechnical criteria for a project.

All decisions about the geotechnical aspects of a project begin with a thor-

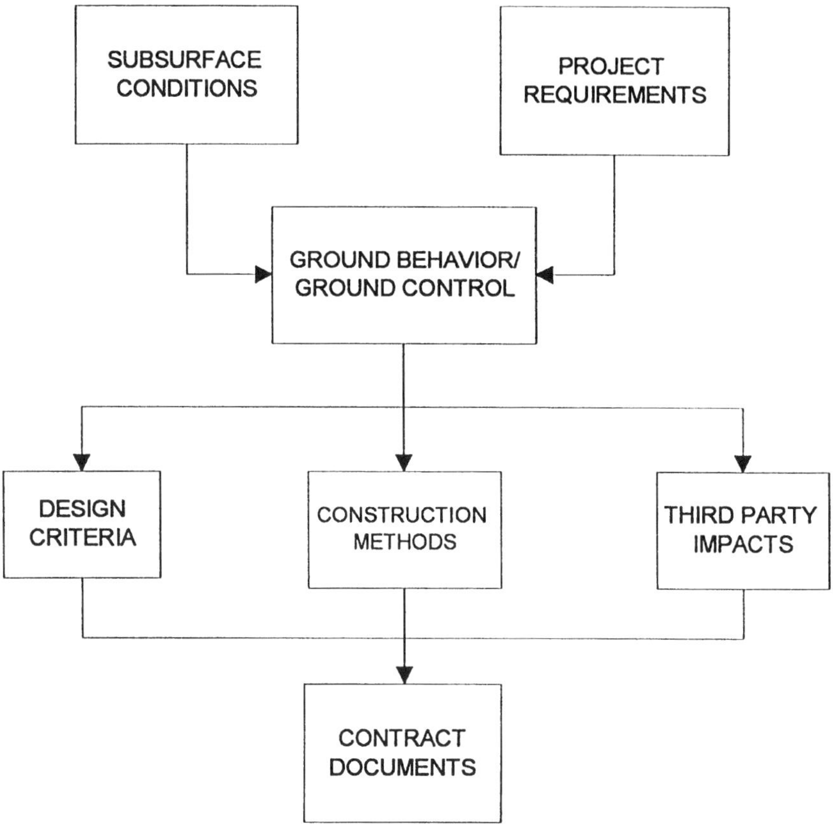

Figure 3.2 Flowchart for evaluating geotechnical issues

ough evaluation of subsurface conditions. Beginning with the subsurface data, the geotechnical engineer develops a reliable model of the subsurface which, to the greatest possible extent, accurately describes the in-situ condition of all soil and rock deposits and the superimposed groundwater regime. This process almost always begins with the plotting of one or more subsurface profiles or contour maps based on site-specific information and on the critical evaluation of other available information.

3.5.1 Subsurface Profiles

Subsurface profiles are an indispensable part of interpreting subsurface information. The profiles should be plotted with important field and laboratory test results shown on the drawing and to a scale that minimizes horizontal-to-vertical exaggeration. At least once during this process, the geotechnical engineer should plot the data to a true scale so that at least one true representation of the scope of the test boring program is available prior to making final conclusions. Otherwise, all data and other input from geological or mapping studies and/or other investigations, either at the subject site or at nearby sites, should be incorporated into this interpretive drawing. When this drawing is finished, geotechnical "thinking" begins.

The first thought process involves the drawing of lines between the borings; some knowledge of the geologic environment is essential to this undertaking. The number and variety of geologic environments that result in soil deposits is large, and environment is associated with stratigraphic sequences that must be incorporated into the profile. For instance, soil deposits can be created by glaciers, by stationary or running water, by the wind, or as a result of in-situ weathering. With respect to running water, an array of both short-term or long-term phenomena are possible, involving streams, rivers, lakes, and/or oceans. With respect to in-situ weathering, probably the single most important variable is whether or not the site has been glaciated. Glaciers remove most, but not all, of the effects of in-situ weathering, leaving a condition at the soil/rock interface that is dramatically different from unglaciated terrain. Windblown deposits are generally characterized by fine-grained and highly uniform deposits of relatively low density.

Sometimes, the best way to develop a subsurface profile is to start at both the top and the bottom and work your way into the middle. If a layer of fill is present, then that layer should be isolated and described. If rock is present, then the top of what might be called "good" rock should be identified. Following this identification, the nature of the soil/rock contact is established so that the bottom of the soil portion of the profile can be determined. With the top and bottom of the natural soil identified, the process of dividing this deposit into layers, lenses, seams, and pockets begins. Lake deposits imply "layer-cake" stratigraphy, whereas stream deposits are much more variable. Deltaic deposits can be either uniform or variable, depending on the size of the delta relative to the size of the site. Glacial deposits can be highly variable, both from stratum to stratum and/or within specific strata. Soil deposits that result in boulders or rock fragments need special attention relative to describing the size and anticipated frequency of these almost always troublesome occurrences.

If the primary objective of the profile is the rock, rather than the soil, then an entirely different method of interpretation is necessary. Whereas soil deposits are generally interpreted relative to their horizontal distribution in space, rock masses represent a three-dimensional snapshot of everything that has happened to a particular volume of ground over a period of tens or even hundreds of millions of years. Every lithology, every stress regime, every weathering epoch, and/or every faulting event is reflected in the rock in every conceivable orientation. Most distressing is the possibility that vertical test borings might not be the best way to investigate a particular rock mass, as the most important characteristics of the rock might also be oriented vertically. Hence, the data derived from geologic mapping, geophysical investigations, or other forms of site investigation might be even more important than the test boring results relative to the interpretation of rock masses.

Superimposed on all of the above is the groundwater regime. Nothing affects soil and rock properties more than being submerged in water, and nothing affects subsurface construction operations more than the encountering of "unanticipated" groundwater events. Deciding on the true character of the groundwater condition is one of the most challenging aspects of geotechnical engineering, including both the pressure distribution of the water in the ground and its potential flow characteristics. Complex soil profiles must be evaluated relative to the possibility for perched water, confined and unconfined aquifers, excess hydrostatic pressure, and water pressure phenomena existing both within and to some distance outside the envelope of proposed construction operations. With respect to groundwater flow characteristics, no other naturally occurring material has a greater range of properties than does the coefficient of permeability of naturally occurring soil or rock deposits. Values of permeability ranging from 10^2 to 10^{-7} cm/sec are common, representing a range of 10 orders of magnitude. Many engineers feel fortunate if the coefficient of permeability can be established within one order of magnitude, but imagine if the same situation were true for concrete or steel. How difficult would it be for an engineer to design a structure if he or she did not know whether the concrete had an allowable strength of 400 or 4,000 psi? This is the challenge of evaluating problems associated with the flow of water through the ground.

Despite all of the above, the geotechnical engineer must create some model, some description, and/or some characterization of the ground that can be carried forward into design. Reasonable and appropriate assumptions must be made about soil stratigraphy, soil and rock properties, and groundwater characteristics such that a good representation of the ground is possible. In the absence of specific information, reasonable assumptions based on previous experience must be made about the thickness of a deposit, the character of a deposit of fill, or the occurrence of boulders in a glacial till. Hopefully, the subsurface investigation will have been sufficiently thorough that the use of such "assumptions" can be kept to a minimum, but the development of the subsurface condition portion of a report is not and cannot be based entirely on scientific principle. At best, this work is the output of a process involving judgment, knowledge, experience, and maybe a little luck, which results in a good working model for, but not an absolutely perfect description of, the ground.

3.5.2 Project Requirements

In addition to an evaluation of the ground, the geotechnical engineer must clearly describe what the project owner wants to accomplish with the project in a section called "*Project Requirements*." Usually, the owner is interested only in the long-term aspects of project completion, but the geotechnical engineer must also be worried about the short-term aspects of project construction and/or third-party impact. Both sets of project requirements are important, both sets can have important cost and scheduling impacts on design, and both sets must be given serious consideration by the geotechnical engineer.

Under long-term (permanent) project requirements, the owner is usually concerned about four major factors:

1. The proper size, shape and location of project facilities;
2. Structural integrity;
3. Satisfactory utilization and service life; and
4. Cost-effective and rapid implementation.

Most owners want a project that costs no more than is necessary, that can be built in the shortest possible time, that works well and lasts a long time, and that fulfills the owner's needs and desires for wanting to build the project in the first place. Even with respect to this issue, however, underground construction is different from aboveground work, as many of the facilities are a means to an end rather than final, useful product. For instance, how does an owner feel about the need for a pile foundation? Surely, such a foundation can be crucial to success of the work, but is it really necessary? Is all of the time, money, and effort associated with the installation of a pile foundation truly required, or is it being provided merely to minimize the geotechnical engineer's liability? Similarly, with many other aspects of underground design such as waterproofing, drainage, seismic forces, lateral restraint, or allowances for long-term deterioration, where does good judgment stop and overly conservative design begin? Somehow, the geotechnical engineer and the client working together must resolve these issues and develop a plan for the acceptable, long-term performance of the completed project.

When the above work is finished, the geotechnical engineer must evaluate all of the temporary facilities that are required in order to build the final facilities, most of which involve making a hole in the ground. All foundations, shafts, tunnels, walls, dams, slopes, and other such facilities involve the excavation and support of soil and/or rock, usually in the presence of groundwater. During this process of excavation and support, it is necessary to control the ground so that no more ground is removed than is represented by the volume of required excavation. Lost ground or excessive water infiltration can add greatly to the cost of or the time required to do the work, and can result in detrimental impact to adjacent properties. Ground excavation, ground control, and ground support represent important aspects of the geotechnical part of the project and form an important part of the description of project requirements.

3.5.3 Ground Behavior/Ground Control Scenarios

Referring back to the flowchart, the geotechnical engineer is now equipped with a good model of ground conditions and a good working description of both the long-term and short-term project requirements that must be fulfilled. With this information, he or she is now ready to evaluate possible ground behavior/ground control scenarios for the work. One of the keys to success for geotechnical design is to anticipate how the ground will behave when it is disturbed and to develop the most straightforward and cost-effective method for controlling that behavior. Scores of examples exist as to how this process works. Take, for instance, a simple spread footing. When the footing is loaded, the soil will tend to move away from below the footing. Hence, the footing pressure must be controlled such that the soil below the footing does not fail and/or settle more than the structure can tolerate. As the soil becomes weaker and/or more compressible, design criteria for the footing may become so onerous that a mat foundation or piers or piles become more feasible. Sometimes, the deep piles become so expensive that the engineer may point out to the owner that an additional level of basement space might result in sufficient load reduction that a floating mat foundation could be used, resulting in a change in project requirements. Or, maybe, the ground could be drained, grouted, or compacted, resulting in a positive improvement of the subsurface condition and allowing the use of a higher-capacity foundation at a higher elevation. Each design scenario requires an iterative evaluation of subsurface conditions, project requirements, and/or ground behavior/ground control responses. Each evaluation also results in some output that may need to be discussed in the Geotechnical Interpretive Report.

Going back to the example above, suppose that the project owner accepts the suggestion for a second level of basement with a floating foundation, but the deeper excavation now makes construction dewatering necessary. Dewatering means that additional soil parameters must be evaluated and additional consideration given to third-party impacts. Suppose the site is underlain at shallow depth by a thick deposit of highly pervious sand, or it becomes known that a service station located one block away has allowed gasoline to infiltrate into the ground. Does the shallow foundation still make sense in the face of these costs/risks? Each option is evaluated for various ground improvement scenarios and/or possible changes in project requirements, design criteria, construction methods, or third-party impact until one or more possible solutions to the problem are developed. During this time, it might be beneficial to talk with contractors about local practice, to prepare comparative cost estimates or construction schedules, and/or to try to quantify specific project risks for the preferred alternatives. Eventually, one approach will be chosen for presentation in the report, although favorable options to that approach could also be discussed. The purpose of the exercise of geotechnical interpretation is to provide the input, the thoughts, and the recommendations that will be presented in the Geotechnical Interpretive Report, as discussed below.

3.6 GEOTECHNICAL INTERPRETIVE REPORT

The Geotechnical Interpretive Report (GIR) is written by the geotechnical engineer for use primarily by the project designer, and seeks to fulfill three important objectives:

1. The description of all relevant subsurface soil, rock, and groundwater conditions at a subject site;
2. The recommendation of the best technically feasible and most cost-effective design criteria for each subsurface project element; and
3. The identification of and recommendations for dealing with and/or mitigating important construction considerations and third-party impacts.

Given below is a proposed table of contents for a typical GIR. Following the table of contents are comments about what should be included in various sections of the report.

An executive summary is appropriate for the GIR, wherein all major thoughts, opinions, observations, conclusions, and recommendations developed as a result of the work can be summarized. This section should be written only when the remainder of the report is finalized; it should contain a specific admonition about the need to refer to both the GDR and the entire text of the GIR for more detailed information. It is also very important not to introduce any new thoughts or new conclusions into an executive summary, but to make certain that it is *just* a summary of ideas expressed in the body of the document.

Under "*Introduction*," the same basic subheadings are used to introduce the same basic thoughts as was done for the GDR. Under "*General*," the major players are introduced, the project is identified, reference is made to the "Project Locus," and the dates of specific scoping documents, meetings, discussions, and the like are provided. Under "*Purpose and Scope*," both the purpose and scope of the entire study and of the GIR are described; some discussion about how the GIR is organized is provided under "*Report Organization*." Finally, significant limitations are established under "*Report Limitations*," as discussed below.

3.6.1 Limitations

Under "*Limitations*," it is important to note, as was done for the GDR, that the GIR is prepared for a specific client, at a specific site, and for a specific purpose and that no other "use" of the document is authorized. Also, the degree to which the GIR is intended to cover the issue of site contamination must be clearly and definitively established at this point in the report. As was stated for the GDR, it is not the intention of this chapter to discuss the issue of "contaminated" sites; reference should be made to other sources for that information. Finally, the writer of the GIR must explain his or her access to and understanding of the data on which the interpretive report is based. Hopefully, all of the data were accumulated by persons employed by the same firm or by firms in which the writer has a high degree of confidence. If this is not the case, then some reference to that fact must be made.

Unlike the GDR, however, there is a long list of other possible limitation statements that may or may not apply to interpretive reports. Quite often, these statements are referred to in a derogatory manner, as if the geotechnical engi-

- Executive Summary
- Introduction
  - General
  - Purpose and Scope
  - Report Organization
  - Report Limitations
- Project Requirements
  - Project Status
  - Proposed Facilities
  - Third-Party Facilities
- Site Conditions
  - General
  - Site Reconnaissance
  - Geologic Setting
  - Subsurface Profile
  - Groundwater Conditions
- Design Recommendations
  - General
  - Foundation Design
  - Earthwork Requirements
  - Third-Party Impacts
  - Construction Monitoring
- Construction Considerations
  - General
  - Excavation
  - Construction Dewatering
  - Excavation Support
  - Third-Party Impacts
- References
- Tables
  - Summary of Field Investigations
  - Summary of Laboratory Investigations
- Figures
  - Project Locus
  - Subsurface Exploration Plan
  - Subsurface Profile—General
  - Subsurface Profile—Specific
  - Project Layout Drawings
  - Design Details
- Appendices
  - Glossary of Technical Terminology
  - Design Investigations by Others
  - Recommended Technical Specifications

Figure 3.3 Proposed Table of Contents—Geotechnical Interpretive Report

neer is somehow trying to avoid responsibility for his or her services; but nothing could be further from the truth. What has happened is that, over time, the courts have ruled that geotechnical reports could be used in ways for which they were not intended unless specific limitation clauses were included in the text. Hence, various limitation statements, as presented below, have been developed for use in a GIR for the protection of both the geotechnical engineer and his or her client from unwarranted claims by the contractor or by any number of other parties who may obtain copies of the GIR for various reasons. (Note that other versions of these and other such statements are also available from ASFE.)

Subsurface Information

The boring logs and related information depict subsurface conditions only at the specific exploration location and at the particular time designated on the logs. Subsurface conditions at other locations may differ from conditions occurring at the boring locations. Also, the passage of the time may result in a change in the subsurface condition at a boring location.

Water level readings have been made in drill holes and observation wells at times and under conditions stated in the boring logs and observation well reports. These data have been reviewed and interpretations made in the text of this report. However, it must be noted that fluctuations in the level of the groundwater may occur due to variations in season, rainfall, tides, temperature, and other factors not evident at the time measurements were made and reported herein.

The stratification lines designating the boundary between soil and/or rock types on the logs of borings and on the geological profiles represent approximate boundaries. The transition between materials may be gradual, and one or more strata may be absent at specific locations.

Foundation Design

This foundation investigation has taken into consideration the building layout, column loadings and structural configurations as shown on Drawings "______________," prepared by "______________," entitled "________________________," dated "_____." When and if further information is developed by the Architect and/or Structural Engineer concerning final design for this structure, then the recommendations and design criteria presented herein should be reviewed by the geotechnical engineer for continued applicability. In addition, it is strongly recommended that the geotechnical engineer be given the opportunity of reviewing final project plans and specifications in order to check that earthwork and foundation recommendations are properly interpreted and implemented in the contract document.

Cost estimates have been prepared solely for the purpose of assisting the project owner in comparing foundation alternatives and assisting with the establishment of design criteria. No other use of this information is authorized. In particular, cost estimates prepared by prospective contractors for the actual construction of proposed facilities must be based on factors such as labor rates, material costs, productivity assumptions and overall market conditions that are applicable only at the time of construction.

Construction Considerations

The following sections of the report include comments related to excavation, dewatering, lateral support, foundation construction, earthwork and other geotechnical engineering aspects of proposed construction. The section is written primarily for the engineer responsible for the preparation of plans and specifications. Since this section identifies potential construction problems related to foundations and earthwork, it may also be of use to owner's representatives involved in the monitoring of construction activity.

The following sections of the report include comments related to excavation, dewatering, lateral support, foundation construction, earthwork and other geotechnical engineering aspects of proposed construction. The section is written primarily for the engineer responsible for the preparation of plans and specifications. Since this section identifies potential construction problems related to foundations and earthwork, it may also be of use to owner's representatives involved in the monitoring of construction activity.

Prospective contractors for the project must evaluate construction-related issues on the basis of their own knowledge and experience in the (local area), or on the basis of similar projects in other localities, taking into account their own proposed construction methods and procedures. In addition, construction activities related to excavation and lateral earth support must conform to the safety requirements of OSHA and other applicable municipal and state regulatory agencies.

Foundation recommendations provided herein are based, in part, on subsurface conditions inferred to exist between the actual locations of subsurface explorations. If changes become evident during construction that differ materially from those described herein, then it may be necessary to re-evaluate these recommendations.

It is recommended that geotechnical engineering personnel be present during construction to provide monitoring services such as observing compliance with design concepts, specifications, and recommendations and assisting with the evaluation of construction changes that may be needed in the event that subsurface conditions are discovered that differ materially from those anticipated prior to the start of construction. The rapid and forthright response to a perceived "differed site condition" can be a crucial element in the successful resolution of a possible claim.

Technical Terminology

Not all readers of geotechnical reports are familiar with the geological and/or engineering words used in these reports. Please refer to Appendix A for a listing of words used in the documents with associated definitions. If you, as the reader, are confused about or would like additional clarification of any aspect of this document, then please do not hesitate to make the necessary contact.

Clearly, not all of the above paragraphs need to be included in every report, but they do make it obvious that a GIR is not a statement of fact, but, rather, an accumulation of professional thoughts, opinions, judgments, and recommendations about the subsurface aspects of a project. It is a professional document, prepared on the basis of professional judgment and experience, and provided for the benefit of a specific client with a specifically defined purpose in mind.

Under "*Project Requirements*," the geotechnical engineer must clearly establish the status of the project at the time of writing the report, clearly describe his or her understanding of the facilities proposed for construction, and clearly express his or her knowledge of adjacent third-party facilities. Some projects undergo major changes *after* completion of the geotechnical report. Building heights are increased, shafts moved, vertical or horizontal tunnel alignments altered, or slopes changed, and the impact of those changes must be identified relative to the geotechnical input and associated recommendations. A conscientious evaluation of nearby buildings and utilities is now also an important part of the subsurface investigation and the associated geotechnical interpretation.

Under "*Site Conditions*," reference must be made to the Geotechnical Data Report and all other sources of information used to develop as complete a picture as possible of the subsurface conditions. A brief write-up is provided about what is observable as a result of a site visit, and as much information as possible is provided about geologic history, stratigraphy, and groundwater conditions. Following this introductory material, detailed discussions are provided about the subsurface profile(s) and the results of specific observations of, and/or the results of tests conducted on, the groundwater regime. In addition to general information about the site, large projects frequently include several project facilities that should be carefully and thoroughly described independently of the rest of the project. Examples of these facilities would be large entrance shafts on a tunneling project, or high-rise towers associated with a low-rise structure.

3.6.2 Design Recommendations

Given all of the above as background information, it is now necessary for the geotechnical engineer to set forth the "*Design Recommendations*" for the project. What type of foundation is needed, and what are the important design criteria associated with that foundation? What design criteria are appropriate for floor slabs, pavements, both long-term and short-term excavation support, and/or permanent or temporary cut or fill slopes? Each of these issues must be presented and discussed so that the project designer can utilize appropriate criteria for design activities. Quite often, the project designer will request associated technical specifications, which could be included in appendices of the report. If the geotechnical engineer uses subconsultants to help with specific aspects of the project, such as seismic design or the design of deep shaft linings, then the results of those studies would also be included in appendices. The provision of accurate and appropriate criteria for the design of all proposed below-ground facilities is the single most important output of the Geotechnical Interpretive Report.

Once all of the design criteria for the proposed project have been established, the impact that construction might have on adjacent, existing third-party facilities must be studied. To do this, it is necessary to invest some effort in learning about these facilities and establishing their sensitivity to possible ground movement. Old utilities with long histories of distress are obvious red flags. Likewise, structures with masonry foundations or shallow foundations immediately adjacent to the property boundary need special consideration. Some of the most troublesome facilities can be large sewer or water lines that are difficult to shut off and can have an impact on a large area outside the immediate vicinity of the site.

Since the site might experience changes relative to what is reported in the GIR, and because subsurface construction can be highly sensitive to construction means and methods (see Chapter 6), it is generally necessary to recommend some type of monitoring program that is compatible with known subsurface conditions and proposed construction activities. These monitoring programs can vary from simple observation of construction activities to highly sophisticated instrumentation programs. The key to success for instrumented observations is to learn how the soil or rock is responding to construction activities and to control that response *before* it causes problems for the site or for third parties. Almost always, if the purpose of the program is only to observe actual problems, then it is usually too late to make meaningful improvements in the work. Observations of actual utilities or structures should be used primarily to prove that they did *not* move or were *not* adversely affected by construction.

Unfortunately, it is sometimes difficult for the geotechnical engineer to convince a project owner of the value of field observations, but an appropriate recommendation for such work must still be made in the report. Then, even if the owner decides not to do the work, he or she would be unable to claim that no recommendation was provided. More serious subsurface problems have been avoided by skilled observations in the field than by any other input to the process of geotechnical engineering.

3.6.3 Construction Considerations

As has been stated elsewhere in this chapter, the geotechnical work does not stop with the provision of design recommendations including suggestions for the monitoring program. In order to build an underground facility, it is necessary to excavate the ground, to control the ground during the process of excavation, and to support the ground during construction of the final structure. Each of these issues also includes important geotechnical considerations, and each must be addressed to some degree in the "*Construction Considerations*" section of the geotechnical report.

Under excavation, the type of ground at the site is characterized relative to its ability to be excavated and its sensitivity to the excavation process. Dense glacial till with boulders will obviously require different equipment to excavate than will loose, silty sand, but the sand will be much more sensitive to disturbance. For really sensitive soils, concrete "mudmats" or graded filters may be necessary to protect the bearing surface. Near the top of rock, highly irregular surfaces are possible, with the potential need to remove rock. If blasting is required, then another entire set of requirements associated with blasting and blasting vibrations must be addressed.

Construction dewatering has been the foremost method for stabilizing soil below the water table, primarily because it provides a demonstrable improvement in soil behavior at relatively low cost. The possible detrimental impact of soil settlement or ground contamination caused by dewatering, however, has resulted in construction dewatering frequently being thought of as too risky. Sometimes, it is better to control the ground by preinstalling cutoff walls or by grouting the ground to improve its behavior. A large number of these "ground improvement" technologies, in addition to construction dewatering, are now available for use under appropriate circumstances.

As was discussed earlier, the geotechnical engineer must also think about the need for excavation support. This type of support must be strong enough to provide a safe working environment and stiff enough to prevent detrimental ground movements. For simple sites, it is possible for the contractor to provide his or her own design for excavation support in accordance with safety regulations. As the opportunity for real damage or risk goes up, however, then the need for the project owner to become involved with this process does also. One of the best reasons to provide design criteria for excavation support in the GIR is to make certain that the contractor has enough money in the bid to do this work. After award, it is still possible to negotiate with the contractor for the installation of a different, but still satisfactory, type of excavation support, if necessary.

Although it is necessary for the geotechnical engineer to discuss important construction-related considerations in a typical GIR, it is equally necessary for the geotechnical engineer not to go too far and provide detailed recommendations regarding construction means and methods. As discussed in Chapter 6, the traditional approach is for the contractor to be solely responsible for selection of the most appropriate means and methods for constructing the project. This issue goes to the heart of the contractor's responsibility for site safety and the right to generate as much profit as possible based on his or her efficient implementation of the work. Once again, however, the issue of construction means and methods assumes an entirely different emphasis for subsurface projects as compared to aboveground work.

As the ground conditions become more difficult, the construction operations more complex, and the third-party impacts more serious, then the need for the owner to exercise more control over the project also increases. As this happens, then it becomes increasingly necessary for the owner to possibly abandon the writing of a traditional GIR for the writing of a more comprehensive type of report, referred to as a Geotechnical Design Summary Report (GDSR). This change in emphasis and the basic format for a GDSR are discussed in Section 3.8. With respect to this discussion of GIRs, however, it is assumed that no attempt will be made in the GIR to discuss the specific means and methods of doing the work, even though some comments are provided about important construction-related ground conditions at the site.

The demands made on a geotechnical engineer for the accurate and thorough interpretation of subsurface information—even as the sites become more difficult, the designs more complex, the construction operations more sophisticated, and the third-party impacts more sensitive—have consistently increased over time. Despite the very real uncertainties that exist about underground conditions and the incessant pressure to decrease budgets at all costs, it is still somehow expected that the GIR will accurately characterize the ground, recommend the best technically feasible and most cost-effective design, discuss which underground construction scenarios may or may not be appropriate for the work, and fully identify and make recommendations for avoiding detrimental third-party impacts.

One of the primary functions of geotechnical interpretation is to check for possible site problems that might not be obvious based on a quick review of subsurface conditions. Some of the worst geotechnical "disasters" occur when two or more "unrelated" events occur at a time or in a manner that, although not obvious at the time, might have been avoided with a little more thought. Situations can occur such as a quick condition-causing settlement, which breaks a gas main, which causes a fire, which—you know the story. One of the most important services provided by the geotechnical engineer is, therefore, detailed *thinking* about the project—thinking about the site, thinking about design, thinking about construction, and thinking about third parties. This thinking must be based on experience, good data, valid theory, and a solid knowledge of local construction practices.

With respect to risk, the geotechnical engineer's foremost objectives must be the identification of realistic project risk, the formulation of appropriate strategies for avoiding as much of that risk as possible, and/or for adequately discussing all remaining risk in the GIR. The best way for a geotechnical engineer to solve a problem at a site is to *avoid* it; this solution must be used as often as possible. When a problem cannot be avoided, such as by moving a structure or by modifying the ground, then it must be identified, assessed, and addressed in the contract document. Obviously, such a statement is easier to make than it is to accomplish, but the best way for a geotechnical engineer to avoid negative impact to his or her practice is to try as hard as possible to avoid negative impact to the project.

In summary, the GIR is written by the geotechnical engineer primarily for use by the project designer. The primary objectives of a GIR are to clearly and forthrightly describe subsurface soil, rock, and groundwater conditions, and to provide cost-effective and technically satisfactory design recommendations for the work. A very important secondary objective of a GIR is to point out and make recommendations for avoiding serious detrimental construction and/or third-party impacts at a site without providing detailed recommendations for specific construction means and methods. Throughout report preparation, the geotechnical engineer's most important input to the process of "design" is comprehensive thinking about the work such that all major subsurface problems are identified and addressed. Most adverse consequences at a construction site do not result from poor solutions, but rather from problems that were never identified and for which no solution was even proposed. Insofar as the geotechnical engineer is able to avoid that "problem," then his or her reports will be satisfactory.

3.7 INFORMATION DISSEMINATION

Having finished a discussion of subsurface investigation and report preparation, it is necessary to revisit the issue of how this information should be made available to prospective contractors. Depending on the size and complexity of the project and owner preferences, the owner is now in possession of a single report, called simply the Geotechnical Report, or, for larger, more complicated projects, two reports, entitled the Geotechnical Data Report (GDR) and the Geotechnical Interpretive Report (GIR). As a minimum, it is recommended that this information, in whatever format, be made available for inspection by prospective contractors for the work. This is the case whether or not the contract contains a differing site conditions clause and whether or not the owner tries to disclaim the information in some way. Courts have ruled time and time again that the owner has an obligation to make the contractor at least aware of all information that the owner has in his or her possession that could affect the contractor's cost or time requirement for doing the work. No long-term benefit can possibly accrue to the owner by trying to hide or otherwise limit distribution of information about subsurface conditions that exist at a site intended for construction. Full disclosure of all such relevant information is always the best course of action.

The next issue that must be addressed, however, is whether or not some of this information should be given the status of a "contract document." If the contract does contain a differing site conditions clause, then it must be assumed that the project owner, for whatever reason, has decided to "share the risk" of subsurface conditions with the contractor and, in doing so, has assumed some responsibility for describing those conditions. Under this scenario, the preferred approach is that the owner should make the GDR part of the contract documents, without reservation and without trying to disclaim responsibility for the accuracy of information included in that document. If the owner wishes to share the risk of subsurface conditions, primarily for the purpose of avoiding large subsurface contingencies in the bid, then he or she must be prepared to produce a GDR in which he or she has sufficient confidence that it can also be shared with the contractor.

3.7.1 Contracting Practices

In order to put the above recommendation into perspective, it is necessary to digress for a moment to a discussion of contracting practices in general. The most basic definition of an ideal contract is a document that clearly and unambiguously establishes each party's obligations and responsibilities to the other party. Hence, the primary objective of a contract is to be definitive and precise and to avoid, to the greatest possible extent, opportunities for misunderstanding and/or controversy. A second important concept is that a contract document is almost always written by one party in an attempt to obtain something from another party. Hence, when all is said and done, one party ends up with something (a long-term benefit) as a result of the activities (usually short-term) of the other party. Following this theme, it is the responsibility of the receiving party to write the contract, and it is that party's responsibility for making the contract definitive and precise. If, at some time in the future, a judge or some other body of dispute resolvers finds room in a contract for interpretation, then the responsibility for and the consequences of that interpretation will be assigned to the party that wrote the document.

If one tries to apply the above two concepts to underground construction, one is immediately faced with a serious problem: the fact that it is simply not possible, under most circumstances, to be definitive and precise when it comes to a description of subsurface conditions and/or of the consequences that those conditions might have on project delivery. Hence, the differing site conditions clause offers a source of contractual relief that addresses this uncertainty; this discussion is about how one goes about limiting the impact of that clause without eliminating it entirely. The question, therefore, is what contracting practices should be implemented in order to adequately share the risk of differing site conditions without one party to the contract obtaining some undue advantage in the process.

Recognizing the above, the owner must retain persons or firms in whom he or she has the utmost confidence to investigate the ground and from whom he or she fully expects to obtain a body of "factual" information that can be transmitted to prospective contractors. If this is done, then the contractor will be under a contractual obligation to take that information into account in the preparation of the bid and in planning the means and methods for the work. If the provided information is reliable, that fact will go a long way toward limiting the owner's exposure to claims of "differing site conditions" since, in order to be successful in pursuing a claim of differing site conditions, the contractor must demonstrate that the discovered subsurface condition differed *materially* from what was described in the GDR and that difference had a *material* impact on the work. Hence, reliable geotechnical data should be a part of any construction contract that also contains a differing site conditions clause.

3.7.2 The GIR and the Contract Document

Opinion is not so universal, however, with respect to the sharing of the interpretive portion of the geotechnical information. Some persons believe that all of the opinions and recommendations in the GIR should also be made part of the contract documents, but it must be remembered that the GIR was prepared by the geotechnical engineer primarily for the use by the project designer. As such, not all of the geotechnical engineer's recommendations will necessarily be accepted by the designer; the geotechnical engineer may not be allowed to participate in the preparation of the contract documents; and/or the geotechnical engineer may not be allowed to visit the site during construction. To simply include the GIR as a contract document, without any of the above follow-through activities by the geotechnical engineer, could be troublesome and could result in increased potential for claims of differing site conditions based primarily on "confusion" in the contract document rather than because of any real problem with the subsurface data or the geotechnical report. Hence, the decision to include or not to include the GIR in the contract document depends on many variables that are not susceptible to simple yes or no answers. There are basically two options for use of the GIR relative to the contract document:

1. The GIR can be "made available" to prospective contractors but not made a part of the contract documents. In this case, the GIR exists as a document that has been brought to the contractor's attention, but is restricted in its scope and use by appropriate limitation clauses used in the report itself; or
2. The GIR can be made a part of the contract documents just like the GDR.

In light of all of the discussion and all the concepts that have been introduced so far in this chapter, and in light of the risks and responsibilities assumed by the owner as a result of including a differing site conditions clause in the contract document, it is recommended that the owner most often implement the first option above, which is to make the GIR *available* to the contractor but to *not* make it part of the contract document. In this way, the owner:

1. Makes the contractor aware of his or her project-specific interpretation of subsurface conditions.
2. Makes the contractor aware of how the data were utilized for the purpose of project design, even though the contractor has no direct, vested interest in that information. At least, and as a minimum, the contractor will now be informed about why certain facilities and/or project elements are required for construction.
3. Makes the contractor clearly aware of the construction considerations and third-party impacts that the owner feels are important or that need to be taken into account during construction. Hopefully, and to the best of the owner's ability, these considerations and/or concerns will have also been addressed in some appropriate manner in the specifications and drawings portions of the contract document.

Under the majority of circumstances, the above is what the owner should accomplish. The owner has provided a reliable package of data about subsurface conditions; he or she has informed the contractor in writing about how that information was taken into account in project design; and he or she has informed the contractor (again in writing) about other possible problems and/or situations that the owner believes "should" be taken into account by the contractor during bidding without tying the contractor's hands, without eliminating the contractor's opportunity to use innovative or cost effective solutions, and/or without telling the contractor specifically how to control the means and methods for doing the work. If all of the above are combined with a sincere desire on the part of the owner to work together with the contractor to do the work and with a good approach to the schedule of payments such that some flexibility is provided for compensating the contractor for truly unknown subsurface conditions, then the owner has done about the best job he or she can for bringing the subsurface aspects of the project in on time and within budget.

3.8 GEOTECHNICAL DESIGN SUMMARY REPORT

As was mentioned at the end of Section 3.2, it is necessary to discuss two additional formats for geotechnical reports before ending this chapter, the Geotechnical Design Summary Report (GDSR), and the Geotechnical Baseline Report (GBR). Sometimes, project owners have a need to assume greater control over the construction means and methods portion of a project, as compared to the conventional approach discussed in the preceding sections of this chapter. When this happens, it is also necessary for the owner to adopt a more comprehensive approach to reporting on the results of the subsurface investigation; this is the Geotechnical Design Summary Report. Probably the owners agency that has the most experience with this method of approach is the Washington Metropolitan Area Transit Authority (WMATA). WMATA began using GDSRs as part of its contract documents for tunneling projects around 1970, and many of the ideas and concepts discussed here have been obtained from articles and speeches presented by WMATA about this topic. Specifically, WMATA has attempted to establish the geotechnical basis for design by producing a report that discusses available subsurface information and sets forth many of the important subsurface design and construction considerations formulated as a result of that information. During the last 25 years, WMATA's approach to using GDSRs has been re-evaluated and re-assessed, but always with the following four objectives in mind:

1. Openly share available information about the geology and about subsurface conditions as derived from the subsurface investigation.
2. Explain the geotechnical basis for design and the underlying approach used for preparation of the contract specifications and drawings.
3. Provide a "level geotechnical playing field" for the preparation of bids by avoiding overly optimistic assumptions about ground behavior and by openly discussing the advantages and disadvantages of various methods of construction for evaluation by prospective contractors.
4. Provide a package of both factual subsurface information *and* the interpretations derived from that information that can be used to help evaluate the validity of claims of differing site conditions during construction.

Hence, whereas the GIR is written by the geotechnical engineer for use primarily by the project designer, a GDSR represents a *collaborative* writing effort by the geotechnical engineer, the project designer, and representatives of the project owner, working together to produce a document that is intended to be read primarily by the *project contractor*. And, whereas the concept of a GIR is based on a belief that the contractor will read the document and make use of it in some reasonable and prudent manner relative to preparing a proposal for doing the work, the GDSR is a contract document that is *mandated* for use by the contractor.

3.8.1 GIR versus GDSR

Although most, if not all, of the issues discussed above for the preparation of a GIR are applicable to a GDSR, there are aspects of a GDSR that require a completely different emphasis as compared to a GIR. With respect to the geotechnical interpretation needed for preparation of a GDSR, the use of the same flowchart and all of the techniques and procedures used to write a GIR are still appropriate and, in fact, the same input relative to subsurface conditions and project requirements would be utilized. A completely different emphasis is necessary, however, relative to establishment of the ground behavior/ground control scenarios that seem appropriate for design and for evaluation of the construction methods and third-party impacts that result from those scenarios. For instance, it is highly likely that experienced construction professionals (e.g., retired professionals who spent all or most of their careers working for major contractors) will be brought in to discuss the project and to provide detailed input relative to the advantages and disadvantages of using various construction methods for doing the work. Hence, brainstorming sessions could be used to truly understand how the project should be constructed and to assemble contract documents that are highly compatible with the preferred construction alternative(s).

Another major difference between a GIR and a GDSR is that, even though some of the above discussion about construction means and methods might take place during the preparation of a GIR, not all of the results of those discussions would be included in the GIR text. Often, a discussion of possible construction methods is helpful in deciding which design recommendation makes the most sense, but the primary output of a GIR is the design recommendation, not a discussion of construction activities. Exactly the opposite is true for a GDSR. The primary output of the GDSR is a summary of construction-related issues with respect to both project design and construction. In essence, it should be possible to use the GDSR in conjunction with all the other project specifications and contract drawings to make certain that all important geotechnical design recommendations are incorporated into a message delivered directly to the contractor. For instance, certain "baseline" statements about how much water the contractor could anticipate flowing into the excavation would be included in the GDSR, together with a description of how the contractor would be compensated for flows in excess of those amounts. If this is done, then language in the dewatering specification must be made compatible with the GDSR and a pay item established in the contract documents that is also compatible with the GDSR's proposed method of compensation. Such is the way that a variety of issues concerning excavation, ground control, excavation support, third-party impacts, and many other important construction aspects of the work would be handled on a project utilizing the GDSR concept.

It must be emphasized, however, that this method of approach to contracting the work is a departure from how most construction projects are undertaken in the United States, and that it requires a highly focused and intensive effort to correctly produce many of the important details of the contract documents. To begin with, it is important to have design and construction professionals associated with the project who really are familiar with the site geology and who really do know a lot about the best way to design and construct important project facilities. It is also important for the owner to satisfy himself or herself that this team has assembled a contract package that will "meet the test of time" and will provide a satisfactory finished facility at least cost and in the shortest possible time.

A second point is that the owner must be certain that all effort expended on his or her behalf by the project team, including the geotechnical engineer, the project designer, construction consultants, and designated owner's representatives, really is a collaborative effort and that the output from that effort is accurately and thoroughly discussed in the GDSR. One of the most important parts of this collaboration is a coordinated effort to make certain that all parts of the contract documents, including the GDSR, the specifications, and the drawings really are coordinated in an attempt to avoid confusion and/or ambiguity about what the owner expects from the contractor, both in terms of preparing the proposal and in terms of actually constructing the project. Sometimes, the biggest problem with this type of approach is that some aspect of the project is discussed in the various reports or the contract documents in such a confusing, contradictory, or ambiguous manner that the cost of or the time for doing the work cannot be determined. Statements such as the soil *may* squeeze, or the rock *may or may not* stand up, or large amounts of groundwater *might* flow into the excavation, are examples of this type of problem. Or, the GDSR may say one thing about a problem, the technical specification something else, and the notes on a drawing something else again. Instances of this type of confusion should be avoided if it is intended to use the GDSR approach to contracting the work.

Third, it is necessary for the owner to insist on good follow-through by the project team into construction such that accurate and complete records of construction activities are produced for comparison with contract indications. One of the most difficult aspects of construction monitoring for subsurface projects is the accurate recording of cause-and-effect situations relative to ground reactions that take place during construction. One of the most common misunderstandings that develop between owners and contractors on subsurface projects is whether or not the contractor is using proper techniques in dealing with known ground conditions. Many claims of differing site conditions revolve around assertions made by the owner that the contractor is not using the equipment in a manner that is compatible with statements made about the ground in the contract documents versus assertions made by the contractor that the ground is not behaving as he or she was led to believe by contract "indications." In general, a complete set of accurate project records is needed to evaluate these conflicting sets of assertions and, hopefully, to obtain some resolution of a claim short of a full-blown dispute.

Last, the GDSR approach very often implies the use of an entire array of other contracting practices for a project. Some of the more common of these practices might be partnering, escrow bid documents, value engineering, alternative dispute resolution, contractor prequalification, negotiated bid pricing, or other practices that clearly imply that the owner wants to work closely together with the contractor to build the best quality project at least cost and in the shortest possible time. Hence, the ground is viewed as an opportunity for cooperation rather than as a source of antagonism, and the owner creates an environment for working together with the contractor to produce great projects. Such is the promise of using both a

GDSR and a cooperative approach to building projects, and, in fact, this can be a very satisfactory way of proceeding with some projects.

Probably the biggest downside to this method of approach, especially for the design professionals involved with a project, is the possibility that, despite all best efforts on the part of all participants, some major problem or major incident still takes place on a project during construction. The fact of the matter is that "unknown" subsurface conditions really are unknown, and something completely unanticipated can be uncovered at a site. Or, some combination of events at a site may result in a major cost or scheduling impact or a third-party problem that simply goes beyond the ability of project participants to obtain resolution on the basis of contractual provisions. If this happens, then it is probably only a matter of time before someone asserts that statements made about the project in the geotechnical report constitute some type of "implied warranty" about what the various parties had a right to expect at the site. At that time, the geotechnical engineer's best line of defense will be the written agreement for the work and the various (hopefully appropriate) limitation statements provided in the reports themselves.

In conclusion, the GDSR is a more comprehensive version of a GIR, which tries harder to take into account the behavioral characteristics of the ground and the construction consequences of those behaviors. Most often, this approach is used for more complex projects that involve higher-risk construction operations and/or third-party impacts. GDSRs are written as a collaborative effort by the geotechnical engineer, the project designer, and the owner's representatives, with strong input from experienced construction professionals and with the expressed purpose of not only providing a satisfactory finished facility for the owner, but also of controlling the means and methods of construction in a manner that the owner believes is necessary to limit his or her liability relative to undertaking the work. Some of the more important aspects of proceeding with this method of approach include a highly coordinated effort relative to the preparation of contract specifications and drawings, a strong desire on the part of the owner to work together with the contractor, and contractual mechanisms for the quick and nonconfrontational resolution of claims of differing site conditions that arise during construction.

As a final note, some projects that have been released for bidding include both a GIR and a GDSR. The belief here is that the geotechnical engineer would prepare a GIR, which would then be given to and used by the project designer as input to the writing of a GDSR. This approach to the work is *not* recommended, for the following reasons:

1. The preparation of two documents creates additional opportunity for confusion regarding the subsurface information.
2. The contract status of the GIR must also be established in some manner that does not conflict with the GDSR.
3. The GDSR is supposed to be one collaborative effort on the part of the

owner's team trying to provide the contractor with the best possible package of data, interpretations, and recommendations. The writing of more than one report conflicts with this objective and creates additional opportunities for misunderstandings on the part of the owner's team, which must be avoided if there is any hope of dealing forthrightly with the contractor.

3.9 GEOTECHNICAL BASELINE REPORT

As of the writing of the chapter, a small group of individuals in the tunneling business is suggesting the use of a third type of geotechnical report, called a Geotechnical Baseline Report (GBR). Although many of the ideas about how this type of report should be written and used during bidding and construction are still in the formative stage, the most basic concept of a GBR is that it would have the status of a contract "specification" and that it would be used to establish definitive "baselines" of information about certain subsurface conditions to be expected at a site. Using these—and only these—baseline statements, the contractor would then be able to bid the project without having to provide an independent interpretation of subsurface information, and with a higher degree of confidence that claims of differing site conditions would be more easily resolved by comparison between the encountered conditions and what was said in the baseline statement regarding these conditions. Hence, the bottom line for this method of approach is to use the GBR to provide clear and unmistakable descriptions of "contract indications" for certain subsurface aspects of the work. (See the discussion of baseline reports in Chapter 4, Section 4.3.9. See also Essex, 1997.)

Although the preparation of a GBR may or may not make sense for a particular type of project, there is substantial controversy at present about how this concept will work in practice. Some of the more controversial aspects of this concept are discussed below.

1. Some persons in the industry have suggested that baseline statements do not have to adhere to geotechnical "reality," but can be used to reflect a willingness of the part of the owner to accept more or less risk on a project. For example, the baseline statement might tell the contractor to expect a larger or smaller number of boulders in glacial till compared to what could reasonably be expected as a result of the subsurface investigation. This might be done in an effort to "protect" the owner from a claim (say, more boulders), or in an effort to receive a lower bid for the work (say, fewer boulders). It is the opinion of this author that all geotechnical baseline statements must reflect geotechnical reality, as revealed by the subsurface investigation, and that any attempt to adjust "risk" in the above manner will be unsuccessful.

2. No one knows what the professional liability is for geotechnical engineers who write baseline statements. Ground behavior during construction is a function of both existing ground conditions and construction means and methods. When preparing baseline statements, the geotechnical engineer should have a fairly good idea of the subsurface conditions and some idea of the kinds of construction methods that will be used at the site, but the details of how those construction operations will be performed in the field are difficult to evaluate (to say the least) before the contractor has even been selected to do the work. How is it possible to write a "definitive" statement of anticipated ground behavior without having a definitive understanding of the details of construction operations?

   Once a contract is signed, the contractor has enormous influence over what happens at a site and on how that work will be performed. There is, unfortunately, still no law or contract provision that precludes something from going wrong at a site, and when that happens, there is always controversy about why it happened, who is responsible, and what the contractor should have known or should have done in an effort to have avoided the problem. It is the opinion of this author that the GBR concept will provide the contractor with a stronger case for stating that almost anything that happens at a site should have been foreseen in a baseline statement and specifically warned against in the GBR. Some geotechnical professionals have gone so far as to state that no geotechnical engineer should participate in the preparation of baseline statements without a specific disclaimer of responsibility from the project owner.

3. Given the novelty of the GBR concept, there is no known case law associated with the use of baseline statements as part of the contract documents. Thus, if something does go wrong, then the case will be adjudicated under existing legal precedents established on the basis of the more conventional method of approach to construction. Questions, therefore, will immediately be raised about whether the baseline statement was reasonably based on available data or about whether each statement went far enough or too far in describing a certain aspect of the work. (See Hatem, 1996.) For instance, suppose a baseline statement tells the contractor to anticipate 300 boulders at a site where, say, only 100 boulders might reasonably be anticipated as a result of the subsurface investigation. In anticipation of 300 boulders, the contractor chooses a piece of equipment that is able to handle the boulders, but which, as a result, advances more slowly in the nonbouldery ground. Upon project completion, it is discovered that only 50 boulders were actually encountered on the project. Can the contractor now file a claim for the lost production he suffered in the nonbouldery ground? Other aspects of this problem might include a need to describe whether the boulders occur individually or in nests, or if the boulders are held firmly in the soil matrix such that they can be cut by the tunneling

machine or easily dislodged and cause problems at the face. Probably the worst type of boulder is one that is only partially exposed in the face such that it deflects the shield or otherwise damages the machine without actually being exposed in the tunnel. How, then, and to what degree can or should all of these aspects of just this one issue be discussed in a baseline statement?

These and numerous other issues associated with this topic—such as value engineering, risk allocation, site safety, the increased possibility of owner "credit" claims due to actual conditions encountered being more favorable than the described baseline conditions, engineering estimates, contract compatibility, and third-party impacts—strongly suggest that geotechnical engineers should approach any request to write a "baseline" statement for a subject project with considerable trepidation. Any effort to do this must be preceded by a high-quality subsurface investigation and then undertaken only for project owners who are highly experienced in dealing with the intricacies of subsurface construction. Otherwise, it is the opinion of this author that a more conventional approach to reporting on the results of subsurface investigations, as described earlier in this chapter, be considered for use on a project.

3.10 CONCLUSION

A considerable amount of information has been provided in this chapter about the topics of subsurface investigation, geotechnical interpretation, and geotechnical report preparation. In general, those activities can be quite straightforward, but they can also be highly difficult and risky as the ground becomes poorer, the project more complex, and the third-party impacts more troublesome. Nothing seems to test the ability of a team of design professionals working together with an owner to bring a project in on time and within budget more completely than does the encountering of a "differing site condition." How best to anticipate and to avoid this problem, if at all possible, are two of the most challenging aspects of the practice of geotechnical engineering.

Despite an attempt on the part of this author to provide a clear and comprehensive discussion of the topic of geotechnical report preparation, there is still more left unsaid about this topic than was covered here. The number of project types, ground conditions, and third-party impacts and the number of owner/contractor teams involved with underground projects is so large that it is simply not possible to discuss every possible combination of problem that might be encountered at a site. Competent geotechnical engineering is a function of theoretical knowledge, practical experience, and professional judgment that can be accumulated only over years of effort. Many writers about this topic have characterized geotechnical engineering as more of an art than a science, and a strong argument can be made in support of that statement. Trying to second-guess Mother Nature on a regular basis while, at the same time, accurately assessing all manner of design and construction options for a site can only be characterized as one of the greatest challenges of civil engineering.

REFERENCES

American Society of Civil Engineers "Avoiding and Resolving Disputes During Construction," Underground Technology Research Council, 1991.

American Society of Civil Engineers "Avoiding and Resolving Disputes in Underground Construction," Underground Technology Research Council, 1989.

Bohlke, Brenda Myers, ed., *Geotechnical Reports in Underground Construction*, American Society of Civil Engineers, Underground Technology Research Council (Washington, D.C., April 1996).

Brierley, Gary S. and T. J. Lyman, "The Geotechnical Aspects of Rock Tunnel Design," *Proceedings of the 1989 Rapid Excavation and Tunneling Conference* (Los Angeles, June 1989).

Brierley, Gary S., Michael J. Robinson, and M. H. Kobler, "Tunnel Planning in Urban Areas," *Proceedings of 10th Annual Canadian Tunneling Conference* (Banff, Alberta, Canada, October 1992).

Brierley, Gary S. and Robert J. Smith, "Contracting Practices for Tunneling Projects," *Proceedings of North American Tunneling Conference and Exhibition*, Boston, MA, October 1992.

"Controlling Tunnel-Lining Costs—Tips for Owners and Designers," *Civil Engineering*. Vol. 48, No., 7, July 1978, pp. 51–53.

Essex, Randall J., "Means of Avoiding and Resolving Disputes During Construction," *Tunneling and Underground Space Technology*, Vol. 11. No. 1, Elsevier Science Ltd., 1996, pp. 27–31.

Essex, R. J., ed., "Geotechnical Baseline Reports for Underground Construction—Guidelines and Practices," prepared by the Technical Committee of the Underground Technology Research Council (American Society of Civil Engineers, 1997).

"Geotech Design Reports Get a Litmus Test," *Civil Engineering*, December 1996, pp. 47, 49.

Gould, James P., "Geotechnology in Dispute Resolution," *Journal of Geotechnical Engineering*, Vol. 121, No. 7, July 1995.

Hatem, David J., "Geotechnical Baselines and Geotechnical Reality; One and the Same, Similar or Not Even Close—Professional Liability Implications," *The CA/T Professional Liability Reporter*, Vol. 2, No. 2 (Boston: Burns & Levinson LLP, December 1996), pp. 1–8.

Matyas, R. M., et al., *Construction Dispute Review Board Manual* (New York: McGraw-Hill Companies, Inc., 1996), p. 163.

O'Rourke, T. D., ed. *Guidelines for Tunnel Lining Design*, Underground Technology Research Council (American Society of Civil Engineers, 1984), p. 82.

"The Risks Associated with Tunneling Projects," *USNC/TT Newsletter*, June 1987, pp.

1–9. A discussion of this paper was published in the December 1988 edition of the *USNC/TT Newsletter.*

"Underground Contracts for the 21st Century," *Civil Engineering*, December 1996, pp. 54, 56, 57.

U.S. National Committee on Tunneling Technology, "Better Contracting for Underground Construction" (Washington, D.C.: National Academy of Science, 1974) (also reprinted as Report No. DOT-TST-76-48, U.S. Department of Transportation).

U.S. National Committee on Tunneling Technology, "Overview of Practice and Legal Issues, Evaluation of Cases, Conclusions and Recommendations," *Geotechnical Site Inspections for Underground Projects* (Washington, D.C.: National Academy Press, Vol. 1, 1984).

4

PREPARATION OF CONTRACT DOCUMENTS FOR SUBSURFACE PROJECTS

Eldon L. Abbott, P.E.
Parsons Brinckerhoff Quade & Douglas, Inc.

4.0 INTRODUCTION

This chapter discusses the methods and procedures available to the design and construction management professional to aid in his or her role as the professional retained by the owner to assist in preparing the contract documents for use by the contractor in constructing a subsurface project. More specifically, this chapter will explore the importance of risk identification, informing and advising the owner concerning risk allocation, and the preparation of contract documents consistent with the owner's decisions regarding risk allocation. In addition, several specific topics relating to contract document preparation are discussed, including the use of disclaimers, differing site conditions clauses, the development of "baseline" reports, and the need for detailed specifications regarding the specific contents of the escrow bid documents.

Subsurface Conditions: Risk Management for Design and Construction Management Professionals, edited by David J. Hatem
ISBN 0471-15607-8

4.1 PRELIMINARY CONSIDERATIONS

4.1.1 Identifying and Understanding Risk Exposure

Chapter 2 discussed various risk management issues associated with the terms of the professional's agreement with the client. This chapter will focus on risk management considerations related to the terms of the contract documents—that is, the various documents that constitute the agreement between the owner and the construction contractor. Although the design and/or construction management professional is not an actual party to the contract documents, the professional ordinarily prepares the contract documents. The extent to which risk has been identified and the manner in which it is allocated in the contract documents not only has a material impact on the allocation of risk among the owner and the contractor, but also affects the professional liability exposure for the professional (see discussion, Chapter 10, Section 10.2.4).

The preparation of contract documents must be based on an understanding of certain risk management principles. Effective risk management starts with each party identifying and understanding its risk exposure and then determining what means are available to allocate, control, and manage those risks. In some circumstances, a professional who has identified the risk may choose not to proceed with any further involvement in that project. As a project moves through the various stages, from planning to conceptual design, contract documents, and finally into construction, the project risks may change or be re-allocated (consciously or not) from one party to another. The professional must understand this evolving concept of risk identification and allocation in order for any risk management program to be effective, and to be in a position to make appropriate recommendations to the project owner regarding allocation of risk in the contract documents.

Good contract documents are an essential component of such a program and the preparation of those documents starts with a well-defined plan by the professional. The old saying, "Plan your work, then work your plan," is very much the rule of effective risk management in preparation of good contract documents. David B. Ashley (1981) states that project risks can be broadly grouped into three categories: (1) performance (i.e., does the project function up to the level of owner expectation?); (2) completion (i.e., can the project be completed for a cost and on a schedule that are economically viable?); and (3) liability risk (i.e., professional design liability, on-site construction safety, and other third-party exposures). This chapter is concerned with the management of risk associated with the development of contract documents prepared or assembled by the owner's professional for construction of the subsurface project. As demonstrated elsewhere (see Chapter 10, Section 10.2.4), there is a direct correlation between the degree to which risk is fairly and effectively allocated in the contract documents and the degree of potential liability exposure of the professional.

David Ashley recommends several principles when considering risk management strategies for a project. First, risk transfer on a project should not be at the expense of removing the incentive for managing the risk. Good risk allocation, says Ashley, is really risk and incentive allocation. Second, the party bearing the risk should be capable of managing the risk. For example, it serves no real or constructive purpose to transfer nonmanageable risks to the contractor since, at a minimum, it forces the contractor to include allowances or bid contingencies for that risk and, in all probability, will lead to the significant potential for disgruntled economic expectations, and the resultant project disruption and claims. Project risk management, in its most general sense, is described by Ashley as an owner-determined, -monitored, and -controlled management program designed to reduce the total negative impact of project risks (which result when risk is inadequately defined and/or unfairly allocated). This reduction in total negative impact on a project can be accomplished through a combination of risk retention by the owner, risk transfer to the contractor, and/or partial risk transfer or retention between the owner and contractor.[1] For subsurface projects in particular, the professional has an important role in assisting the owner in identifying project risks and advising the owner concerning the most feasible, reasonable and realistic approach to allocate and manage those risks.

The process begins with the professional developing an understanding of the owner's risk tolerance (i.e., the owner's attitude about retaining and transferring risk) with the goal of reducing the total cost and negative impact on the project. The owner has many risks with which to contend, including:

- Risk of loss of development capital
- Risk of a cost increase for the project
- Risk of loss of the project's value due to inadequate performance
- Risk of adverse publicity from the press or the public
- Risk to the partially completed project of damage during construction
- Risk of differing site conditions and related claims
- Risk of losing eligibility of grant or other funding sources due to mismanagement of the project[2]

Similarly, the contractor must also assess and identify the potential risk on a subsurface project. The contractor has to evaluate the contract documents before submitting a bid to determine how the risk of constructing the work has been allocated by the owner (typically, with some advice from the owner's legal counsel and other professionals). On large subsurface projects, it is common for contractors, as a normal step in preparing a bid, to have their legal counsel prepare their own risk assessment of the contract terms and other clauses in the contract to determine how the project risk has been allocated. If the contractor senses that the risk of constructing the work has been shifted too far in its direction, the contractor must then decide whether to pass on the project, submit a bid, and, in the latter situation, whether to include some contingency for perceived excessive risk assumption. The contractor's risk may be summarized as:

- Risk of not making a profit on the project or even losing the contractor's own capital on the project through underestimating the complexity of the work, i.e., underestimating the effort required to complete the work
- Risk of damage to adjacent property due to construction work activities, especially construction means and methods
- Risk of construction delays, causing the owner to impose liquidated damages for late completion
- Risk of injury to workers due to construction activities or job site conditions
- Risk of the impact of the "unknowns," especially those relating to subsurface conditions on the project, or the risk of incompletely or inaccurately assessing the cost or time impact of certain events (e.g., differing site conditions), especially in the absence of contract provisions allowing for equitable adjustment
- Risk of negative recommendations and associated reputational damage (and consequent loss of revenue and possibly bonding capacity) in the event of adverse project experience due to claims, disputes, and associated litigation

Risk management for the design and construction management professional is a continuous process, which begins with the decision of whether to pursue the project and then follows on through the proposal, contract negotiations, and project execution, including the quality control program. Risk management for the professional is an all-encompassing program of internal quality control (within the design team) and external management of project risk in accordance with the owner's defined risk management avoidance/tolerance perspective for a specific project. This risk management program ideally is accomplished through the identification, assessment, and allocation of project risks and the careful preparation of unambiguous contract specifications and a good design, and the clear allocation of risks in accordance with the owner's risk management objectives.

Professionals involved in a subsurface project should prepare their own risk assessment for the project. Actually, it is recommended that the professional prepare two risk assessments for the project. One risk assessment analysis would cover the professional's perceived risk in undertaking the project. Such an assessment would identify the risks and possible mitigation measures for eliminating, managing, or minimizing their effect on the professional.

The second risk assessment analysis is the all-important one addressed to the owner, since it is the one that seeks to identify the risks of proceeding with the project. For this analysis, it is suggested that very senior, experienced owner representatives, professionals, construction managers, and constructability experts identify, evaluate, and brainstorm the risk issues on the project. It is important for this level of risk assessment to be very specific to the level of such issues as groundwater inflows, rock jointing, boulders, running ground,

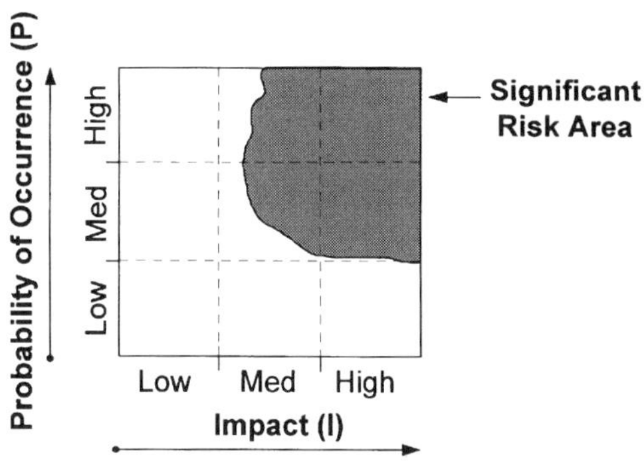

Figure 4.1 Risk probability and impact analysis (Kuesel, 1979)

and so forth, so that the perceived risks can be assessed and determinations can be made by the owner as to how these risks are to be allocated in the contract documents. It is often helpful when performing this risk assessment to visualize and rank the risks (see Figure 4.1) by probability of occurrence, as opposed to potential for adversely affecting the project. An alternative to this visual risk classification would be to chart the risks using a standard spreadsheet program, which can then be used to rank, sort, and track the resolution and allocation of the identified risk throughout the contract document preparation phase of the project. In order for this risk assessment to be accomplished, it is critical that the owner has retained the professional to undertake an adequate site investigation for the purpose of identifying the risk potential existing in the project. (See Chapter 2, Section 2.1.3.1; see generally Chapter 3, for further details on this subject.)

Project execution for the professional can mean many different things to different people. In this instance, we are focusing on project execution as related to the methods or contractual clauses available to the professional to assist the owner in fairly and clearly allocating risk. These methods or clauses should, when fairly applied, produce contract documents with risks clearly allocated to those entities that are best able to manage and control them.

4.1.2 Knowing the Client

Knowing the client and his or her "risk tolerance" is an important first step. Some clients do not want to accept any more risk than absolutely necessary, and desire to transfer as much risk to the construction contractor as possible through the contract documents. Owners who have a low threshold for risk often choose to specify more costly, conservative designs that can be constructed within known bounds of risk, cost and time. The subject of client selection as it relates to professional liability risk is discussed further in Chapter 1.

Some clients understand that clear contract documents, combined with a fair allocation of risk and with effective, timely and decisive management during construction, will ultimately result in lower constructed cost, higher quality for the completed work, and significantly decreased incidence of disruptive claims and disputes. Public-sector owners, however, may face different realities and constraints in how they are required to treat risk on a project. Whether they can manage unknowns within the context of project funding constraints, and whether they are able to carry contingency budgets within the project budget, may not always be within the control of the owner's staff actually managing the project, but may be controlled by a higher level of agency administration or advisory-level board. The inability to carry sufficient contingency in a project funding budget may eventually determine how readily the contractor can be reimbursed if a differing site condition is encountered. Furthermore, it is not unusual for the owner's project manager to be evaluated on the basis of the completed project coming in at or below the original project budget carried in the funding application for the public sector's budgetary process and constraints.

For these reasons, the public sector is often very challenging for the professional to navigate through and to be able to complete contract documents that fairly allocate the project construction risk. The more knowledgeable the professional is in understanding what drives and motivates the owner's decision making, funding availability, and risk management philosophy and program, the easier it will be to end up with a completed project that meets the owner's objectives and expectations, and, hence, produce a satisfied client. Of course, the owner must be willing to accept the advice and recommendations of the professional on these subjects.

4.1.3 Understanding the Owner's Risk Management Program

Just as important as knowing the client is developing an understanding of the owner's risk management attitude or philosophy, especially as embodied in any risk management program. It is not unusual that an owner may not have an established risk management program and, in that case, it is essential (and in the professional's best interest) to prepare a risk assessment report to help the owner clearly define and understand the risks associated with the project. A project risk assessment report may be written from several perspectives: i.e., the professional's risk, the owner's risk, or the construction contractor's risk. However, the risk assessment report referred to in this instance is a report that is prepared after one or more project team (owner, design team, construction manager, legal counsel) brainstorming sessions have occurred, which have reviewed and detailed the project risks and their probability of occurrence, assessed the impact of the identified risk on either the owner, contractor, or other project participant, and identified a possible risk mitigation action plan for assigning the risk to that entity best able to manage and control the identified risk. The results of this risk assessment should be summarized in a report identifying risks, possible consequences to the project, risk mitigation actions available to manage the risks, and a recommended plan of action. This report would be submitted to the owner for concurrence and direction to the professional. The owner's concurrence and/or direction on each of the specific risk assessments and action items contained in the report may then become the professional's work plan for risk allocation in preparation of the final contract documents.

As previously mentioned, it is important for the professional to understand what drives the owner's risk decision-making process. Is it accountability to the "rate paying" public? Does the owner have the resources to assume some of the financial risk with the goal of obtaining a lower bid price and, eventually, a less costly constructed project? Is the owner using public funds, private financing, or his or her own resources? If a project is reimbursably funded from a state or federal agency, what are the criteria that must be met for changes or changed conditions encountered during construction to be eligible for reimbursement to the owner? The answers to these questions will have an important influence on the recommendations that the professional develops for the owner's consideration about assigning risks and in preparing the provisions of the contract documents relating to risk allocation (which reflect the owner's decisions with respect to such matters).

Finally, the professional needs to understand the owner's contractual method for handling hazardous materials and contaminated soils and/or groundwater. How will the owner handle these materials, if encountered? Does the owner have a specific "hazardous materials" contractor or professional response team already under contract? What is the contractor to do if hazardous materials are encountered during construction? Is the contractor responsible for sampling and testing of the material to determine if it is actually contaminated? Answers to each of these questions must be factored into the professional's risk assessment report and in the development of a checklist of what needs to be included in the contract documents. The owner's clear directions on how each of these issues is handled in the contract documents helps both bidders and the successful contractor to assess the risk that is allocated to each party by the contract documents.

4.2 IDENTIFYING, ASSESSING, AND ALLOCATING RISK

Good contract documents start with planning the work and developing a "mind set" or thought process that is carried throughout the development of the documents. An important starting point in developing the work plan is to read and understand the general conditions of the owner's construction contract that will be used during execution of the construction project.[3] The professional, having an understanding of the risks to be managed, should review the general conditions to determine the extent to which they accurately reflect the owner's informal and project-specific decisions regarding risk allocation, especially as they relate to subsurface conditions. Some of the questions to be researched during review of the general conditions are:

- Does the contract contain a differing site conditions clause?
- Does the contract contain a site investigation clause?
- Does the contract contain a provision for expedited or alternative dispute resolution (e.g., a dispute review board)?

- How does the contract deal with contract schedule, contract completion date, and liquidated damages?
- Does the contract contain a "no damages for delay" clause?
- What is the order of precedence for the contract documents? Which document governs another in the event of a conflict?
- How are contaminated soils addressed? What are the statutory requirements?
- Will the boring logs, data reports, and/or geotechnical interpretive reports be part of the contract documents? Are they given only for information?
- Will the contract contain a geotechnical baseline report? If so, what is the contract status of the baseline report?
- Does the contract disclaim the accuracy or completeness of any reports, data, or other information (or any portion thereof) made available to the contractor?

In dealing with an owner who already has a set of "general conditions," the following is a checklist of some of the specific issues to be addressed in the "special provisions" (after a careful review of the owner's front-end documents):

1. How are risks allocated in the owner's front-end documents? Is a differing site conditions clause incorporated into the documents? Has the owner made it clear that payment will be made if physical site conditions materially different from those indicated in the contract documents are encountered, or will the risk of those conditions be entirely imposed on the contractor?
2. Do these clauses need to be clarified or modified? (Clear allocation of risk is key.)
3. Identify disclaimers in front-end documents; decide if they need to be modified, qualified, deleted, or added.
4. Are boring logs or other data to be included in documents? Are geotechnical reports part of the contract documents? Are they provided for information only? What disclaimers are in these documents? Are they consistent with those in the front-end documents?
5. Were the documents prepared and reviewed by the owner's legal counsel, knowledgeable in construction law, by the owner's procurement department, or both? Is it possible to make changes to these documents, or will changes have to be made through the "special provisions"?
6. Is the contractor told to visit the site or make his or her own subsurface investigation?
7. What are the notice requirements in the contract documents regarding variations in anticipated conditions—differing site conditions—and what is the required notice to be given by the contractor if a claim is to be based

on alleged errors/omissions by the design and/or construction management professional?

8. Is there a "no damages for delay" clause included in the contract between the owner and contractor? (If enforceable under governing state law, this clause can pass the financial risk of delays for any reason on to the contractor unless the contractor typically can prove that one or more of the parties have engaged in bad faith or willful, malicious, or grossly negligent conduct. In other words, unless the contractor can demonstrate that the delays were caused by the professional or owner acting in one of the above manners—i.e., exercising bad faith, negligent conduct, and the like—then the contractor will not be able to collect compensation and the sole remedy will be a time extension.)
9. Are designs prepared by the contractor to be reviewed by the professional and, if so, for what purpose? Are the contractor's designs required to be stamped by a registered professional? Do the contract documents completely assign responsibility for the adequacy and constructibility of contractor-furnished design to the contractor?
10. Will the contractor be allowed to propose "value engineering" (VE) design changes or alternative designs? Are these designs to be stamped by a registered professional engaged by the contractor? Will the contractor's designer be obligated to provide professional liability insurance? Is it clear in the documents that the professional is reviewing the VE proposal only for its conformance with the final design intent and performance, and that the review does not extend to the details of the design and/or the contractor's means and methods of construction? Who has responsibility for coordinating the contractor's value engineering or alternate design (if accepted) with the overall project design? Will the contractor and/or its designer be obligated to indemnify the owner and professional for liability arising out of the VE design?

4.2.1 Assessing the Risk to Be Managed

As described earlier, each project has specific, identifiable risks that can be assigned to one or more of the project participants to be managed. Some risks are obvious and, through the standard general conditions, are automatically assigned to either the owner or the contractor. Some examples are contract completion within a specified time limit, responsibility for construction means and methods, construction safety, performance of certain elements of work within an allowance item, performance of work at bid unit or lump sum prices, weather delays, "acts of God," and insurance requirements, to name just a few.

For any subsurface project, it is the uncertainty of the subsurface conditions that results in the greatest risk exposure to both the owner and contractor, for increased time and cost requests from the contractor. This uncertainty of subsurface conditions results in claims from the contractor for such things as excess water, unsuitable soils, bedrock within excavation limits, ground movements due to excavation or other contractor operations, and damage to nearby utilities, subsurface structures, or buildings. Albritton, Hatheway, and Underwood, in their 1983 RETC paper titled "Improvements in Geotechnical Aspects of Contract Documents for Underground Construction," states:

> Geologic and ground water conditions are a major consideration in underground construction. The more complex these conditions, the greater are their influences on the various elements of construction. In general, these conditions can be grouped as follows:
>
> - Engineering properties of soil, rock, or weak rock;
> - Structural characteristics of rock or weak rock;
> - Anomalous or frequently changing geologic conditions;
> - Hydrostatic pressure, location, and volume of water; and
> - In situ geologic stresses.
>
> The geologic conditions, in turn, exert a primary influence on excavation stability, as well as on temporary and permanent ground stabilization systems. The important first derivative of these conditions and factors is, of course, the rate of excavation.
>
> Generally speaking, the relative hardness of earth materials controls the rate of excavation. The degree of ground deformation that may occur after excavation is a function of material properties, discontinuities, ground water conditions, and the stress field surrounding the excavated opening. These factors, in varying ways, affect the standup time, a critical factor in maintenance of contractor excavation progress.

For a tunnel project, a list of unique specific subsurface risks can be identified. Soft ground tunnels can have risks associated with granular soils below the water table, yielding large water inflows through the tunnel face, causing face stability problems. Water inflows causing drawdown of the groundwater table that results in ground settlements due to changes in effective stresses within the soil mass, and cobbles and/or boulders within the tunnel regime, can cause significant delays (due to the time required to negotiate and execute the necessary change order to proceed with the work) and increased cost when encountered with no provisions to handle such occurrences within the contract. Rock tunnels can have similar risks such as water inflows, fault zones within the tunnel alignment, unstable rock conditions, changing rock characteristics such as rock strength, hardness, jointing, and so on.

For subsurface projects, identifying the level of risk associated with any one of the identifiable, usual risk problems requires a carefully planned and executed field exploration and laboratory testing program. Sampling and testing using well-established industry procedures is an important element of the pro-

High Financial Impact		High water inflows Hard rock Unstable tunnel face Mixed face tunneling conditions
Low Financial Impact		
	Controllable	**Uncontrollable**

Figure 4.2 Financial impact/controllability matrix (after Ashley, 1983)

fessional's quality assurance/quality control (QA/QC) responsibility (see Section 3.3 of Chapter 3 for further details on this subject).

Another useful way of looking at specific project related risk is presented, again by Ashley (1983), in his article entitled "Financial Risk Management for Underground Construction." Ashley presents "Financial Impact/Controllability Matrix" (Figure 4.2) as a tool to identify which elements of the project risk have the potential for greatest cost impact. This helps to determine which management action can best be applied to change a threat that, in turn, will establish control over the remaining risk.

The two axes on the table are "Financial Impact" and "Controllability." A combination of high financial impact and uncontrollable risk should be a "red flag" to the owner and professional to concentrate on managing that set of risks. Actions on the risk items shown in a matrix should concentrate on reducing the financial impacts, as well as establishing a mechanism through the contract to exercise some control over the remaining risk. Ashley points out in his article that if one cannot control the risk of a cost overrun source, then one should compensate by lowering or removing its impact. By this same reasoning, gaining more control over a risk dilutes the need for eliminating its financial consequences and, of course, the best of all possible worlds is gained by completely controlling a risk and removing its impact as well.

4.2.2 Risk Elimination or Allocation?

4.2.2.1 Elimination

Elimination of risk on a subsurface project is not a realistic goal. Risk is inherent in subsurface work, given the uncertainties in the actual site conditions to be encountered and the uncertainties of how site conditions will behave during the construction process. Although an adequate scope of subsurface investigation can reduce the risk of these uncertainties, no amount of investigation can reasonably be expected to reveal all ground conditions to be encountered.

Thomas Kuesel, in his 1979 RETC article titled, "Allocation of Risks," states:

> Unnecessary risk can also be eliminated in the area of unknown physical conditions. It is not sufficient for the owner to accept risk under a "changed conditions" clause if the extent of site condition information made available to bidders does not give convincing evidence of probable conditions. Bidders will still add substantial contingencies for undefined or unexplored site conditions.

Thus, on many large-scale and complex subsurface projects, it is not uncommon to see, as a part of the early contracts, a full-scale pilot tunnel or pumping test, or some other full-scale field test that will give prospective bidders the actual field data that they need to significantly reduce the amount of money set aside to cover the risk in their bid contingency (see generally Chapters 3).

4.2.2.2 Mitigation

Mitigation, or risk reduction, is the more realistic risk management approach on subsurface projects. There are measures that can be taken to mitigate risk on subsurface projects in addition to conducting an adequate preconstruction subsurface investigation. One area in which owners may practice initiative in mitigating risk for certain projects is by setting up an active preconstruction planning program to reduce construction restrictions before they become a project problem or risk. Items falling into this category would include early site work contracts, preselection of disposal areas for excavation material, and prepurchase of long-lead items.

Another initiative, which is being implemented in the Massachusetts Highway Department's (MHD) Central Artery/Third Harbor Tunnel (CA/T) project, is a variation of the contractor's "prebid" meeting. All prospective bidders are invited to attend a "prebid" meeting, at which they are encouraged to ask questions concerning the contract documents currently being advertised for bid. Prior to convening the prebid meeting, MHD, along with its section design consultants and management consultant, selects the factors and areas of project risk that are felt to have the greatest potential impact on the project. At the prebid meeting, a presentation is made of those factors and risks, their uncertainty, and the baseline conditions for those factors and risks that MHD requires the contractors to carry in their bids. This presentation is carefully presented and recorded, and minutes are sent out to all bidders by addendum before the bids are accepted for the work. Through this "prebid" meeting, the owner establishes a record that, at a minimum, the bidders were aware of the basis of uncertainty in the contract documents and were thereby in a position to make minimum assumptions regarding what should be carried in the bid for the uncertainties on the project. In many respects, this procedure (in appropriate circumstances) may serve in establishing a "baseline" for selected, high impact/high risk segments of the CA/T project.

As documented by the prebid meeting approval, it is possible to develop creative ways to mitigate some types of risk in a construction contract. It is the professional's role to think creatively and assist the owner in identifying and initiating new ways to manage project risks.

4.2.2.3 Allocation

Kuesel (1979), in his paper titled "Allocation of Risks," concludes with some general principles for allocation of unavoidable construction risks:

1. The contractor should bear all risks over which he can exercise reasonable control. These include all matters relating to selection of construction methods and equipment and prosecution of the work, except as this control is impaired by the action of third parties.
2. Truly unpredictable risks (natural disasters, force majeure) are properly allocated to insurers. The owner may in some cases choose to be a self-insurer.
3. Site conditions should be defined as a basis for contracting. The contractor should bear the risk of variations that might reasonably be anticipated, but the owner should bear the costs of substantial differences. "Reasonably" and "substantial" are necessarily contentious terms, but the area is too complex to admit black-and-white definitions.
4. In the area of third-party effects, risks should be allocated to those best able to deal with the third party. This principle would assign to the owner the risks related to government agency regulations and to agreements with adjacent property owners. Risks associated with labor and subcontractor agreements and disputes should be assigned to the contractor.
5. Construction safety should be the responsibility of the contractor, although financial risk with respect to third parties is properly allocable to insurers (either the contractor's or the owner's).
6. The allocation of risks due to general economic factors (material and labor price escalation, foreign exchange rates, general strikes) depends on the duration and location of the work. For domestic projects expected to take less than two years to complete, it may be reasonable to ask bidders to price these risks. For projects with overseas or inaccessible locations, and for those with long completion schedules, the owner may properly assume part of these risks through "rise and fall of costs" clauses and other relief provisions.

Once the extent and nature of the risks associated with a specific project have been identified and prioritized as to their level of financial impact and potential for control, the professional is on the track to make recommendations to the owner regarding the fair and appropriate allocation of those risks (with proper incentives) to the party best able to contractually manage those risks during the construction phase. It is the professional's role, based on investigation, experience, research, and an understanding of the owner's "risk tolerance" attitude, to make *recommendations* to the owner about risk allocation. These recommendations should be based on the principles of ability to control risk and fairness; however, it is the *owner* who *makes* the *decisions* about how risk is to be allocated. Once those decisions have been made *by the owner*, it is the obligation of the professional to implement those decisions in the preparation of the contract documents.

4.2.3 Identifying and Allocating Risk for Design Adequacy and Construction Means/Methods

4.2.3.1 Design Adequacy

Owners and construction contractors have certain obligations to each other by virtue of their contractual relationships and traditional legal standards that are implied or derived from those relationships. The legal implications of those obligations are discussed in Section 10.2.2.1, Chapter 10.

The owner impliedly warrants to a construction contractor the accuracy, completeness, and constructability of design-related information and construction documents furnished to the contractor. The rationale for this implied warranty assumes that since the owner (typically through a design professional) undertakes to prepare and furnish detailed design and construction documents to the contractor, the owner should be responsible for design adequacy (see Section 10.2.2.1(A) of Chapter 10).

The contractor has the obligation to exercise due diligence in bidding the project—i.e., if a contractor is presented with an obvious omission, inconsistency, or discrepancy in the design or construction documents, the contractor must, at a minimum, request clarification. Further, the contractor has the obligation to construct the project in accordance with the design and the contract documents furnished. Generally, the contractor is under no obligation to check and verify the accuracy, completeness, and constructability of design and contract documents furnished by the owner to the contractor.

4.2.3.2 Construction Means and Methods

The basic principle of risk allocation between owner and construction contractor is that the construction contractor is in the best position—by virtue of experience, qualification, expertise, and ability to control—to be responsible for construction means, methods and techniques, sequences, and procedures, including safety precautions and programs. This assumes that neither the owner nor the professional has prescribed and/or required specific construction means and methods in the design and/or contract documents, and that neither the owner nor the professional has given advice, instruction, direction or supervision, or exercised control over, construction means and methods.[4]

4.3 TRANSLATING RISK ALLOCATION DECISIONS INTO CONTRACT DOCUMENT PROVISIONS

Up to this point, the professional has been researching and gathering facts about the owner's risk management program, its "risk tolerance" attitude, insurance requirements and/or programs for the project, and, specifically, how the owner wants to assign the risk for the specific project. Once risk has been identified and assessed, the professional has made appropriate recommendations regarding risk allocation, and the owner has made decisions, the professional may then proceed with the preparation or revision of the contract documents to reflect those decisions. Execution of this phase will require the professional to modify specific aspects of contract documents, remembering that, for the professional, the goal is to provide clear, accurate documents that fairly assign the risk to that entity best equipped and able to manage and control the risk. (See Figure 4.3 for a list of risks subject to allocation.)

4.3.1 Differing Site Conditions Clause

Legitimate differing site conditions represent physical site conditions that, when encountered, are materially different from those indicated in the contract documents and/or materially different from those normally encountered on a project of the character under construction. (Refer to Chapter 10, Section 10.2.1 for more details.) Typically, when these conditions are encountered, additional time and cost for the completion of the project result. Even in circumstances in which the contractor is entitled to additional time and/or compensation on the basis of a valid differing site conditions claim, an owner often may have a difficult time accepting any obligation to pay or to grant a time extension request associated with the claim. The owner's reaction is understandable (at least in the first instance), in light of the fact that payment of the differing site conditions claim will frequently increase the cost and/or time for performance of the construction work, but will rarely, if ever, increase the value of the completed project in the traditional sense represented by a scope addition or quality enhancement.

The primary issue to be addressed whenever a differing site condition is actually encountered is who—the owner or the contractor—bears the economic or performance risks associated with the encountering of the condition. Under the general "common law" rule, the contractor bears the risk that performance of his or her obligations may be more time-consuming or costly than it had anticipated when he or she submitted the bid. (For a more complete discussion of the common law rule, see Section 10.2.1 of Chapter 10.) In the absence of a specific provision in the contract documents providing the contractor some relief (either by way of a time extension or additional compensation) from the general common law rule, the contractor must generally bear the time and cost impacts associated with the discovery of differing site conditions.

The most generally accepted variation to the common law rule is accomplished through the inclusion of a differing site conditions clause in the contract documents, which recognizes the right of the contractor and the owner under certain limited circumstances, and subject to various procedural requirements, to an equitable adjustment in time and price when materially different physical conditions at the site are encountered. In fact, many state and virtually all federal government contracts are required by law to contain a differing

Acts of God (force majeure)
Adequacy of project budget/funding
Adequacy of design
Adverse weather
Ambiguous specifications
Asbestos
Attitudes of other parties
Bidding procedures
Changes in laws and regulations
Changes in pricing
Changes in procedures
Clarity of plans
Competence of cost estimating
Concurrent work
Consequences of delay
Consequences of disruption
Constructability of design
Contractor qualifications
Cost escalation
Delays in decision-making and in addressing and solving problems, approaches
Delays in transmitting drawings and instructions
Delays in presenting problems
Differing site conditions
Embargoes
Environmental compliance
Environmental constraints
Equipment availability
Equipment quality
Equipment suitability
Exchange rates
Existing utilities and underground facilities
Governmental acts
Governmental stability
Ground subsidence
Ground support requirements
Groundwater
Hazardous waste
Labor availability
Labor force substance abuse
Labor productivity
Labor skills
Managerial and supervisory capability/availability
Managerial and supervisory efficiency
Material availability
Material quality
Material shortages
Means, methods, and techniques
Measurement of pay items
Mobilization cost payment methodology
Owner bureaucracy
Owner decision-making process
Owner familiarity with construction
Owner-furnished material and equipment—quality
Owner-furnished material and equipment—timeliness of delivery
Payment cycle
Permits and licenses
Quality of construction management
Quality of inspection and monitoring
Quality of plans
Retention
Safety
Schedule constraints
Site availability
Site access
Site congestion
Site drainage
Site security
Soil stability and strength
Subcontractor availability
Subcontractor qualifications
Subcontractor reliability
Sufficiency of plans
Supplier/vendor competence
Supplier/vendor performance
Third-party impacts
Timeliness of construction management, contract administration and coordination
Timeliness of inspection and monitoring
Union strife and work rules
Unproven designs
Unrealistic performance time
Unrealistic contract terms
Utility availability
Utility dependability
Variations in quantity
Warranty obligations

Figure 4.3 Representative listing of risks, risk groups, and risk sources subject to management/allocation via contract terms or contracting practices (Smith, 1995)

site conditions clause providing (upon satisfaction and proof of varying requirements) for some time and/or cost relief to a contractor that actually encounters differing site conditions.

A typical differing site conditions clause distinguishes between so-called Type I and Type II claims. For instance, under the American Institute of Architects' (AIA's) standard version of a differing site conditions clause, Type I conditions are defined as "subsurface or otherwise concealed physical conditions which differ materially from those indicated in the contract documents," while Type II conditions are defined as "unknown physical conditions of an unusual nature, which differ materially from those ordinarily found to exist and generally recognized as inherent in the construction activities of the character provided for in the contract documents." (See AIA Document A201, General Conditions of the Contract for Construction, 1987 edition, paragraph 4.3.6.)

Another widely used differing site conditions clause is found in the EJCDC Standard General Conditions of Construction Contract, 1997 edition, paragraph 4.02, "Subsurface and Physical Conditions." This clause directs the contractor to those reports specifically identified in the supplementary conditions to the contract for an explanation of subsurface conditions; the clause also directs the contractor to drawings identified in the supplementary conditions that depict physical conditions in or relating to existing surface or subsurface structures, and that were used by the professional in preparing the contract documents. The clause further limits the contractor's reliance on these reports and drawings to the general accuracy of the "technical data" contained in the reports and drawings and excludes these reports and drawings from the definition of "contract documents." Further, the contractor's right to rely on any interpretive information or utility data from other sources not verified by field data is expressly disclaimed. Under paragraph 4.03 of the EJCDC Standard General Conditions of Construction Contract, the contractor can submit a claim for a "differing site condition" if any one of four of the following benchmarks of subsurface or physical condition at or contiguous to the site is uncovered or revealed:

- the condition is of such a nature that it establishes that the "technical data" on which the contractor was entitled to rely is materially inaccurate
- the condition is of a nature that a change is required in the design, i.e., contract documents
- the condition differs materially from that shown or indicated in the contract documents
- the condition is of an unusual nature, and differs materially from conditions ordinarily encountered and generally recognized as inherent in work of the nature required by the contract documents

Inclusion of these types of differing site conditions clauses in contract documents generally results to some degree in a shift or re-allocation of the time and economic risk from the contractor to the owner of the construction project. Under the general common law rule, where the contractor assumes the risk of increased time and cost impacts due to the encountering of differing site conditions, a prudent and cautious contractor typically will include in its bid a contingency to ensure (or, at a minimum, to provide some level of economic comfort) that the work can be profitably performed if differing site conditions are actually encountered. Since, under that common law rule, the contractor has to set the amount of contingency to carry, it is in the contractor's interest to carry a contingency that is on the "high side." This is of particular concern to the owner since the contractor may include in the bid a contingency that is never disclosed or utilized because the contractor never actually encountered the kind of differing condition that prompted the contractor to include the contingency in the bid in the first place. Of even greater concern, and in contrast to contractors who carry this high contingency, is the probably more common low-bid (most optimistic) contractor, who discounts the risk and includes a small (or no) contingency in the bid in an effort to win the contract. To avoid these unnecessary contingencies in bids that are intended by the contractor to guard against the common law rule, many owners include differing site conditions clauses in the contract documents, which, to some degree, shift the risk of additional time and cost impacts to the owner due to the discovery of genuinely unanticipated and materially different site conditions. Owners who have been involved in numerous projects generally have found that they can save money by assuming the risk of certain unforeseen site conditions, and thereby allow contractors to remove the contingencies placed in their bids and lower their bid amounts for the work.

4.3.2 Disclaimers

A related issue is whether disclaimers regarding information made available to the contractor should be included in the contract documents. Decisions concerning this issue implicate a number of related considerations requiring engineering judgment and legal input (from both the owner's counsel and the professional's counsel). As a matter of engineering judgment, the professional needs to make an evaluation as to the reliability of the information or other factors that may affect whether the information should be disclaimed, either in whole or in part; presumably, and as a purely technical matter, the greater the confidence level as to the reliability of that information, the less the need for any disclaimer. Additionally, the professional must consider the relevance of the information to the final design; if the information is not necessarily relevant, then it may be appropriate to disclose it but to disclaim the accuracy and/or completeness of that information, rather than to withhold that information altogether. In addition to engineering judgment, the professional should advise the owner to seek the advice of experienced legal counsel regarding the effect and/or enforceability of such disclaimers, as well as what should be done in the preparation of contract documents to increase the likelihood of enforceability of such disclaimers (see Section 10.2.1.1 of Chapter 10 for further details).

Legal counsel should also consider the extent to which disclaimers will be enforced by the courts when the contract documents also contain a differing site conditions clause. The owner should be advised that the inclusion of disclaimer provisions may result in bidders including a contingency for the risk associated with the disclaimed information, thereby increasing the dollar amount of bids and the contract price, irrespective of whether those risks are ever encountered or materialized and despite the inclusion of the differing site conditions clause.

Disclaimer-type contractual provisions may take several forms, including provisions that:

- Disclaim responsibility for the completeness and/or accuracy of site information;
- State that information regarding site conditions is not part of the contract documents and is furnished solely for the information and convenience of the contractor, and is not guaranteed;
- State that the contractor should make its own independent examination of site conditions;
- State that estimated quantities are furnished for bid comparison purposes only and are not guaranteed; and
- State that information regarding site conditions contained in the bid and contract documents are not a part of the contract documents and were prepared solely to assist the owner in connection with the design of the project.

As a general guideline, disclaimers, when used, should: (1) not be general blanket statements, but, rather, should relate to specific information contained in the documents or reports; (2) be located in close proximity to the information that is being disclaimed; and (3) clearly allocate to the contractor all risk for the data that are being disclaimed. In addition, the professional should advise the owner that there is a risk that the disclaimer clause may not be enforceable under the governing law (the owner should consult legal counsel to determine the enforceability of the disclaimer clauses). Finally, if disclaimer clauses are going to be inserted into the construction contract, the bidders should be given sufficient time to study the information being disclaimed and instructed to make their own investigation of the site, as necessary, to assume the risk of the disclaimed information. The indiscriminate use of disclaimers could lead to unfair risk allocation between the owner and contractor and, as such, may increase the risk of professional liability (see Section 10.2.4 of Chapter 10). For that reason, the professional should consult with its own legal counsel regarding the appropriateness and phraseology of disclaimers.

4.3.3 Site Investigation Clauses

Courts have ruled that a contractor has an implied obligation to make a minimal inspection of the site to familiarize himself or herself with the site. Most contracts include an express "site inspection clause," obligating contractors to inspect and familiarize themselves with the site and conditions at the site. AIA

Document A201 and the EJCDC Standard General Conditions of Construction Contract (EJCDC No. 1910-8) both refer to such site inspections by the contractor, though their approaches differ slightly.

AIA Document 201 references the contractor's site inspection under subparagraph 1.2.2, which informs the contractor that the act of signing the contract is considered an acknowledgment that the contractor has visited the site, become familiar with the local conditions under which the work is to be performed, and correlated personal observations with the contract requirements. Under this contract provision, even though the contractor may not have actually visited the site during the bid period, the act of signing the contract for construction is an acknowledgment that he or she has done so.

The EJCDC document, in sub-paragraph 4.03(C)(2)(b), precludes a contractor's claim for an adjustment in the contract price or time if the claimed differing site condition could reasonably have been discovered by the contractor as a result of any examination, investigation, exploration, test, or study of the site and contiguous areas required by the bidding requirements or contract documents to be undertaken prior to the signing of the contract. In the case of the EJCDC wording, the contractor is explicitly informed that no additional compensation will be permitted for any condition that could have been reasonably found by the contractor during the prebid investigatory period required by the bidding and contract document requirements.

The application of a differing site conditions clause must consider the presence and effect of the site inspection clause. When a site inspection clause is included in the contract documents and the contractor fails to perform the site inspection, the contractor may lose the right to recover for a differing site condition that reasonably could have been disclosed or inferred from such a site inspection.

Unless qualified, disclaimed, and/or precluded by the terms of the contract, the contractor is entitled to rely on information received from the owner, especially when he or she does not have sufficient time or opportunity to conduct a thorough investigation or when such an investigation would not be reasonably expected.[5]

4.3.4 Design Versus Performance Specifications

The professional should be careful, when preparing the contract specifications, to consider whether each specification section is a "design" specification or a "performance" specification. Most contract documents contain both types of specifications; it is important for the professional to understand the potential risk allocation consequences of the distinction. Design specifications explicitly detail the design of the work and, typically, do not permit deviations. Design specifications carry with them an implied warranty from the owner that the specifications are complete, accurate and constructable. In the context of risk allocation, the owner who issues a design specification bears the risk of design adequacy or, in other words, bears the risk of the accuracy, completeness, and constructability of the design (see Section 10.2.2.2 of Chapter 10 for further information on this subject).

Performance specifications, on the other hand, specify the results or objectives to be achieved, and/or the criteria on which the design and/or construction will be based; the contractor uses experience and ingenuity to develop the design details and the means of achieving the desired end result. The contractor is given discretion to develop the details of implementing the design, consistent with the stated performance criteria. The risk of failure to perform under a performance specification is allocated to the contractor. Generally, for a contractor to successfully pursue a claim based on defective performance specifications, the contractor must prove that performance was either impossible or commercially impracticable.

If a project is to be bid through the public bidding process, most public procurement statutes and regulations (typically absent special legislation, such as design-build legislation) preclude an item of work that is permanent in character and part of the completed project (e.g., a permanent slurry wall system) to be specified by a performance specification, since the work typically must be competitively bid based on a detailed design contained in the contract documents.

4.3.5 Specification of Construction Means and Methods

Subsurface construction carries with it the risk of ground deformations that can be detrimental to nearby underground utilities and/or adjacent structures. Generally, ground deformations are dependent on the method of construction selected by the contractor and the timeliness of the contractor's work. The design professional (in developing a state-of-the-art design) often will consider and prescribe some portion of the contractor's means and methods as a way to specify the timeliness of the contractor's work or the actual method of construction. Before undertaking the prescription of means and methods specifications, the reader is encouraged to read Section 7.6 of Chapter 7 and Section 10.2.2.2(B) of Chapter 10 with regard to the liability implications of prescribing means and methods in contract documents.

On many large public-sector projects, the owner must negotiate third-party agreements as part of the project agreements. These agreements may require that means and methods of construction be specified in the contract documents. When it is necessary to specify construction means and methods, the professional should first review the professional services agreement with the owner to see that appropriate clauses are included to clarify and carefully draw the lines of responsibility for the means and methods provided in the contract documents. However, in regard to the preparation of contract documents, the professional and the owner should carefully review the general conditions, with the view that there should be a specific provision included within the general conditions or the special provisions that specifically defines, allocates, and limits the extent of the owner's, the professional's, and the contractor's respective responsibilities for the design of particular means and methods contained in the documents, and disclaims responsibility of the owner and professional from any other means and methods of construction for the project.[6]

4.3.6 Unit Price Versus Lump-Sum Bidding Documents

Although not always looked upon as a risk management technique, the way in which the professional assembles the contract schedule of bid prices (the bid proposal form, including its associated measurement and payment provisions for the respective items of work) can be an effective tool in allocating risk, either by multiplication of unit-priced items by an estimated quantity or by some percentage of a lump sum for work completed. For the work elements of a project that a "financial impact/controllability matrix" (or some other method) has identified as risks to be managed, breaking out a lump sum item, which cannot be physically or financially quantified at the time of contract bid and acceptance, into a unit price item with an estimated quantity contained in the bid form can make it easier for a contractor to handle the unknown risk.

However, the downside to this is that a small-quantity item for which a unit price is requested from the contractor in the bid can be unbalanced by the bidder to a very high unit cost with the hopes of a higher quantity of the item bringing in extra profits under a bid price. The professional cannot always prevent this from happening, but the best defense against this is a solid estimation of the worst case financial impact of the item with a quantity set to match the owner's and professional's best estimate of the actual quantity. This "best estimate" actual quantity can then be inserted into the bid form with a fixed minimum cost inserted—the (so-called minimum-cost approach)—by the owner as the unit cost of that item to be carried by all bidders on the work. Usually, this minimum cost is fixed below the estimated cost to perform the work to prevent a contractor from reaping a windfall if there is an overrun on the item. By using less than the actual cost, the minimum cost allows for competition by responsible bidders and also ensures that a contractor will be paid at least part of the cost when the item of work must be performed. See *Boenning v. Brick Township Municipal Utilities Authority*, 374 A.2d 1214 (N.J. Super. Ct. App. Div. 1977) (upholding owner's requirement that contractor carry a minimum cost to reduce possibility of "penny" or nominal unbalanced bids), *cert. denied*, 384 A.2d 516 (1977).

This "best estimate" quantity should be developed when the final design level quantities are taken off the final bid set of contract documents. The professional's quantity takeoff staff should perform the quantity takeoff in the same manner that the contractor's estimators would when preparing the bid estimate. The quantity takeoff should be accurate and complete, and the professional should use this opportunity to make a final verification that all work in the project is included within the measurement and payment clauses of the contract documents. The quantities developed should represent a true assessment of the work required and should be of a level of reasonable accuracy that will stand up under a critical review. If a contractor discovers a gross error in the quantities of work estimated in the documents, he or she may conclude that the bid prices should be unbalanced. This could provide a windfall profit for the contractor and added costs for the owner if the actual quantities of an item significantly differ from the quantity given in the bid list. Conversely, if the bid item underruns significantly, the contractor could claim for loss of overhead and/or loss of anticipated profit. All of this notwithstanding, it is not always possible in subsurface construction for the professional to accurately estimate the quantities of each item that may be needed to accomplish the construction of the completed project. As with other topics discussed in this chapter, the professional's role in the development of estimates may have professional liability implications (see Section 10.2.1.1 of Chapter 10; see also Hatem, 1996).

Another alternative way of handling these "guesstimate" items is to provide in the contract documents for an allowance item unit cost to cover payment for the item in question with instructions to the bidder that any additional cost over the allowance given must be included by the contractor in the other items of the contract, and that no separate payment will be made for work under that item. The problem with the allowance approach, however, is that it shifts the burden to the owner (and/or professional) to develop cost estimates for performing specific types of components or work and, hence, to anticipate or assume specific construction means and methods that will be required in the performance of that work. These assumptions, on which the owner (and/or the owner's professional) will predicate the allowance cost, may later be problematic in defending contractor claims on the theory that the anticipated construction means and methods and/or allowance costs constituted implied warranties that the work could be accomplished in the manner and/or at that cost.

Several issues should be considered when utilizing unit price provisions as a means of administering the differing site conditions clause in the contract documents. Most obvious is that the contract should compute the contractor's compensation for given items on a unit price, rather than a lump-sum, basis. The contract should also specify that any increase or decrease in quantities is not a change in the contract, i.e., not subject to renegotation of the unit price due to a change in quantity. If significant changes in quantity are possible, the owner may want to set unit prices for different ranges of quantities. Unit prices bid by the contractor should have clear payment provisions, which tell the contractor that the unit prices bid contain all markups, and constitute the full payment for all contractor costs associated with the item of work. Of particular importance for administering the differing site conditions clause, a procedure should exist and be identified for measuring and auditing the actual quantities of unit price items installed or removed. It may also be advisable to specify the method of calculation of the unit quantity so that the contractor fully understands what he or she is to price in the unit price item to be bid.

4.3.7 Variation in Quantity Provisions within Contract Documents

A common differing site conditions claim involves allegations of variations in quantities. For instance, a contractor may assert that the quantity of material anticipated to be installed or excavated on the project (based on the contractor's reasonable interpretation of the contract documents) differed from the quantity of material actually installed or excavated.

The potential for differing site conditions claims over quantity variances may be minimized or avoided by including limiting language in the contract documents. For example, language could be added to a unit price contract to provide that the quantities expressed are *estimates* only, and that a variation in the final quantities encountered will not form the basis for a differing site conditions claim unless the final quantities actually encountered exceed a certain percentage from the estimated quantities. Similarly, in a fixed-price contract, lists of unit prices can be developed for various quantities of subsurface materials expected to be encountered (i.e., the excavation of rock or clay). The fixed unit prices will then be adjusted upward or downward should the actual quantities encountered differ materially from the quantities indicated in the contract documents. In any event, the contract documents must specify how and under what circumstances unit prices will be adjusted if the quantities actually encountered differ from those indicated in the contract documents.

Taking proactive measures during the contract drafting and negotiation stage to carefully define and qualify the quantities expressed in the contract documents, and to expressly provide for unit price adjustments can reduce the time spent by project participants negotiating if and when a dispute arises over a differing site condition based on quantity variations, and can help to prevent such a dispute from evolving into a formal differing site conditions claim. (For further discussion on this subject, see Darroch and Sieracki, 1992.)

4.3.8 Inclusion of Reports and Studies in Contract Documents (Disclosure of Information to Bidders)

One of the most challenging decisions that the owner will have to make for a subsurface project is whether to include reports and studies as part of the contract documents, or to provide these materials to the contractor only as reference material, with appropriate disclaimers. This question is one that is intensely debated throughout the subsurface design and construction industry and is probably best answered on a project-by-project basis, and only after the owner receives advice from qualified legal counsel.

On large subsurface projects, it is becoming standard practice to include a "baseline report" as part of the contract documents, and to include other pertinent reports, such as interpretative reports, as reference documents. (See discussion in next section.) This reduces the possibility of a statement in the geotechnical interpretative report unknowingly carrying the force of contract behind it, when it may have been intended only as an amplification of a thought being presented in interpreting the data for the report at hand. It is important that the appropriate disclaimer be attached to define the status of the other reference information, and their priority and relevance in relation to the baseline report. However, the professional should consider that not all physical conditions can or will be addressed by a baseline definition and, therefore, the inclusion of the data report and/or interpretive report as part of the information for the contractor's reliance may be appropriate. When this is done, care must be exercised to give the data report and/or interpretive report the proper status within the framework of the contract document hierarchy.

Care should be exercised in selecting which studies and reports are to be made available to the contractor for review at the time of bid. The concern is that misleading or otherwise unreliable information from an earlier report that has been superseded or changed during the course of design iteration and development will end up confusing or distracting the bidders, rather than enhancing bid competition through more informed bidders.[7] The important points are: (1) carefully selecting and defining what information and reports are to be included in the contract documents; (2) with respect to information disclosed but not made a part of the contract documents, whether it should be disclaimed or not; (3) with respect to information made a part of the contract documents, making certain to achieve consistency in the information, and to establish the order of priority; and (4) with respect to the use of baselines in the reports and other documents, achieving clarity and measurability in the articulation of those baselines.

4.3.9 Baseline Reports

The primary purpose of the baseline concept is to provide a clear and objectively defined mechanism—the "baseline"– for the contractual allocation of risk for certain site conditions and, as such, to clearly and objectively establish the "basis" in the contract price for the degree of risk and associated financial cost for the baseline conditions. These reports also facilitate the administration of the differing site conditions clause in the event of claims that seek adjustments in the contract price or time. The baseline(s) should be clearly, concisely, and consistently defined in the contract documents. The baseline need not necessarily reflect, represent, describe, or portray actual or factual geotechnical conditions. (Essex, 1997) Different baselines may result from the same geotechnical data, depending on the owner's degree or level of risk tolerance/adversity. When the owner sets baseline(s) for a project, it is recommended that the baseline report have a statement that reinforces the distinction between geotechnical fact and the contractual baseline, such as the following:

> ... [T]he provision of a baseline in the Contract is not a guaranty, representation, nor a warranty or indication, express or implied, that the baseline conditions will, in fact and actually, be encountered; rather, the baseline is primarily intended to define, for contractual purposes, those conditions which, if determined to be more adverse than defined, may result in an equitable adjustment of the contract price or time. The baseline is not a guaranty or warranty that the baseline conditions actually will be encountered in the performance of the work; nor may the contractor exclusively rely upon the baseline for the planning or performance of any aspect of its work, including without limitation the selection, design, or imple-

> mentation of the means, methods, techniques, sequences and procedures of construction to be employed by the contractor, and safety precautions and programs incident thereto.

The inclusion of a baseline will certainly facilitate the determination of entitlement to an equitable adjustment (assuming, of course, satisfaction of other prerequisites, such as reasonable reliance, adequate and timely notice, and a causal connection between any claimed time and cost impact and physical conditions materially or substantially different from the baseline). There may well be professional liability implications for the professional in connection with the use of baselines (see Section 10.2.5 of Chapter 10; Hatem, 1996; and Hatem, 1997).

The preparation of clear, concise baseline reports is an evolving and much discussed and debated topic at present-day gatherings of subsurface design professionals, contractors, dispute review board members, owners, and the industry as a whole. The reader is encouraged to read Chapter 3, on the preparation of reports and studies, for a more detailed discussion of the topic. See also Essex, 1997.

It is important that the professional and the owner work together to set the contractual baselines that will be included in the contract documents. The owner must understand the implication on the contract bid price of establishing a baseline that is either too optimistic or too conservative (when judged in relation to geotechnical reality). A baseline can be established differently depending on the "level of risk adversity" chosen by the owner (with the advice of the professional). This level of risk adversity should reflect the owner's attitudes and decisions as to whether the risk and costs of potential differing site conditions should be fixed and built into the front-end construction costs, or whether those risks and costs should be addressed by the claims or change order process as and if potential differing conditions are uncovered. The professional has the responsibility to properly and accurately translate the owner's attitudes and decisions into the baseline definition.

4.3.10 Escrow Bid Documents

Also recommended for complex subsurface projects is the inclusion of a clause in the contract documents that allows for the placement of the contractor's bid documents "in escrow" for the duration of the project, and allows for the consultation of those documents should a dispute arise regarding what the contractor included in the bid. The caution to be used here is the need to be specific about what the professional and owner require to be included in the escrow bid documents. Typically, some of the most important elements of a contractor's bid are covered by one or two lines in the documents, which provide no clear conclusion as to whether the contractor included the disputed item in the bid price or not. The contract documents, as much as is reasonable, should contain some specific guidelines about what details or backup data the owner wants to be able to track in the escrow documents. Another issue that often is the subject of discussion and dispute especially during a subsurface project pertains to how the means and methods of construction that are actually used by the contractor relate to those the contractor assumed when putting the bid together for the project and whether the latter assumptions were reasonable. Therefore, it is suggested that the contractor be required to give some level of detail in the escrow bid documents about the anticipated means and methods of construction that were considered at the time of bid, and how they might be affected by a change in the contract indications and/or in the baseline given in the documents. Many professionals believe it is not a realistic expectation that contractors will indicate meaningful information about assumed means and methods of construction in the escrow bid documents, but with a clear set of requirements about the kind of information to be included in the escrow bid documents, a pre-award check of those documents for compliance and completeness will demonstrate the bid responsiveness, as well as provide the basis for facilitating post-award disputes and claims. The key is a well-thought-out guideline for what information should, as a minimum, be included in the escrow bid documents. In developing that guideline, some consideration should be given as to what types of information would be relevant in the resolution of disputes and/or claims relating to differing site conditions encountered during construction.

4.4 RISK MANAGEMENT IMPLICATIONS FOR THE PROFESSIONAL INVOLVED IN THE PREPARATION OF CONTRACT DOCUMENTS

4.4.1 How Review Comments Are Documented

It is the professional's responsibility—and definitely in the owner's best "risk management" interest—to document and respond in writing to every review comment given by any source during the design process. Once again, in the age of today's "freedom of information" access to all project records, comments by any party during the design process can be used by the contractor to assist in proving a case of "defective design" or "defective specifications." Written responses of resolution of comments should be required by the owner on every project, and should also be a standard quality control measure for a professional's firm. These procedures are necessary to provide a clear record of the evolution of the contract documents up through bidding of the project, and will help to reconstruct the thought process of the project if a question comes up during construction of the project.

4.4.2 The Designer's Quality Control Program and Its Role in Risk Allocation

Most professionals have an operating policy within their respective firms covering project quality assurance/quality control activities. Many insurance carriers strongly encourage written assurance of a firm's quality control efforts through publication of a firmwide "quality manual" available to all employees. Complex projects, however, require a more deliberate and specifically dedicated program of quality reviews to maintain an acceptable standard of risk management. Quality should be looked upon by the professional as a deliberate and continuous process that starts with a project specific set of procedures about which every member of the project team is aware and has a role to play in the execution. Quality control formalizes the traditional elements of the design process, such as coordinating, checking, examining, reviewing, and supervising. Once a firm or project team has developed a quality manual, it is imperative that the procedures be adhered to rigorously throughout all phases of project execution. Effective quality control programs require that:

- Well-qualified professionals are assigned to perform the project tasks;
- Qualified specialists are assigned to oversee the design services;
- The technical discipline leaders are instructed to supervise the checking of each design element by a responsible professional;
- Peer and constructability reviews are conducted by qualified personnel experienced in construction and the contractor's view of bidding for work;
- Multidisciplinary, coordinated reviews of multidisciplinary projects occur; and
- The quality control process is documented, to the degree appropriate to each project.

A professional's reputation is of utmost importance to the firm. Therefore, achievement of quality professional services should be, first and last, an attitude of mind that must be a part of all personnel on the design team and within a firm.

4.4.3 Computer-Aided Design (CADD) Quality Control

As the use of computers increases in the professional's preparation of contract documents, legal issues relating to the ownership, use, and sharing of computer-generated data, software error liability, and legality of electronic data storage are all adding to the complexity of designer quality control and risk management. The following are general guidelines taken from a paper written by Michael Ingardia and John Hill (1992), from which a professional can frame a project specific CADD quality control program:

1. Do not assume that the software program that has been selected for use on the project is the right or best one without first reading all program documentation carefully. If questions arise that cannot be answered from the documentation, call the software vendor.
2. Once the professional has decided that the software is appropriate, read

and check the documentation and formulae used by the software for validity to the problem at hand. Discuss with past users the program's assumptions and limitations. Are they valid for the problem at hand?

3. Assess the margin of error in the results that the software program generates. Not all computer-generated results give answers to the precision of three decimal places.
4. Make sure that the professional that sets up the problem for analysis and the professional who checks the data input and output have both received adequate training on the use of the software program.
5. Thoroughly check all data input and output.
6. Check for data loss or errors when translating computer data from one system to another. Computer translation programs are seldom 100 percent accurate.
7. Limit access to data files through the use of password encryption.

4.4.4 Transmission of Data to Clients or Subconsultants

When submitting computer files or data to clients or subconsultants, Mark Copilot (1996) suggests that the professional must always be concerned about unauthorized changes to the files. The professional can be held liable for unauthorized changes to a data file simply because it is believed that the entire work product belongs to the professional. The professional may want to consider including a disclaimer in the electronic computer file that is being transmitted in a form similar to the following:

> Acceptance of this digital CADD information (the "materials") indicates your understanding of and agreement with the following terms and conditions. If you do not accept these terms and conditions, you must promptly return this information to ______________________ and not use any of the information contained therein.
>
> Any modification to or reuse of the materials for other than the specific purposes intended without ______________________'s prior written authorization, by **Outside Firm**, its agents, or others shall be at the user's sole risk and without liability or legal exposure to ______________________. Furthermore, **Outside Firm** agrees to defend, indemnify, and hold ______________________ and its employees harmless from all claims, injuries, damages, losses, expenses, and attorney's fees arising out of such reuse or modification to the materials.
>
> The submitted data files are intended to work only as described. These files are compatible only with XXX program, version ____ operation on an IBM-compatible PC under DOS ____. ______________________ makes no warranty as to the compatibility of these files beyond the specified release of the above-stated software or on older versions of the software. ______________________ is not responsible for uses of the material outside of or beyond the scope intended.

For CADD, it is important that the professional's contract specify the file format, including:

- Drawing levels/layers
- Symbology
- Reference file structure
- Plotting pen table requirements, etc.

If the professional's contract does not contain the details concerning CADD file format, it is recommended that this be clarified in writing before any design is started to avoid rework, which is often not reimbursable.

4.4.5 The Professional's Role During the Construction Phase

Particularly in subsurface projects, it is essential that the owner retain professional consultants to observe the work during construction for conformance with the contract documents and to provide other important administrative duties, such as the review of shop drawings and other contractor submittals and the investigation and evaluation of differing site conditions claims. (See Section 2.4 of Chapter 2 of this book for more information on the role of the professional during the construction phase.)

In general, shop drawings are drawings, diagrams, schedules, and other data specifically prepared for the project by a contractor, specialty trade contractor, manufacturer, supplier, or fabricator, which shows some portion of the work. These shop drawings are typically more detailed than the design drawings. In reviewing shop drawings, the owner and the professional should maintain their focus on whether the work, when completed, will conform to the contract documents; the contractor is to determine how the work will be executed and should be solely responsible for construction means and methods. The professional should review shop drawings only for the limited purpose of checking for conformance with the information given in the contract documents and the design concept. The professional's and/or owner's actions in reviewing shop drawings or similar submittals during the construction phase of a project can significantly increase their risk exposure on a project if they begin to cross over into the contractor's responsibility for construction means and methods. It is sometimes hard to resist the temptation to get involved in these matters.

In regard to temporary structures, the role of the professional should be to evaluate the effect of the temporary structures on the integrity of the completed structure. This may mean that the professional may want to indicate in the contract drawings how long such elements of temporary work, such as temporary bracing, shore-up, reshoring, and the like, need to remain in place, such as in the case of bracing of partially completed structural walls during backfilling. The professional usually should not review temporary construction shop drawings except where necessary to determine compatibility with the design of the completed structure (see Section 6.5.2 of Chapter 6 for further details on this subject).

4.5 DESIGN-BUILD ISSUES

The design-build method of project delivery is becoming increasingly more attractive to project sponsors and owners today, as they look for alternate delivery systems that will reduce project costs and risks from their standpoint, and the potential for claims and litigation. From the perspective of the owner, design-build provides the owner with single-point responsibility for both the design and construction of the project. Before a professional enters into a design-build agreement, however, there are several questions that should be asked and documented:

1. Who is the client in the design-build project? If the professional is a subconsultant to the contractor, then the client is the contractor. What happens if the contractor does not build what has been shown in the design documents? A professional should be particularly vigilant when a contractor does not want a professional to perform all of the standard basic construction phase services for a design-build project.
2. Is the professional working for the contractor? Is the contractor a subcontractor to the professional? Who is responsible for on-site safety and construction means and methods, particularly when the professional is prime or in a joint venture with the contractor?

Design-build projects can appear to be attractive ways for expanding one's client base and to gain entry into new areas of professional practice. One should approach such a venture with caution, with appropriate qualifications and experience, and only after having thoroughly researched the teaming partners and owner for whom the project is to be performed. Depending on how the design-build project team is set up, a professional can assume significantly more risk than that found in the more traditional design/bid/build/delivery method of construction, and may involve itself in areas beyond its qualification or expertise. (See Section 10.2.3 of Chapter 10.)

4.6 CONCLUSION

Management of risks for a subsurface project requires application of all of the owner's and professional's experience and resources coupled with clear, definitive contract documents, which contain fair and equitable allocation of project risks between the owner and contractor. When all these factors come together, the result is a constructed project at the least cost to the owner, with an appropriate profit to the contractor and professional. The professional, in the role of advising the owner on the content of the contract documents, has the responsibility for identifying, assessing, and describing the risk to the owner, making fair and realistic recommendations regarding risk allocation, gaining the owner's concurrence with those recommendations, and implementing the owner's risk allocation decisions through the preparation and inclusion of appropriate clauses in the contract documents, which clearly and fairly allocate risk according to the respective project participants' ability to manage and control the risk. The execution of this risk management program must begin in the earliest stages of the project and continue in one form or another until the project is completed and accepted by the end user. To be effective, the professional must make risk management as much a regular part of his or her "mind set," as most professionals think of quality control, checking of calculations and other routine services provided by every professional working on the project.

NOTES

1. Other chapters of this book have addressed risk allocation between the owner and professional. See, generally, Chapter 2 and Section 10.2.2.1(A) of Chapter 10.
2. The professional, likewise, has risks that must be recognized and managed throughout the design and construction process. The professional's risk (assuming that the professional has carefully understood and defined its responsibilities in the professional services agreement) includes:
   - Lost design fee
   - Claims of damage due to professional negligence
   - Loss of reputation
3. In many instances, those general conditions, as well as any supplementary and/or overriding conditions, are prepared and/or reviewed by the owner's legal counsel. In any event, it is strongly advised that the professional recommend that the owner retain qualified legal counsel to review these conditions.
4. This does not mean that the professional cannot, or should never, specify construction means and methods in the contract documents; however, if the intent is to specify means and methods, then it should be done selectively, knowingly, and with great care. If the professional knows that he or she will have some involvement in the design of particular means and methods at the time of the negotiation of the owner/professional agreement, the nature and extent of that involvement, and each party's respective obligations, should be clearly identified in the professional services agreement. In addition, clarification should be made that the professional's involvement in the specific means and methods in the contract does not create or imply a broader involvement in the contractor's means and methods for the entire contract. It is also advisable for the professional, under these circumstances, to provide on-site observational services for the purpose of determining compliance by the contractor with the means and methods specified (see Chapter 6, Section 6.5; and Chapter 10, Section 10.2.2 for additional details concerning this subject).

   Where the professional is not aware of the need to become involved in construction means and methods at the time of the negotiation of the professional services agreement, as soon as the professional becomes aware of that need, an amendment should be written to the professional services agreement to cover that specific means and methods involvement. In addition, under this circumstance, a change order is needed under the construction contract to modify the contractor's responsibility from "total" to "specifically limited."

5. Subsurface construction involves inherent risks of the unknown. The more effort that owners are willing to invest in defining the baselines of site conditions, material properties, and expected behavior of material properties, the lower the contingency for uncertain subsurface conditions that contractors have to include in their bids. Performance of an adequate site investigation and making all of the subsurface information available to contractors with a right to rely on the information reduces a contractor's bid contingency, especially when the investigation and information are clearly given as a baseline condition, and a differing site conditions clause which can be fairly administered during construction is included in the contract documents. Some professionals and owners feel that giving the geotechnical information to the contractor without disclaiming it will only lead to claims if the information is wrong. This attitude misses the point. By including the information and establishing clear, measurable baselines from which differing site conditions can be evaluated, fairer allocations of risks are established between the owner and contractor and the contractor knows that if materially different site conditions are encountered, then he or she will be entitled to an equitable adjustment.
6. If the professional does become involved in the design of particular construction means and methods, the professional should assume responsibility for that design and should have a continued involvement in the development and implementation of that design, including the review of shop drawings and other submissions regarding the design of the particular construction means and methods. The professional should also observe (for limited purposes) the contractor's implementation of the construction means and methods.
7. However, the professional should also be aware that, in today's "freedom of information" era, all studies and reports generated on a project will be available to the contractor after the bid and contract signing. Therefore, if a report or study has been prepared very early in a project that appears to contradict the final design documents, the professional should be certain that the reasons for variance with the earlier study have been documented in the project record files. This will assist in defending against possible allegations of "defective designs" or "superior knowledge" later on in the construction of the project, if problems arise in execution of the contract by the contractor.

REFERENCES

American Institute of Architects, "General Conditions of the Contract for Construction," AIA Document A201 (1987 ed.).

Albritton, John A., et al., "Improvements in Geotechnical Aspects of Contract Documents for Underground Construction," *Proceedings*, RETC, Vol. 2, 1983, pp. 1139–54.

Ashcraft, Howard W., Jr., Esq., "Avoiding and Managing Risk of Differing Site Conditions Claims," in R. F. Cushman and D. R. Tortorello, eds., *Differing Site Conditions Claims* (New York: John Wiley & Sons, Inc., 1992).

Ashley, David B., "Construction Project Risk: Mitigation and Management," *Proceedings*, PMI/Internet Joint Symposium (Drexel Hill, PA: Project Management Institute, 1981), pp. 331–40.

Ashley, David B., "Financial Risk Management for Underground Construction" *Proceedings*, RETC, Vol. 2, 1983, pp. 1203–17.

Cohen, Daniel E., "Minimizing Conflicts Between Design Professionals and Contractors," *Construction Briefings*, 2nd series, October 1995.

Copilot, Mark J., "Legal Issues: Ownership and Sharing of Computer-Generated Data," PB Network (New York: Parsons Brinckerhoff, Inc., 1996).

Darroch, B. R. and R. J. Sierack, "Proving and Pricing Damages," in R. F. Cushman and D. R. Tortorello, eds., *Differing Site Conditions Claims* (New York: John Wiley & Sons, 1992).

Engineers Joint Contract Documents Committee, "Standard General Conditions of the Construction Contract," EJCDC No. 1910-8 (Alexandria, VA: 1996 ed.).

Essex, Randall J., ed., "Geotechnical Baseline Reports for Underground Construction—Guidelines and Practices," prepared by the Technical Committee of the Underground Technology Research Council (ASCE, 1997).

Gould, James P., "Geotechnology in Dispute Resolution," *Journal of Geotechnical Engineering*, July 1995, Vol. 121, No. 7, pp. 523–34.

Hatem, David J., Esq., "Differing Site Conditions: Potential Liability and Precautions for Design Professionals," *Defense Counsel Journal*, October 1994, pp. 565–72.

Hatem, David J., Esq., "Geotechnical Baselines and Geotechnical Reality: One and the Same, Similar or Not Even Close—Professional Liability Implications," *The CA/T Professional Liability Reporter*, Vol. 2, No. 2 (Boston: Burns & Levinson LLP, December 1996).

Hatem, David J., Esq., "Geotechnical Baselines: Professional Liability Implications," *The CA/T Professional Liability Reporter*, Vol. 3, No. 1 (Boston: Burns & Levinson LLP, October 1997).

Ingardia, Michael P., and Hill, John, "CADD Risk Management For Design Firms, CADD Contracting Guidelines for Design Firms" (Washington: American Consulting Engineers Council, 1992).

Kuesel, Thomas R., "Allocation of Risks," *Proceedings*, RETC, Vol. 2, 1979, pp. 1713–24.

Parsons Brinckerhoff Quade & Douglas, Inc., *Parsons Brinckerhoff Project Management Policies and Procedures* (New York: 1996).

Smith, Robert J., Esq., "Risk Identification and Allocation: Saving Money by Improving Contracts and Contracting Practices," *The International Construction Law Review* (Lloyds of London Press, Ltd., 1995), pp. 40–71.

Thomas, Randolph H., et al., "Understanding Defective Specifications," *Journal of Construction Engineering and Management*, Vol. 121, No. 1, March 1995, pp. 40–71.

5

RISK MANAGEMENT CONSIDERATIONS FOR COMPLEX SUBSURFACE PROJECTS

Thom L. Neff, Ph.D., P.E.
Parsons Brinckerhoff Quade & Douglas, Inc.

5.0 INTRODUCTION

This chapter will continue the general theme of this book, "how to best define and then fairly allocate the inherent risks for subsurface engineering and construction," but with particular emphasis on the "extra" care required for complex subsurface projects.

Section 5.1 will define a complex subsurface project, and note that size/cost is not necessarily the best indicator of complexity. Section 5.2 will discuss the phases through which a typical complex subsurface project must pass, emphasizing the importance of coordination and communication as the project moves from conception to reality. Section 5.3 describes the roles and responsibilities of the various players on these complex subsurface projects, with suggestions regarding alternative relationships. Sources of uncertainty and risk are presented in Section 5.4, along with optional means to define and control them. Section 5.5 briefly reviews the more common environmen-

Subsurface Conditions: Risk Management for Design and Construction Management Professionals, edited by David J. Hatem
ISBN 0471-15607-8

tal constraints or impacts likely to be encountered and offers several possible mitigation measures. The risk management process can be enhanced and improved in numerous ways, which are described in Section 5.6, to include formal systems, such as total quality management (TQM), and less formal aids, such as incentives and computer software. Section 5.7 presents a framework, the STEPS Approach to Risk Management, that is both inclusive and flexible enough to improve the risk management process for complex projects.

Much of this chapter will consist of formalizing a reasonable approach to risk management, using lessons learned from actual experience on such projects. The critical importance of communication and coordination, beginning at the earliest possible time in the planning process, will be emphasized.

5.1 DEFINITION OF A COMPLEX SUBSURFACE PROJECT

An average or typical project should not require an unusual approach to risk management. However, now that we have clearly entered the information age, with laptop computers, the Internet, voluminous and complex state and federal regulations, soil-structure interaction finite element analytical programs, and the like, we sometimes think all projects are complex. But in reality, a $10 million parking garage being constructed in horizontally-bedded, dense sediments in the Midwest will likely present fewer complex civil and geotechnical issues than a $100 million combined sewer outfall tunnel system in a glaciated northeastern U.S. city.

Size or cost alone, however, do not represent adequate criteria for determining whether or not a project is "complex." In this age of mega-projects, any project costing more than $1 billion would automatically fall into the complex category. It is difficult, however, to pick a dollar value above which one might call a project complex and thus employ unusual risk management techniques. A cursory review of the current mix of U.S. projects that have a significant geotechnical element suggests that $25 million is a reasonable figure to select for subsurface projects. If the project in question has a total value of $25 million or more (in 1997 U.S. dollars), it is likely to present risk issues sufficient to warrant extra consideration of risk management and allocation.

A second important criterion to consider is the percentage of the total project cost that comprises geotechnical factors. Of all the disciplines that comprise civil engineering, geotechnical elements pose the largest single source of uncertainty and risk. A significant number of claims and lawsuits result from the geotechnical or subsurface portions of projects. It appears reasonable to select 25 percent as a guide number in this case, i.e., if the cost of the geotechnical or subsurface portions of the project exceeds 25 percent of the total project cost, the project will likely require some extraordinary risk management and mitigation measures. Another more subtle issue relates to the actual complexity of the site geology at the project location. A deep excavation in a homogeneous, granular, dry material in the southwestern United States presents far fewer potential problems than the same excavation in downtown Boston with soft, compressible clays and high groundwater levels. In general, complex geology means greater levels of uncertainty and risk. It is important to know local history and precedent in this regard, but a 25 percent (geotechnical portion of the total) figure appears a reasonable guide.

A third criterion to consider when assessing the complexity of a project is the actual number of project interfaces that must be dealt with during the process of planning, designing, and constructing the project. An "interface" is defined as a relationship between two entities involved with the project that is formal enough to require two-way communication over time, and that this communication and resulting correspondence involves proposing, defining, and meeting certain commitments that may arise from existing legislation, regulations, agreements, or contracts.

Figure 5.1 lists a number of interface categories. Obviously, specific projects could have more interfaces than listed here, but the important point is that the greater number of interfaces, the more complex will be the execution of the project. Certain compromises between the various entities will require time and other "considerations" that may slow the process and add to the total project cost and schedule.

In summary, the "25/25/25 Rule of Complexity" states:

- If the project will cost more than $25 million,
- If the project has more than 25 percent of the total cost dedicated to geotechnical elements, and
- If the project has more than 25 interfaces,

- Abutters (facility owners, with their consultants, attorneys)
  - — Buildings
  - — Utilities
  - — Subways, railroads, tunnels
  - — Streets, highways
  - — Waterways
- Regulating agencies (local, state, federal, international)
- Insurance company(ies)
- Financiers
- Adjacent projects (in design or construction)
- Public
- Client(s)
- Contractors and subcontractors

Figure 5.1 Project interfaces

1. MWRA—Boston Harbor Cleanup—Outfall Tunnel ($300 million)
2. Central Artery/Tunnel Project (Boston) ($8.5 billion)
3. Narragansett Bay Commission Combined Sewer Overflow Project (RI) ($400 million)
4. Los Angeles Metro Project ($8.8 billion)
5. North Station Transportation Improvement Project (Boston) ($80 million)
6. 63rd Street Subway (New York City) ($420 million)
7. West Side Light Rail Transits Tunnel (Portland, OR) ($150 million)
8. Post Office Square Parking Garage (Boston) ($80 million)
9. Fall River (MA) Combined Sewer Overflow System ($100 million)
10. Seattle Bus Tunnels ($44 million)

Figure 5.2 Complex subsurface projects

then, by definition, it is a complex subsurface project and some extraordinary risk management and allocation measures probably need to be considered. Figure 5.2 lists current and recent U.S. subsurface projects that fall into the complex category.

Figure 5.3 shows the completed two-level 7th Street/METRO Center Station, for the Los Angeles METRO.

Figure 5.3 Two-level 7th Street/METRO Center Station, Los Angeles METRO

5.2 PROJECT PHASES

Projects don't just "happen." Most major projects possess a long, complex history that has preceeded the beginning of their actual "life" period, i.e., when significant funding is in place and some serious feasibility evaluation or design service work is under way. A partial list of the most frequently utilized project phases includes:

1. Concept/Idea; Preplanning
2. Pre-Feasibility
3. Feasibility/Planning
4. Concept Design
5. Financing/Insurance
6. Preliminary Design
7. Permitting/Memoranda of Understanding
8. Final Design
9. Construction Management/Construction
10. Operation/Maintenance
11. Modification/Rehabilitation

The sequence of these phases will vary somewhat, and other phases could be added.

With few exceptions, most such projects begin with owner/client perceived needs. Owner/client in-house capability to carry out one or more of the 11 noted project phases varies between the extremes of a small water and sewer department that may contract out virtually all phases after some initial concept ideas and preplanning efforts, to a large agency that can perform all project phases with its own forces. Most major projects stay in the planning and feasibility phases for extended periods of time, frequently as long as 10 to 20 years, while political, social, and economic issues are defined, debated, and generally compromised, before actual design and construction begin. It is rare that technical issues dominate in the early phases, although in recent years, environmental compatibility issues and selected advancements in the state-of-the-art (design and construction) have resulted in exceptions. The comprehensive justification for any major project must include five key aspects:

1. **S** (social)
2. **T** (technical)
3. **E** (economic)
4. **P** (political)
5. **S** (synergy)

Note: The acronym STEPS will be employed in the text that follows.

To get a project through phases 1 through 5, i.e., acceptable feasibility and concept design studies and firm financing in place, the owner/client must create a comprehensive and fair justification for the project that includes the five STEPS aspects. At this point, the proponents have proven that the project offers social and economic benefits without undue environmental impacts, the local and higher political entities support both the project and the means to fund it, and all agree that it can be accomplished with known technology. Many assumptions regarding project uncertainty and risk are always made at this time, and some of them may have a fairly high probability of occurring. Project opponents use the same brand of logic, but with different interpretations of the risk and uncertainties, while trying to prove that the "idea" is a gigantic waste of taxpayers' dollars. It remains important to clearly define the risk, uncertainties, and associated assumptions at this point for two important reasons:

1. The assumptions and interpretations may have been unduly unconservative in order to breathe life (funding) into the project, and
2. Frequently, the owner/client brings a new engineering firm (which will likely have a different view of risk and uncertainty) on board at this time to carry out preliminary and final design studies, and prepare contract documents for bid.

At all phases, the importance of communication and coordination between the owner/client and his or her agents, *i.e.*, engineers, consultants, lawyers, and so on—a true "open" dialogue—cannot be overemphasized. It's the owner's/client's project (and money) and he or she should get what he or she wants. The dialogue will ensure that the owner/client has information on which to base his or her decisions. This dialogue needs to constantly focus on a number of key elements:

1. Overall project objectives
2. Efficiency and cost-effectiveness of various processes, e.g., planning, design, permitting, construction
3. Standards to which the client's agents are to be held accountable
4. Specific client political, economic, and/or social needs and objectives
5. Level of technology desired by client related to initial, capital, and O&M costs
6. Definition and fair allocation of risk (the uncertain elements) on the project, for design and construction phases

The financing and insurance for a project are usually undertaken by the owner/client with assistance from the engineer and others. Financing can frequently drive the project schedule. For large, complex projects, the insurance schemes need to be adequate for the associated risk, but there are various policies and approaches that can do the job. The selection and execution of the appropriate plan requires careful attention by the owner/client and his or her agents, especially with respect to attitudes toward risk tolerance/adversity. Frequently, the financing elements of the project will demand involvement in insurance decisions.

Wrap-up insurance policies, both for construction (contractors) and professional liability (designers/engineers) are not uncommon on large projects. These wrap-up policies provide a common focus for all project participants. They help emphasize that risks are, to some extent, shared by all, and that the project participants have a vested interest in doing their particular portion of the job well. These policies typically provide coverage for the entire project. Rates are generally lower than for individual policies, and the premium is typically fixed over the duration of the project. These policies often include educational programs and, sometimes, loss prevention programs, usually advanced by the insurance companies, but a prudent owner/client will demand that his or her staff, as well as key people from all project participants, take an active part in these programs.

As a project moves into the active design phase, the owner/client must make many critical decisions that will set the tone for the entire project. There is, unfortunately, no single approach that is embraced by all, nor any approach that will guarantee success. Local precedent should play a key role in the decisions, but not to the detriment of the overall project objectives. It is not uncommon for a very large project (a subway, for example) to constitute a once-in-a-lifetime event for a community; therefore, a new "structure" may need to be created to carry out the work. The new subway systems in Atlanta and Washington, D.C., come to mind, wherein "authorities" (MARTA and WMATA, respectively) were created and staffed to do the work. Obviously, politics would play a key role in such a case. The authority, in effect, becomes the client. A partial list of early project decisions would include:

1. Will preliminary design and final design be done by the same entity?
2. Is the project large enough to have multiple designers? If so, who will coordinate the various design contracts? How?
3. Who will be responsible for permits, and who will coordinate between the various requirements of local, state, and federal agencies?
4. Is the project large enough to warrant special attention for mitigation of abutter concerns, as well as associated agencies and entities? Will memoranda of understanding be required? Will force accounts be required? Funding? Coordination?
5. How will hazardous and contaminated materials be handled?
6. Who will serve as the construction manager? Will there be multiple construction contracts? What will the client's "philosophy" be regarding the processing of claims and changes? For credits?

7. What role will the designers/engineers play during construction?
8. How will disputes be handled during construction? Dispute review boards? Partnering? Arbitration? Other?
9. Will the geotechnical aspects be handled by single or multiple entities for borings, testing, analyses, design, construction recommendations, Geotechnical Design Summary (baseline) Reports?
10. What local, state, and federal design codes will apply to the project?
11. How will potential impacts to historic structures and archeological aspects be handled? By whom?
12. How will industrial relations be handled with the numerous unions that may staff the various contractor operations?
13. What standards and systems will be utilized for CADD services? How will they be enforced and updated?
14. What pre-qualification criteria will be required for design/engineer firms and construction contractors? (Note: Pre-qualification criteria concerns documentation of the successful completion of similar previous work.)
15. What criteria will be utilized for the integration of minority business enterprises and disadvantaged business enterprises into the project?
16. What safety standards will be employed on the project, and who will be the primary enforcer?
17. How much leeway will the construction contractors be permitted in varying from the contract documents through value engineering cost proposals (VECPs) or "unique" means and methods?
18. What level of detail will be required in the submittal process for both design and construction to ensure that the client obtains the required product at an agreed price, and on schedule?

There is a wide variety of operating structures that can work effectively to manage and control complex projects. The first decision the owner/client needs to make is: how much of the design will be done in-house? It is rare, on a complex subsurface project, for the client/owner to do all design and/or to coordinate all design done by different independent design consultants. To release some of the design to others can take the form of a single design firm doing all design, multiple firms handling parts of the project design with coordination and management by the client/owner, or hiring another entity frequently called the design manager (DM). The client/owner will also have to decide who will carry out the construction management function. Often a single entity called a managing (or management) consultant (MC) manages both design and construction, or a separate design manager and construction manager can carry out the work—but this requires someone to coordinate the two entities. Most large projects will require multiple construction contracts, demanding careful and comprehensive interface coordination to ensure project success.

5.3 PROJECT PLAYERS

The potential number of firms, agencies, and individuals that can be involved in a complex project is high. As mentioned earlier in this chapter, the number of interfaces will dictate the number of players, to a large extent, but numerous other factors contribute to both the nature of the players and their total number. It is important to distinguish between primary (principal) players and secondary and tertiary ones. A primary player would have at least the following criteria:

1. Long and continuous connection to the project;
2. A major financial or professional "stake" in the project and/or direct accountability to some funding source or agency;
3. Making important project decisions over a long period in the project life in one of the following areas:
   —Social
   —Technical
   —Economic
   —Political
4. Making major contributions to the overall tone/philosophy for the project and more or less control the project outcome.

Noting the above criteria, it appears logical to classify the following as primary players:

1. Owner/client
2. Design engineer/manager
3. Construction manager
4. Prime contractors
5. Funding agency or entity

The owner/client often includes several funding agencies, especially if the project involves federal funding.

A list of secondary players that serve important roles, but roles that are generally defined by the primary players, would include:

1. Insurance firms, brokers
2. Legal firms, lawyers independently retained by the various primary players
3. Permitting agencies (depending on the project, could be primary players)
4. Consultants, or subconsultants working for owner/client, design engineer/manager, construction manager, or prime contractor

5. Subcontractors working for owner/client, construction manager, or prime contractor
6. Affected political entities—local, state and federal

A tertiary group of players, ones that have some stake in the outcome of the project, or that have a contractual or legal connection to one of the primary or secondary players, would include:

1. Abutting property owners
2. Affected utility companies
3. Suppliers
4. Residents and pedestrians in affected areas
5. Academic or other research entities

A critical look at these three lists of players emphasizes the potential complexity of the system required to coordinate and communicate with all project players in ways that respect the rights and responsibilities of each. While a dictatorship can be efficient, such a management mode is impossible in our open and democratic society. A participatory democracy is a more realistic project model, but one with focus, strong leadership, and effective planning and management controls. It makes sense that the "primary" players are largely responsible for setting overall project criteria, and for establishing, in a formal way, how the project and its participants will deal with risk and uncertainty. A simple and effective means to do this is to begin the "dialogue" at the earliest possible time, i.e., the preplanning phase, and use (from the beginning) the concept of a "contingency fund" to acknowledge that project uncertainties do exist, and that they must be dealt with in a realistic and cost-effective manner.

The project requires a formal program or system to be in place as early as possible to clearly define the uncertain elements of the project, as well as how the elements will be dealt with as the project unfolds. Decisions regarding this aspect of the project will have ramifications in every other project area; thus, they should be made only after careful consideration, and with a serious attempt to predict how the decisions will affect costs and schedules, design philosophies, contract formation, bidding strategies, and the like. The owner/client (along with the funding agencies) should have the most input here—it's their money and their facility. There will be heavy input from the engineers, since they enter the project early and generally have a long association with it. Legal and insurance advice should be sought through the appropriate mechanisms (independently on behalf of the owner and the engineer or other professional). The program requires constant updating as the project unfolds (and the uncertainty of various elements changes) and requires constant communication to all project participants, especially the "new" ones.

As the project develops, the secondary and tertiary players come on board and must have full awareness of the project approach. The contractor, one of the most important players—a primary player—comes rather late to the "party." The contractor must feel welcome, must feel like a key and primary player, and must fully understand (and accept) the rules of the game to be an effective team player. Suggested means to establish and carry out a risk management program will be discussed in Sections 5.6 and 5.7.

5.4 SOURCES OF UNCERTAINTY/RISK

The initial step in managing risk is to clearly identify and define the risks, along with the sources of uncertainty that lead to risk. Subsurface conditions for complex projects constitutes a general term that defines a wide variety of specific topics and/or situations that have their own unique criteria and range of influence. In attempting to create a workable system to manage such risks, a comprehensive and logical framework is required, and should include at least the following elements:

1. Specific topic titles: clear nomenclature, boundary limits, relationship to other specific topics
2. Source and nature of uncertainty: single or multiple source(s), limited or extensive occurrence(s)
3. Possible means to reduce uncertainty: costs, effectiveness
4. Specific risk(s) that result from the uncertainty: possible penalties—scope expansion, exceeding budget and/or schedule
5. Contingency plan(s) (contingency fund, insurance, other) to cover the cost of the possible occurrence of event(s) that result from the uncertainty
6. Method/system to track and update the project uncertainty/risk as the project moves forward through design and into construction
7. Method/system to coordinate/communicate items 1–6 above with all participants

A critical look at the sources of uncertainty/risk for the subsurface portions of large projects would result in the following partial list:

1. Geologic profile: location of "boundaries" between various soil (naturally occurring and manmade fill) and rock layers within the horizontal and vertical extent of the possible "effects" of the planned project
2. The engineering and chemical properties of the soil and rock layers noted above
3. Groundwater conditions within the soil and rock layers noted above, e.g., amount of water, chemical properties of the water, static and dynamic pressures within the entire groundwater regime, and ability of the water

to flow through the various soil/rock layers when subjected to pressure differences

4. Occurrence and nature of both natural "obstructions" (boulders, peat zones, etc.) and man-made obstructions (rubble fill, old piles, old foundations, wharf structures, etc.) within the soil deposits, or fault zones within rock layers
5. Occurrence and nature of any subsurface hazardous materials or other contamination that would require special mitigation during excavation and subsequent construction
6. Nature of waterproofing details for planned subsurface structures; constructability issues, QA/QC measures, warranties, leakage criteria
7. Nature of details for expansion and/or construction joints for subsurface structures; constructability issues, QA/QC measures, leakage criteria, compatibility with interior finish materials/systems
8. Nature of planned temporary support of excavation (SOE) schemes; constructability issues, QA/QC measures
9. Effect of planned excavation and subsequent construction on various environmental aspects of the construction area: traffic, noise, dust, vibrations; and the effect of climatic conditions on planned excavation and construction sequence(s) and schedule(s): extreme swings between heat and cold, and excessive snow or rainfall
10. Compatibility of SOE schemes with planned subsurface structures, compatibility with adjacent existing facilities and utilities (will predicted deformations exceed "permissible" facility and utility movements?)

For each of these 10 items, adverse conditions can develop during the design and construction process that can seriously affect the final project schedule and cost. A prudent owner and his or her engineer will expend adequate resources during the planning and design phase to define, as completely as possible, the details associated with each possible source of uncertainty, and carry out studies to reduce the resulting risks to manageable levels (cannot be zero risk). At the same time, a system should be put in place to track and control actual construction conditions, and an adequate contingency fund must be established. It is noted that this is a partial list of sources of uncertainty and that local conditions and precedent will dictate the appropriate list for a specific project. Figure 5.4 gives an example of how one might define and track selected project uncertainties.

5.5 ENVIRONMENTAL CONSTRAINTS/IMPACTS

The impact of recent environmental legislation and regulation on the subsurface construction industry has been quite significant. As noted in Section 5.4, environmental issues present a major source of uncertainty and risk for this work.

Topic	Source(s) of Uncertainty	Possible Mitigation Measures
1. Geologic conditions	Variations in boundaries and layer geometry	Additional borings, test pits, field investigations, testing evaluations, analyses
2. Soil and rock engineering properties		
3. Groundwater conditions		
4. Subsurface obstructions		
5. Subsurface hazardous materials		
6. Waterproofing requirements		
7. Expansion or construction		
8. Support or excavation requirements		
9. Environmental Issues: noise, dust, traffic, vibrations, weather		
10. Response of adjacent facilities to planned construction		

Figure 5.4 Sources of uncertainty/risk

This section will explore, in some additional detail, the potential for creating constraints and impacts, and discuss what approaches and specific programs may be required to reduce them to a manageable level and permit a successful project conclusion. The list below partially covers major elements generally encountered:

1. Groundwater management (dewatering, disposal, treatment, or recharge)
2. Groundwater pollution
3. Buried, contaminated, or hazardous materials
4. Contaminated soil or rock (to be excavated and disposed)
5. Dust (which may be contaminated)
6. Noise
7. Traffic and pedestrian congestion, or restraints

8. Vibrations
9. Special worker health and safety precautions

Recall that we are limiting our concern to issues related to subsurface conditions. Items 1 through 4 above are obvious, while items 5 through 9 will occur from the excavation, support of excavation, removal of excavated material (or groundwater), and subsequent construction of underground facilities.

These issues generally receive serious attention in the planning stages of the project because they must be addressed in the project environmental impact statement/report (EIS/R). Also, numerous local, state, and federal agencies must review and approve this document, which generally occurs somewhere in the preliminary design process. Thus, the owner/client and his or her selected engineer have the most input into the success of getting environmental approvals within a reasonable time frame, and without having to accept expensive and time-consuming mitigation measures.

The public (informed and vocal) is involved to a great extent, via hearings and neighborhood meetings, and when combined with the involvement of the agencies, the process can be slow, contentious, and exceedingly stressful on all participants. It is critical that the owner/client and engineer work as a team during this process, with emphasis on clear communication and close coordination. Once completed, the EIS/R and related memoranda of understanding (MOUs) with various agencies and abutters will set the tone for many critical aspects of design and construction. Working together, the owner/client and engineer need to jointly decide on the most efficient and cost-effective means to meet required regulations and provide acceptable mitigation measures where necessary. These decisions, coupled with an agreement on the general nature of the project, i.e., a state-of-the-art "monument" versus a "serviceable and affordable" facility, need to be made, recorded, and jointly implemented.

It is not uncommon for the commitments required to complete a successful EIS/R to add 20 percent to 40 percent to the total cost of a project, particularly in an urban area. Unfortunately, the level of uncertainty and resulting risk (potential cost and schedule impact) is significant for the nine items mentioned above. Thus, it remains critically important that this process start early (in the preplanning stage) and that the owner /client fully understand the social, technical, economic, and political issues and constraints that exist in the project area. The owner /client will typically have input and advice from numerous entities during these early phases and must be proactive in selecting an approach that is both feasible and affordable. If a project is properly packaged and presented to the public, abutters, and regulating agencies, it can make a significant difference in the time and cost of getting it moved from the preliminary design stage through construction, operation, and maintenance. When the engineer becomes part of the project team (i.e., a key player), the engineer and the owner/client must form a single voice and operating instrument in order to carry out the planned work.

5.6 MANAGEMENT ENHANCEMENTS

5.6.1 Incentive System

In any endeavor, good performance is considered desirable. From a philosophical viewpoint, few would argue with stated objectives: We want to do a good job, we want to "win," we want to excel above the "norm." Unfortunately, there is little agreement on how to consistently achieve good performance. Observation of the world around us reveals many examples of success in a wide variety of fields, such as professional sports, entertainment, manufacturing, and the focus of this book—the heavy (subsurface) construction industry. We can easily know more than a little about the reasons the Chicago Bulls, Tom Cruise, the Gillette Company, and Peter Kiewit Sons are consistent "winners." A listing of several obvious strategies often cited for consistent success would include:

1. Hire well-educated, experienced staff, i.e., talent.
2. Provide strong, creative leadership.
3. Provide continuous, focused training for all levels of employees.
4. Provide good pay, benefits, working conditions, and opportunities for growth and advancement.
5. Provide creative, real motivational assistance to employees at all levels within the organization: incentives, bonuses.
6. Have a strong, creative, continuous strategic planning and general planning activity that constantly looks ahead to where the firm (entity) or project is headed relative to direct and indirect competition in the larger arena(s) in which the entity operates.
7. Have a strong management and control system in place to monitor and modify (as needed) the results/achievements of items 1 through 6.

Of these seven items, management experts generally agree that item 5—motivation through incentives—is the key ingredient in achieving, over time, above-average performance. Bill Gates and Warren Buffett are exceptions; the rest of us "human beings" both need and respond to real (direct) incentives. Those charged with creating a large, complex subsurface project must be constantly aware of the value of incentives in helping to achieve desired performance. The owner/client and engineer team need to work together to set the tone for these activities. Management literature is full of good ideas that can easily apply to the subject at hand. A brief example will illustrate the general idea.

Peter Kiewit Sons, an Omaha-based international contracting firm and owner of other business units, earned \$244 million on revenues of \$2.9 billion in 1995. Over the last five years, Kiewit stock has risen nearly 300 percent—and a key ingredient in this remarkable success is employee ownership. More than 40 percent of the company's 2,900 salaried employees hold stock in the business and typically incur heavy margin debt to make their purchases. Only employees deemed to have made a significant contribution to the success of the company are given the chance to participate. At Kiewit, there is no safety net in the form of a pension or profit-sharing plan for shareholders; only their stock stands between them and financial oblivion. There is a *direct* and *easily evaluated* link between the worker and the desired performance objective. Similar lessons abound in any area of endeavor, for those who wish to do a little research. Thus, starting with the owner/client, and extending through each member of the created project "team," the most critical performance objectives need to be identified and clearly defined. Then, to the greatest extent possible, the key players who can contribute to achieving those performance goals should have some direct and easily evaluated incentives built into the role they play. This approach will then provide a sound basis for possible success.

Once the owner/client has a preliminary design and firm cost estimate in place, he or she may choose to provide incentives to the final designers in the form of a portion of their fee being earned via the meeting of selected project schedule milestones and/or achieving acceptable design costs and construction costs (bid prices and completed costs). These incentives, of course, need to address the roles of the design manager, the final designer, and the construction manager.

Incentives for construction contractors can take the form of bonuses for early completion and the encouragement of a fairly administered value engineering (VE) program, whereby the contractor can reap a savings for creative, workable ideas to build the project.

5.6.2 Partnering

The concept of "partnering" has a relatively short history in the construction industry. The term, as used in this chapter, refers to the process developed by the U.S. Army Corps of Engineers (COE) and now used by numerous entities in the construction industry.

Partnering is a way of doing business that helps the provider of services and the recipients of the services work together to achieve both their mutual and individual goals and objectives. It is particularly relevant in the current construction industry climate, in which economic, technological, and legal considerations make every project a challenge to complete on time, within budget, and to the recipient's satisfaction.

The following definition comes from the COE pamphlet, "Partnering" (Alternative Dispute Resolution Series Pamphlet £#4/TWR Pamphlet-91ADR-P-4):

> Partnering is the creation of an owner-contractor relationship that promotes the achievement of mutually beneficial goals. It involves an agreement in principle to share the risks involved in completing the project, and to establish and promote a nurturing partnership environment.

Partnering is not a contractual agreement, however, nor does it create any legally enforceable rights or duties. Rather, partnering seeks to create a new cooperative attitude in completing construction contracts. To create this attitude, each party must seek to understand the goals, objectives, and needs of the other.

Successful partnering depends on a positive attitude. The essence of partnering is the courage to change old attitudes. Important supporting elements include:

- *Commitment.* All participants, particularly management, must support the intent and the goals of the relationship. Partners should enter into the relationship with a winning attitude.
- *Trust.* Teamwork depends on mutual trust. Trustworthiness stems from words and actions being clear and consistent, based on competency and character, and substantiated by demonstrated efforts to follow through.
- *Understanding.* Stakeholders must respect each other's feelings, responsibilities, expectations, and limitations. They also must expect and accept honest differences of view.
- *Excellence.* Partners strive to create the best possible product or deliver a premium service. Partners should expect and encourage a mutually agreeable level of excellence as a standard of the working relationship.

A complete discussion of this approach can be found in "A Project Partnering Guide for Design Professionals," published by the ACEC and the AIA.

5.6.3 Total Quality Management (TQM)

This approach to achieving quality in the workplace was created by Dr. W. Edwards Deming, who, interestingly enough, had helped with Japan's postwar industrial resurgence. He invented a structured approach to creating quality products—a simple, but profound, model.

Deming was one of the first to visualize a "system" that included all the processes and interactions associated with the planning, design, and production of a product. At its worst, a "bad" system contains parties that are uninformed and do not cooperate—there is little or no measurement of quality, no baseline for improvement, no way to know if the process is getting better or worse. When a system is working at its best, the opposite is true; there is a sense of teamwork, inspections produce qualitative and quantitative information that is used to improve both the product and the "process" of creating the product. The model emphasizes communication, teamwork, and process improvement, and requires specialized training in teamwork, structural problem solving, measurement, an understanding of variation (uncertainty and risk), and appropriate technical training to make it happen. In the early 1980s, the move toward quality began in the United States with the petrochemical, electronics, and automotive industries. The movement has grown rapidly in other industries to where it is now common in engineering consulting firms and construction companies.

For complex subsurface projects, the need for extraordinary communication, teamwork, and "process" improvement almost demands some form of quality management. Because most such projects cover a long time period, it remains impossible to create the perfect delivery system at the beginning of the project. The flexibility and willingness to respond effectively to new information is crucial to the overall success of such projects. TQM is not a fad.

To help shed some light on the effectiveness of TQM, Congress asked the General Accounting Office (GAO) to study the impact of TQM on the performance of U.S. companies. In a report issued in May 1991, GAO reviewed the management practices and achievements of 20 companies that had introduced TQM programs. The report provides the following summary:

- Companies that adopted quality management practices experienced an overall improvement in corporate performance. In nearly all cases, companies that used total quality management practices achieved better employee relations, higher productivity, greater customer satisfaction, increased market share, and improved profitability.
- Each of the companies studied developed its practices in a unique environment with its own opportunities and problems. However, there were common features in their quality management systems that were major contributing factors to improved performance.
- Many different kinds of companies benefited from putting specific total quality management practices in place. However, none of these companies reaped those benefits immediately. Allowing sufficient time for results to be achieved was as important as initiating a quality management program. (GAO, 1991)

The GAO study was the first of its kind. Its significance lies in its affirmation of TQM. The study is available, without charge, from the Association of General Contractors of America.

Construction has lagged behind other industries in implementing TQM. Part of the reason for this delay is the perception that TQM is for manufacturing only. With more experience and practice requiring TQM of their traditional suppliers, owners/clients are starting to look at designers and constructors as suppliers. As a result, contractors are being asked to implement TQM to improve the design/construction process. Many projects have been designed and built using a TQM approach.

Thus, TQM presents another option available to key team players to help ensure that planned budget and schedule are achieved, or that variations that do occur are known early and agreed as acceptable in light of "new" information. Like all management tools, its effectiveness is directly proportional to the level of commitment and effort put forth by the participants.

5.6.4 Computer Software

Of the many "tools" available to the managers of projects, computer software presents a powerful, flexible, and dynamic one that can greatly increase the likelihood of success in managing risk, especially on complex projects. We do well to remind ourselves that, as recently as the late 1970s and early 1980s, computers were largely used to speed up the dull, drudge-like aspect of our work, e.g., number crunching and payroll functions. The advent of the personal computer and a few seminal events—e.g., the invention of spreadsheets and databases—started us on an exciting ride that continues without end. The products that give us the potential to be better managers are also seemingly without end. The complexity and power of newer software requires a team approach and very critical leadership to ensure that computer utilization on complex subsurface projects is effectively applied and is, in fact, an enhancement. There are far too many real-life examples of computer systems that actually inhibit optimum performance by failing to provide timely information on the truly critical project variables. Industry leaders like Microsoft, Oracle, Lotus, and dozens of others have given us software packages to keep track of virtually everything on our project, to show how everything is related to everything else, and, most importantly, to explain what happens (over time) to everything else, when we alter, ever so slightly, one of the many project "variables." The programs themselves—be they complex models, frameworks, simple spreadsheets, or bar charts—are enhanced by the use of powerful graphic packages that permit the display of the key variables and their relationships in dazzling colors. Employing such technology is not without potential drawbacks. For every success story of how computer technology has contributed positively to a project, there are at least 10 "horror" stories that claim that the same technology failed miserably. The reasons for this are simple, but not easy to correct.

Every successful computer application involves the integration of two things: (1) intimate knowledge of the complete vagaries of the problem at hand (in our case, all aspects, from initial planning to construction and operation, of a large, complex, subsurface construction project); and (2) intimate knowledge of the "applicable" hardware and software required to "model" the situation. Finding one person who has skills in both areas is rare. Thus, teamwork and creative, flexible integration and coordination are required. Also, top management must lead such an effort, and continually emphasize the importance of overall project objectives.

The project "model" must be set up in the early stages of the project, and grow in size and complexity as the project moves from preplanning through later stages. The minimum project breakdown needs to identify the scope, schedule, and budget for each major task and subtask. Using the critical path method (CPM) or a similar approach, relationships between the critical tasks regarding sequence and priorities need to be created, but this is not likely until the preliminary design phase. For our purposes, we must go one additional step—i.e., analyze each task and subtask for the elements of uncertainty/risk that could prevent either the scope of the task from changing, or the schedule or cost of that subtask from missing the stated objectives. A program is then required to evaluate and control progress toward the stated objectives. Only the most experienced professionals (planning, design, construction, operation, and computer science) can construct and implement such a risk management program. It takes great care to select the truly critical elements, and be able to devise and implement an "early warning" system that reveals when the trend tracking is adverse, and can respond with appropriate actions to get things back on track. The relationship between the owner /client and the engineer is crucial to making this risk management system work. A clear, consistent philosophy, quickly executed when necessary, is the only way this will happen. For large, complex subsurface projects, the interrelationship of the social, technical, economic, and political elements will often present seemingly impossible choices. Frequently, the best technical answer is not acceptable for other, more compelling reasons. Thus, the need for continual dialogue between owner/client and engineer is reinforced.

The following paragraphs describe several software packages currently used on complex projects. Properly utilized, these packages can help to effectively manage and present current, complete project information to the decision makers on a timely basis.

Oracle is a relational database that has been available for use on the personal desktop computer since 1984. Relational databases structure and store data; the structuring is based on relationships within the data. Oracle is used on large and complex projects to store, structure, and retrieve the vast amounts of data input by many different departments. The Oracle server allows rapid access to the data and provides for sharing of data among many departments. This saves electronic space and time as the data are entered into a common storage space. Many users can then access the data, depending on their needs and the rights they are granted. For example, using Oracle, geotechnical department personnel can retrieve data as reports, graphs, and location plans (by linking Oracle and CADD plotting programs) in a matter of minutes. This allows for rapid dissemination of data and timely decisions and actions. Reports can be fashioned to suit specific needs, such as abutters receiving instrumentation data specific to their building. A current drawback to the Oracle system is the need for a computer programmer to append changes and new applications to the system. The changes then need to be tested for technical use by the engineers. This process adds time to go from development to use of new and changed applications. A partial solution to this is to provide some training to the engineers in report and query preparation.

The chief advantages of Oracle are:

- A great amount of data is available to many users, each user having different needs
- After data are entered into the Oracle database, there are no limitations to the reporting formats that can be written to retrieve and distribute the data
- Data integrity is easily provided by triggers and specific rights issued to each user
- Speed of data entry and retrieval makes data available in a timely manner to those who need to interpret the data

Graphic Data System (GDS) is an object-based, computer-aided design-drafting (CADD) software package that is developed and marketed by GDS Corporation of Englewood, Colorado. GDS runs on several workstation hardware platforms, including DEC, Hewlett Packard, and SUN Microsystems. GDS provides the full set of CADD functions normally found in the workstation-based products. These include 2D drafting, use of reference files, plotting, attribute data, 3D modeling, spatial (GIS) modeling, macro programming, BASIC programming, 4GL interfaces, SQL database interfaces, and customized menu creation. Additionally, GDS has the following unique capabilities or features (i.e., not found in other competing CADD products):

1. Object-based approach to the creation, editing, viewing, and management of graphic entities. When implemented correctly, this feature can provide a high level of data segregation of granularity, thereby enabling the establishment of centralized graphic databases that are shared by several groups of users.
2. Area Management System (AMS) module enables the creation of seamless maps across a large geographic area. This feature greatly simplifies CADD file management, provides easy access to central graphic databases, and allows users to work on the same graphic databases concurrently.
3. Dynamic Graphics module provides real-time updates without redrawing or repainting the screen. Objects can change color and position based on internal or external data. This feature is extremely effective for traffic control/congestion management systems.

Geographic Information System (GIS) technology integrates spatial data (stored in CADD systems) with tabular data (stored in relational database systems). Integration of these two data streams makes possible the spatial analysis of events and enhances reporting these events in tabular and graphic formats.

A relational database management system consists of an organized collection of information that uses a set of computer programs to manage, retrieve, and analyze report data. These data are typically known as attribute data, and are stored in tables. One characteristic of a relational database management system is that it tracks relationships between records in different tables. The attribute data are indexed to gain fast access, to ensure that each record (row) in the table does not duplicate other data, and to correlate information from different tables. The major advantage of a relational database is that it can impose integrity checks to ensure the accuracy and consistency of data.

To communicate with the relational database, Structured Query Language (SQL) is used. This international standard language (ISO as well as ANSI Standard) was developed in the early 1970s to communicate with a variety of different types of databases, from simple spreadsheets to relational databases. SQL is easy to learn because it is based on English-like statements.

SQL is used to query, insert, update, delete, and modify data. It allows users to create the database tables, store, and change information in these tables, as well as to maintain the database itself. It also interfaces with additional programming languages, thereby allowing for the processing of many sets of records at a time.

While relational database management system technology is powerful and sufficient to store and analyze attribute data as well as generate reports, it does not incorporate spatially defined graphics, such as map data. However, the attribute data contained in the relational database can be linked to files containing graphics, such as CADD files that contain the location of the various geotechnical instruments.

5.7 STEPS APPROACH TO RISK MANAGEMENT

This section provides a generalized framework to assist in the creation and management of a risk management program for large, complex subsurface construction projects.

There is no magic or mystery about effective risk management on complex subsurface projects. Obviously, effective risk management can result only from a team effort by numerous individuals and entities with different backgrounds. Proactive leadership and full support from top management are also essential ingredients. Since the owner/client initiates pre-project planning, and then, along with the engineer, starts planning and preliminary design, a set of initial, clear, project objectives regarding scope, schedule, and budget must emerge. These objectives must be completely defined with an array of tasks and subtasks required for successful completion. Sections 5.6.1 through 5.6.4 above offer suggestions for optimizing the development of these management tools.

The next step, the real beginning of risk management for the project, takes each task and subtask and analyzes it for potential uncertainties—uncertainties that lead to real risks (i.e., risk of not completing the task on time, nor within budget). This analysis is the crucial step, and often is done in an incomplete manner, or not done at all. Engineers, contractors, and construction managers frequently emphasize only technical and economic issues when performing an analysis or design effort. The political and social aspects of an issue can easily overwhelm the other two, and force a resolution that is neither the most economical nor the most technically viable. The analysis and justification for such a decision can become extremely complex as an owner /client attempts to justify what might initially appear to be a less than ideal economic or technical solution. In these cases, a detailed cost-benefit analysis must assign "value" to technical, political, and social issues, and this process requires careful judgment.

The important point is that in conducting such an exercise, the owner/client and the engineer will decide early on where the primary uncertainty/risk-related issues lie, will have clearly defined them, and will have put in place a tracking management and control system that will detect variations from predicted objectives in time to apply required corrective actions. A second point that is clear—especially for large, complex, subsurface projects—is that all important decisions relating to scope, schedule, and budget factors range beyond the purely technical and economic aspects.

Logic would suggest that the optimum solution would be the one that gives the best technical solution at the lowest price, but experience shows that this rarely happens, because we must function in a real, complex world, where compromise is often required. The STEPS approach, developed by the author, seeks in an informal way to review the important project elements with an eye for five possible aspects that can alter scope, schedule, and budget as the project moves from pre-planning through construction. As mentioned previously, STEPS is an acronym resulting from the first letter of the following five words:

- **S**ocial
- **T**echnical
- **E**conomic
- **P**olitical
- **S**ynergy

Social: All projects get done by, and affect, people: owners, engineers, contractors, construction managers, and the many citizens of the communities in which the projects are designed and constructed. People do not always act logically, nor in a predictable fashion. It is important to prevent an adversarial attitude from developing early in the project, even in the planning stage. Public and client education cannot start too early: stressing and explaining the obvious—there will be some impacts and change! However, a properly conceived project will minimize the impacts and deal effectively with the changes, will come in at a fair and reasonable cost, and will not have major short-term or long-term negative impacts. The more communication that can occur, the more likely that problems can be resolved. Public input to the process is often helpful, since the users of the projects (the ultimate payers) frequently have an efficient way of looking at things that some professionals lack. It is also important to note that the owner and citizens of the community where the project is to be built usually understand local politics, whereas the engineers and contractors may not be local and may need significant education in this regard.

Technical: We generally can control this aspect better than the others. We understand stresses and strains, and deformations; how to design the various structural elements; the appropriate safety factors to employ; and when to inject appropriate conservatism. Uncertainty does exist, but it can usually be dealt with by applying a reasonable amount of analyses.

What is essential is to have engineers and construction staff both involved during the design, to ensure that practicality and constructability have been considered. For most underground structures, one need not worry about the third decimal place in a stress analysis. Some younger engineers, with their vast array of finite element techniques, can get lost in an elegant analysis and design something that is not practical, or perhaps even not buildable. The plans and specifications must reflect constructability reviews by experienced contractors. Design firms must have this expertise in-house, or utilize appropriate consultants in the design process.

Economics: The economic aspects of underground projects are always complex and can easily overwhelm other project factors. It starts with the engineer's estimate. If this misses the mark, the project begins on an uncertain note. If the winning bidder is very low compared to all other bidders, it also puts a cloud on things, even if all bidders are pre-qualified (which they should be for any underground project). If the local economy differs from the national economy, added inconsistencies can develop, depending on the mix of prime and subcontractors. Cost increases can occur during planning/design, and particularly during construction. A properly conceived risk management plan should avoid a complete surprise. The "plan" must anticipate possible problems and have contingency plans and funds in place.

Politics: Often, technical individuals are not fully aware of political issues and the broad ramifications of the projects with which they are associated. Local agency regulations, local union politics, local OSHA enforcement guidelines, and even local politically-based funding and project payment practices can all influence the project. Elections that change local, state, or national politicians just before bidding, or during the construction, can have a dramatic effect on a large, complex project. There are various ways that political entities can create and maintain vested interests in certain projects, and these interests can have significant effects on the project. Politicians may favor certain local unions that have punitive work rules and add unnecessarily to a project cost. Political aspects remain difficult to cover adequately in a set of contract documents. Cooperation and communication among the key entities can help to minimize the problems, but the "plan" must consider and attempt to quantify these aspects.

Synergy: All good projects have synergistic aspects. But, so do bad projects. Synergy, unfortunately, works both ways. A common definition of synergy follows:

> Synergy denotes the combined and simultaneous action of separate parts which, taken together, have greater total effects than the sum of their individual parts, and this action is essentially not predictable given the knowledge of the rules governing the separate parts.

More simply stated, it suggests that we cannot make totally accurate predictions, that risk and uncertainty exist, and that complicated ventures (e.g., complex subsurface projects) will have surprises. Uncertainty leads to risk, and, thus, it remains critical to ensure that all members of the team pull together—and pull in the same direction. We need to think about ways to foster a true spirit of cooperation on such projects to define and manage the risks. If the documents are fair to all, and if they are especially fair in allocating the risk in the project, then a good incentive for cooperation is in place. The concept of synergy forces us to put some energy into mitigation and contingency planning, since it is impossible to get it "right."

In reviewing and critically evaluating the five aspects of the STEPS approach, several common themes emerge:

- The importance of a written detailed plan
- The importance of defining and allocating uncertainty/risk
- The importance of accepting the concept of change and providing adequate contingency (in terms of time and money)
- The importance of two-way communication (dialogue) and coordination (inter- and intra-discipline)
- The importance of clearly defining all major assumptions, and allowing for potential changes in assumptions for social, technical, economic, or political reasons
- The importance of integrating design and construction elements into the process from the beginning
- The importance of providing real incentives for all critical objectives

At the beginning of any complex project, the key players must set out as detailed a plan as possible for the entire project, putting into writing the scope, schedule, and budget for all tasks and subtasks that can reasonably be included at the time. The list of primary tasks shown in Figure 5.5 is a reasonable beginning.

Planning
Conceptual design
Financing/insurance/funding
Permits/regulation/memoranda of understanding
Feasibility studies
Preliminary and final design
Construction management/construction
Bid strategies/contracting methods/procurement
Environmental impacts/mitigation (EIS/R)
Operations/maintenance
Modification/rehabilitation
Claims/changes (design and construction)
Contingencies (design and construction)
Quality assurance/quality control (design and construction)
Public participation
Cost estimating (planning through construction)

Figure 5.5 Project task list

Specific projects could have more or fewer topics. For each primary task (or assignment), a list of relatively detailed subtasks should be then added, as suggested in Figure 5.6.

At this time, each task and subtask should have a cost attached, as well as a duration (with start and stop date included). The detail of this exercise depends on the background and experience of the key team members, and, to some extent, their willingness and ability to take a realistic look at what lies ahead. When completed, this exercise can produce a simple bar chart for the entire project that shows the scope (by task and subtask), schedule, and budget for each task and subtask. While interfaces and contingencies are considered at this time, they are only implicitly included in the bar chart. A partial sample of what the bar chart could look like is shown in Figure 5.7.

The contingency for both time and money lost must be defined clearly, and, at the planning stage, might realistically be set as high as 50 percent. This is especially true if the project is in either an urban area or an isolated area, as both have high uncertainty. As the project moves through the various phases, the overall project contingency (set by the owner/client) should be adjusted in conjunction with the key players with a guide as shown below:

Planning 30–50 percent
Concept Design 20–25 percent
Preliminary Design 15–20 percent
Final Design 5–15 percent

One of the biggest problems in complex subsurface projects stems from the fact that they *are* complex, and they span long periods of time—so there is the real danger of underestimating both interface issues and actual schedules

1. **Task**—Planning
   Subtasks—Define planning procedures—by owner/client, by teams, by subconsultant, by design consultant
   Define project objectives
   Define environmental issues and permit needs, project interfaces, project funding needs, project schedule, project costs
2. **Task**—Concept Design
   Subtasks—Restate project objectives, prepare concept design of required technical disciplines (civil, structural, geotechnical, mechanical and electrical, traffic, architectural, survey right-of-way, environmental), define packaging and interfaces, define schedule and costs of project elements, define project design criteria, etc.

Note: List, in detail, all assumptions.

Figure 5.6 Project subtask list

<table>
<tr><td></td><td colspan="12">1996</td><td colspan="11">1997</td></tr>
<tr><td>Project Task/Subtask</td><td>J</td><td>F</td><td>M</td><td>A</td><td>M</td><td>J</td><td>J</td><td>A</td><td>S</td><td>O</td><td>N</td><td>D</td><td>J</td><td>F</td><td>M</td><td>J</td><td>J</td><td>A</td><td>S</td><td>O</td><td>N</td><td>D</td><td>O</td></tr>
<tr><td>1. Planning</td><td colspan="12">---------------------------?---------------------------</td><td colspan="11">-----------------------------</td></tr>
<tr><td>1.1 Define Procedures</td><td colspan="12">----------------------- ($1.4M) -----------------------
--</td><td colspan="11"></td></tr>
<tr><td>1.2 Define Project Objectives</td><td colspan="12">----------------------- ($1.7M) -----------------------
--</td><td colspan="11"></td></tr>
<tr><td>1.3 Define Environmental Issues, Permits Needed</td><td colspan="12">----------------------- ($2.6M) -----------------------
--</td><td colspan="11">------------</td></tr>
</table>

Note: Contingency for time and cost should be stated for each subtask, and recorded on this chart.

Figure 5.7 Project task schedule and budget

and costs. It remains important, then, to start with as detailed a set of tasks as possible at the outset (with defined schedules and costs), a clearly stated set of assumptions, and a realistic contingency plan (allocated funds and time).

To begin this process, the project team needs to create a table along the guidelines shown in Figure 5.8. A specific project will have more or fewer subtasks, depending on the nature of the project and the makeup of the team. It is important to review and revise the table as the work progresses. This table recognizes the uncertainty and risk, from the beginning, and will help focus on realistic measures to define and then deal with possible adverse results. This table also helps to focus on specific sources of uncertainty and how they may be linked. This exercise is difficult and time-consuming, and can be effectively done only by a multidisciplined team of experienced individuals. Each task and subtask must be analyzed for sources of uncertainty related to one of the five STEPS issues. The synergy issue related primarily to interfaces, those critical areas where the "pieces" must fit together, e.g., traffic planning between adjacent construction contracts, dewatering effects beyond the project boundaries, climatic effects on planned construction sequences/methods, or availability of disposal areas for project-created waste projects. This task analysis strives to determine what aspects of the five critical STEPS issues may change during the project "life," what is the likelihood of the changes actually occurring, and what will be the effects on the current scope, schedule and budget for that task.

As the project moves through additional phases, more detail must be included into the process, and a comprehensive document similar to that shown in Figure 5.9 should become the primary project risk management tool. In this table, the *specific* risks are dealt with via *specific* contingency or mitigation plans (with appropriate scope, schedule, and budget). The risk assessment needs to start with each specific task and subtask, but will eventually expand within a relational data base to cover planning, design, and construction phases, and the complex interface aspects.

Completing a table such as the one shown in Figure 5.8 for all tasks and subtasks will force the key players to assign realistic contingencies, and, more importantly, also suggest effective dialogues among all appropriate project players to mitigate many of the potentially adverse effects of the uncertainty and the resulting risks they present to the project.

The bar chart (Figure 5.7) can be redone after Figure 5.8 is completed, and both documents can be dated and agreed to by the key players. At this point, the created information can be put into an Oracle-type database, or a project management software pack, such as PRIMAVERA, where the contingencies and interfaces can be more formally defined, and control/tracking reports can easily be generated to assist in managing the overall effort.

The keys to the success of the STEPS approach are:

1. Continuous dialogue among key project players

Task Description	Start Date	End Date	Sources of Uncertainty S, T, E, P, S	Risk (Impact) Scope, Schedule, Budget	Possible Mitigation Actions	Controls

Note: The level of detail that goes into this table is directly related to the local, site-specific project issues. The social, technical, economic, and political issues will be unique for every project, as will the synergy (or interface) issues.

Figure 5.8 STEPS matrix for project tasks

Task/Subtask/Topic	Specific Risk(s) Cost	Schedule	Contingency or Mitigation Plan(s)	Control and Communication System(s)
1. Geologic conditions	Extra cost for removing or stabilizing unspecified materials.	Delays during excavations of later period of construction.	Unit prices, with time factor built in, to carry out tasks not in "base" bid.	Careful logging of actual conditions vs. those predicted by bid documents; clear criteria to evaluate potential claims, etc.; system to involve and inform owner and contractor.
2. Soil and rock engineering properties				
3. Groundwater conditions				
4. Subsurface obstructions				
5. Subsurface hazardous materials				
6. Waterproofing requirements				
7. Expansion or construction joints				
8. Support of excavation requirements				
9. Environmental issues: noise, dust, traffic, vibrations, weather				
10. Response of adjacent facilities to planned construction				

Figure 5.9 Sources of uncertainty/risk

2. Written, dated, detailed, and signed (by key players) risk management plans that include all major assumptions
3. Periodic reassessment of sources of uncertainty, as well as reassessment of the risks to scope, schedule, and budget resulting from the uncertainties
4. Tracking of the creation and implementation of all mitigation and contin-

gency efforts, along with a continuous evaluation and implementation of a "lessons learned" program
5. Periodic updates of the plans that reflect the reassessments

5.8 CONCLUSION

This chapter has dealt with risk management considerations for complex subsurface projects. Managing risk for any subsurface project will likely entail a focused effort by experienced staff in order to achieve success. Complex projects, by their nature, require extraordinary efforts in this regard, because the complexity can create adverse situations in surprising and unusual ways.

As a general guideline, if a project meets the following criteria, an "extra" effort should be employed in the area of risk management:

1. Total construction cost exceeds $25 million.
2. The cost of the geotechnical/subsurface portion of the project exceeds 25 percent of the total cost.
3. The number of project interfaces exceeds 25, where an interface is defined as a relationship between two entities involved with the project that is formal enough to require two-way communication over time, and where this communication and resulting correspondence involves proposing, defining, and meeting certain commitments that may arise from existing legislation, regulations, agreements, and/or contracts.

The time required to move through the project phases for complex subsurface projects (preplanning through construction) can easily exceed a decade. Thus, formally tracking the many changes that will ultimately occur in the project objectives, as well as scope, schedule, and budget for the tasks required to meet the objectives, becomes a critical need. The roles of the many players in such projects must be clearly defined. The key players—i.e., owner/client, engineer, contractor, construction manager, and funding agency—will set the tone for the project, but there are a host of secondary and tertiary players (regulatory agencies, subcontractors, legal entities, abutters, and so forth) that must be included in the overall project plan, in a meaningful way, if the project is to achieve success.

A critical step in risk management for complex subsurface projects is the formal identification and definition of potential sources of risk, i.e., uncertainty related to achieving scope, schedule, and budget milestones for all project tasks. This exercise needs to begin in the planning stage, and will grow in detail as the project moves through successive stages. This exercise must assess uncertainty from the following five areas:

1. Social (local precedent, abutters, communication, impacts)

2. Technical (state-of-the-art or "functional")
3. Economic (short-term, long-term, initial cost, operating cost)
4. Political (local precedent, control issues, changes)
5. Synergy (interfaces, inter- and intra-dependency)

Changes will occur throughout the project life, and while it is conventional to concentrate on technical and economic issues, experience emphasizes that one or more of the other three areas can easily overwhelm the project if not adequately addressed on a continuing basis.

Environmental issues frequently play a major role in complex subsurface projects, especially in urban areas. The uncertainty presented by potentially contaminated soil and groundwater, noise and vibration control, traffic and air pollution mitigation, and other environmental issues can add millions to the project cost and add years to the schedule (permits, compliance, and so on). Thus, these issues must be identified, tracked, and controlled very early in the project development.

Maintaining control of the risk management system for a complex subsurface project will require state-of-the-art management techniques. Success in managing risks on these projects will only come with a highly focused effort by a multidisciplined team that recognizes the value of continuous communication between all project participants. In addition, it is strongly recommended that the team employ a variety of proven management techniques that have shown success in related complex situations, e.g., clearly identified incentives attached to the meeting of specific goals, total quality management (TQM), or partnering. One key to making the entire management system work is the successful employment of a computer-based control and tracking system. Such a system must be created and customized to "fit" each specific project. Tools such as relational databases, CADD and GIS systems, spreadsheets, and other software and hardware can be combined in creative ways to enhance the performance of the overall risk management system. It is absolutely critical, however, that the creation and implementation of the system also be directed by an experienced, multidisciplined team. It is extremely rare to find planning, design, construction, software, and hardware skills in a single individual (or even the "typical" group). Thus, this group needs to be carefully formulated with people who have successfully completed similar projects before.

To aid in the development of an effective risk management plan, the STEPS approach can help focus efforts by emphasizing that significant changes in scope, schedule, and budget are likely to happen during the life of the project, but that these changes can be effectively managed and controlled if they are anticipated early enough in the project. The project team must recognize that, in addition to conventional technical and economic considerations, social, political, and synergy (interface management) issues can easily present serious threats to successful project completion, and must be given the same emphasis in order to ensure a successful project outcome.

In reviewing and critically evaluating the five aspects of the STEPS approach, several common, important themes emerge:

- Developing a written, detailed plan
- Defining and allocating uncertainty/risk, and formal written mitigation and/or contingency plans to be employed when evaluations indicate that predictions are varying from initial conditions
- Accepting the concept of change and providing adequate contingency (in terms of time and money)
- Two-way communication (dialogue) and coordination (inter- and intra-discipline)
- Clearly defining all major assumptions, and allowing for potential changes in assumptions for social, technical, economic, or political reasons
- Integrating design and construction elements into the process from the beginning
- Providing real incentives for all critical objectives
- Conducting periodic formal evaluations of the status and progress of all key elements of the risk management plan to include success and progress for mitigation and contingency efforts

6

ENGINEERS' INVOLVEMENT IN CONSTRUCTION MEANS AND METHODS ON SUBSURFACE PROJECTS

Richard M. Simon, Ph.D., P.E.
GZA GeoEnvironmental, Inc.

6.0 INTRODUCTION

This chapter deals with the manner in which the engineer of record, engaged by the owner to design the permanent end result of the subsurface project, may also become implicated in the construction means and methods that are customarily the sole responsibility of the contractor. Excluded from this discussion is the role of the engineer engaged by the contractor directly, a subject covered in Chapter 8.[1]

An engineer engaged in the design of a subsurface project or the subsurface portion of any civil or architectural project is more likely to become implicated in the means and methods of construction of the subsurface works than in other areas of constructed facilities. The engineer has the best opportunity to study the nature of the subsurface, but conditions are never completely known until

Subsurface Conditions: Risk Management for Design and Construction Management Professionals, edited by David J. Hatem
ISBN 0471-15607-8

they are exposed during construction. To reduce some construction contingencies that may arise from the contractor's selected means and methods interacting with unknown soil and rock conditions, the engineer may choose to restrict the contractor in his or her choices of means and methods. The contractor's choice of a method of construction may influence the properties of the soil and rock, changing properties that may be relied on by the engineer for the design of the completed project. Finally, the complexity and cost of the temporary constructed facilities may rival the value of the permanent subsurface works that the engineer may design. Economic considerations may require that the engineer consider the design and construction of the temporary means and methods of construction along with the permanent subsurface structures to achieve an efficient, integrated design.

6.1 THE CUSTOMARY CONSTRUCTION ROLES

John Clark (1993) describes the "Basic Trichotomy" of authority and responsibility in constructed projects delivered under the traditional design-bid-build arrangement. In addition to funding the entire venture, the owner is responsible for developing the functional requirements of the project. The engineer develops a concept to meet the owner's needs, and translates and documents these requirements into a specific design concept. The contractor's role is to assemble the means and methods of construction required to complete the constructed facility in accordance with the end concept developed and documented by the engineer. In the traditional design-bid-build delivery approach, the engineer and contractor function under independent contracts with the owner. The contract documents, in the form of plans and specifications, provide the principal means of communication among the three parties.

The Engineers Joint Contract Document Committee's (EJCDC's) set of Agreements for Engineering Services and Construction Related Documents (1993 ff.) incorporate the customary relationships among owner, engineer, and contractor. The engineer is in charge of developing the design in accordance with the professional standard of care, (see Chapter 10, Section 10.1.1.2), and is responsible for the adequacy of that design to achieve the end result of the project, consistent with the owner's program and the project as a whole. The contractor is charged with designing and implementing a process to carry out the design to achieve the completed end product specified. This process includes the means and methods, sequences, techniques, and procedures of construction, as well as any associated safety precautions and programs, and all incidental or temporary devices required to construct the project (collectively, the "construction means and methods"). The owner relies on both the contractor and engineer to achieve a constructed facility that will satisfy the owner's requirements and criteria communicated at the outset of the project conceptual phase. A principal goal of the respective contracts between owner and engineer and between owner and contractor is to clearly allocate these distinct roles and responsibilities.

The EJCDC "Standard Form of Agreement Between Owner and Engineer for Professional Services" (1996) limits the engineer's role during construction by stating:

> During the Construction Phase, ENGINEER [sic] shall not supervise, direct, or have control over Contractor's work, nor shall ENGINEER have authority over or responsibility for the means, methods, techniques, sequences or procedures of construction selected by Contractor, for safety precautions and programs incident to the Contractor's work in progress, nor for any failure of Contractor to comply with Laws and Regulations applicable to Contractor's furnishing and performing the Work.

The EJCDC "Standard General Conditions of Construction Contract" (1990) allocates the complementary responsibility:

> CONTRACTOR [sic] shall be solely responsible for the means, methods, techniques, sequences and procedures of construction, but CONTRACTOR shall not be responsible for the negligence of OWNER or ENGINEER in the design or specification of a specific means, method, technique, sequence or procedure of construction which is shown or indicated in and expressly required by the Contract Documents.

Similar allocations of roles and responsibilities are recognized in standard contract forms promulgated by other professional societies and contractor organizations, and used by many public agencies engaged in construction. (AIA, 1987a, 1987b). The contractor has the skill and experience to devise the means and methods of construction; he or she is in control of the construction work, and his or her competence in performing that role provides the contractor with the best prospect to control the risks of construction and the opportunity to win bids and reap profits.[2]

6.2 CONSTRUCTION MEANS AND METHODS IN DESIGN ENGINEERING

With the engineer's involvement in and responsibility for construction means and methods ordinarily precluded by his or her own contract of engagement, and responsibility for means and methods assigned exclusively to the contractor in the construction contract documents, why and how does the engineer become involved in construction means and methods?

6.2.1 Engineers' Inadvertent Involvement in Means and Methods

The owner and engineer may plan for the engineer to take substantial responsibility for means and methods, often, if not always, with little or no additional compensation, in all or a portion of the construction for such reasons as described in Section 6.2.2 below. However, there are two common ways in which the engineer's actions may exceed his or her contractual obligations, and he or she may thereby inadvertently inherit construction means and methods responsibility:

1. The contractor may make submittals that contain information that the engineer must review to verify conformance with the contract documents and the engineer's design intent. Frequently, the submittals will also contain additional information for the contractor's use in material fabrication or installation, or sequences or procedures for doing the work, which are the contractor's sole responsibility. If the engineer gratuitously comments or marks up these means and methods elements, or if the engineer fails to explicitly state or otherwise indicate that he or she has excluded those portions of the submittal that relate to means and methods from his or her review, the engineer may acquire some responsibility for the outcome of the contractor's work which he or she may have actually or tacitly reviewed.
2. During the progress of the work, the engineer may suggest (or even direct) a way for the contractor to achieve an end result specified in the contract documents. By this action, the engineer may shift responsibility for construction means and methods from the contractor to the engineer.

Clearly, such inadvertent intrusions should be avoided, and the engineer's field representatives must be properly trained to refrain from such casual assumption of the contractor's responsibility. Should such gratuitous advice work well, it will gladly be accepted by the contractor without acknowledgment. If the contractor's adoption of the advice does not achieve the hoped-for outcome, the contractor may well seek additional compensation from the owner or engineer for interfering with the contractor's means and methods of construction by providing imperfect recommendations. The contractor may even seek extra compensation caused by costs incurred due to the contractor's own improper execution of recommendations that the engineer had no contractual obligation to provide.

6.2.2 Means and Methods Intentionally Incorporated in the Design

The owner and engineer have more than a casual interest in the contractor's execution of the means and methods of construction on subsurface projects. The success of the overall project depends on the success of the contractor in implementing the design and doing so on time. The engineer dealing with subsurface projects may frequently and ordinarily "encroach" on the otherwise exclusive territory of the contractor's development of means and methods in the following situations:

1. The engineer may provide soil and rock design parameters for adoption by the contractor in the temporary structure design. Generally, the engineer has implemented the soil and rock sampling and testing program and has a greater opportunity than the contractor to develop properties of the soil and rock at the site. The engineer is most familiar with the limits and uncertainties in the explorations and characterizations performed under his or her own supervision. Therefore, it may be preferable for the engineer to interpret the data to develop soil and rock parameters for use in both the permanent and temporary design elements. This role of the engineer is sometimes embodied in a Geotechnical Design Summary Report, which will guide contractors in the factors used in development of the permanent design and which the engineer believes will impact the means, methods, and temporary construction and the anticipated contingencies associated with the contractor's work (Gould, 1995). Selection of temporary soil and rock design parameters may either be suggested or required by the construction contract.
2. The engineer may provide the ultimate design or performance criteria for the means and methods designed by the contractor. This might include displacement limits for sheeting and shoring; testing procedures for tiebacks, struts, and rock bolts; treatment criteria for temporary water discharge; and grout take limits for tunnel or similar subsurface construction. Although the contractor retains overall responsibility for the design and ultimate performance of the temporary construction, the engineer has provided end result performance restrictions within which the contractor must carry out his or her responsibilities for the temporary design and construction.
3. The temporary means and methods of construction may be integrated into the completed project. A clear example is a concrete diaphragm wall that will serve both temporary and permanent earth support functions. The engineer may prescribe wall thickness and reinforcing to support the permanent soil loads, while the contractor may have to add reinforcing and other details to sustain different soil pressure and support distributions that may develop during the several stages of a braced or tied-back excavation construction.
4. In order to obtain necessary permits for a project, there may be special construction requirements, or the owner may have entered into certain agreements with neighbors and municipal officials that require imposition of certain means and methods. Sometimes, public hearings are required in advance of bidding. This may occur for deep excavations adjacent to existing buildings, work near railroads, work in public ways, as cut-and-cover tunnels, and other work with traffic disruptions. In order to obtain timely approvals, the engineer may design the temporary structures to obtain approvals prior to the time the construction contract is let to meet the overall project timetable.

5. Large projects covering wide areas, such as subways, tunnels, and highways, may be built by multiple contractors with adjoining contracts meeting at a common interface. Since one contractor has control of only one side of the interface, the engineer may have to design temporary means and methods of construction for both contractors, or restrict both contractors' choices in temporary means and methods to assure compatibility at the interface.

6. Nearly all construction contracts have a target completion date, by which the work must be finished. The engineer must anticipate the work that the contractor must accomplish in order to develop a practicable schedule and time limit for the contract. The means and methods and resulting schedule chosen by the contractor should appear feasible, in light of the engineer's developed schedule, so that there is a level of confidence that the contractor will finish the project on time and within budget. Both the engineer and the owner, not to mention the contractor, must be able to identify that a project is behind schedule soon enough so that the contractor can apply resources to accelerate the work to meet the schedule. The potential consequential damages passing to the owner because of the contractor's failure to complete the project on time may be far greater than the contractor can bear financially. Moreover, the surety company paid to guarantee completion of the work may balk at protecting the owner if neither the owner nor engineer monitors the contractor's progress of the work and provides timely notice of the delay to the surety.

7. The contractor's work may affect neighbors and other third parties. If the project is built by multiple contractors, adjacent contractors' temporary means and methods work must be compatible to avoid interference with independent contractors. If a contractor fails to protect adjoining structures or interferes with adjacent contract work, the project may be delayed or hampered by injunctions, claims from affected adjacent contractors, and other third-party interference. In many instances, owners will indemnify certain third parties for damages resulting from the construction, so the owner and his engineer may have a keen financial interest in how the contractor executes his work. The owner will be left to deal with these third parties long after the contractor's work is complete.

8. The observational method of design in engineering was referred to by Terzaghi (1958), Casagrande (1965), and Peck (1969) as an approach to reduce uncertainty and cost in subsurface engineering. The observational approach requires a close coordination between the design process and the construction process. The goals of adopting this coordinated effort is reduction in contingencies, over-design, and costs to the joint benefit of all the stakeholders in the project—owner, engineer and contractor.

6.3 DISTINCTIONS OF SUBSURFACE ENGINEERING

Engineering and construction in the subsurface are different from other design and construction elements. The ground is dictated by the location of the site. It may be possible to influence the properties of soil or rock for a project by adjusting the project location. Soil or rock properties of a tunnel, for instance, can be influenced by the selection of tunnel route or depth. But once the project location and depth are set, the properties of the soil and rock are fixed. The challenge for the engineer and the contractor is to use the soil and rock given to produce the permanent, end-result constructed facility.

Even though the properties of the materials are fixed, their definition is often the subject of considerable disagreement. Chapter 4 discusses the engineer's challenging assignment of exploring, testing, establishing, and communicating the engineering properties of soils, rocks, and groundwater. Regardless of the effort expended, knowledge of the subsurface is always incomplete and imperfect. Recently, the Underground Technology Research Council debated the optimum way to communicate soil and rock conditions from engineer to contractor in the form of technical reports.

Even if the site is fixed, soil and rock properties can be intentionally changed—for the better or worse. Soil improvement techniques have been developed that can densify loose granular soils (dynamic compaction), stiffen and strengthen soft cohesive soils (surcharging with or without wick drains), lower permeability (grouting and soil mixing), and reinforce soil (with piles, stone columns, and jet grouting). Soil can be cemented with permeation and jet grouting to increase shear strength and reduce compressibility.

Engineering properties chosen for design may be influenced by the manner in which these properties will be used in design. The engineer's choice of soil parameters may be influenced by his or her understanding of the variability of the geologic materials at the project site, as well as the intended use of the parameters in the various analyses and elements of the design for both permanent and temporary structures.

The performance of a particular permanent element or temporary means and method of construction may be more or less sensitive to a particular soil or rock parameter. The engineer's subsurface explorations and laboratory testing are typically described as intended solely for use in design of the project. Nevertheless, the same exploration and testing data are almost invariably used by contractors in bidding the work and in designing temporary means and methods of construction. In presenting design criteria for soil properties to be used by the contractor for temporary design, the engineer must anticipate and understand the several alternative means and methods that reasonably may be adopted by the contractor to construct the project. The engineer may choose to provide several sets of parameters appropriate for the various means and methods options from which the contractor may choose. Anisotropic shear strength parameters are an obvious soil property influenced by loading conditions. More frequently, the engineer will provide the data developed for the permanent design and allow the contractor to adopt appropriate design parameters to use in design of temporary means and methods of construction. This practice may not completely relieve the engineer and owner from some responsibility for the contractor's means and methods development.

The properties of geologic materials are influenced by interaction with the structures placed on and in them. Construction means and methods chosen with insufficient attention to disturbance of the soil and rock can unintentionally alter the geologic materials from their conditions prior to the start of work. Uncontrolled water flow can loosen granular soils and produce voids and sink holes. Some soils and rock masses are strain softening, i.e., their ability to resist shear stress diminishes substantially once they are stressed beyond their peak shear strength. Sensitive clays can drastically lose strength if permitted to strain beyond their failure limit. Rock masses that are permitted to move along discontinuities will no longer mobilize the strength from interlocking interstices of natural tight rock joints. A rigid temporary structure will accept high stresses to limit significant displacements of the supported soil and rock materials. But a flexible structure may allow displacement, which can lead to loss of soil or rock strength if the materials are strain softening, so that the flexible system may require even greater ultimate strength than the rigid system. If the contractor's means and methods permit deterioration of the soil or rock engineering properties, the ground may no longer provide the support on which the engineer is relying to found or support the permanent structure, and the contractor may find an unusual and unanticipated challenge in supporting the temporary soil and rock loads for the temporary construction.

6.4 SPECIAL SUBSURFACE CONSTRUCTION PROBLEMS

Several types of subsurface construction suggest specific types of involvement by the engineer. The contractor's performance in the design and execution of temporary means and methods of construction can have consequences on the final project, sometimes well beyond that which may be insurable. At a subsurface shopping mall project in Massachusetts, for instance, the subcontractor blaster was entirely responsible for the design of blast charges and matting to control vibrations and contain fly rock. An unfortunate arrangement of rock structure, combined with a complex blasting sequence, led to lifting of the mats early in the sequence of shot delays, thereby permitting release of fly rock from later detonations. The release of blast rock struck buildings as far as 1,500 feet from the blast site. A dozen workers on the site received minor to serious injuries, but no one was killed and no member of the public was injured. Although the contractor's insurer paid considerable money for damages, insurance could not completely indemnify the owner for the delay of the project that lay partially completed for nearly two months while the fire marshal waited for results of an insurance investigation and for the contractor's plans for increased containment of the future blasts. The developer/owner incurred more than $1 million in extra interest costs, and his tenants were delayed by an entire shopping season due to the insufficient conservatism in the contractor's method of covering blast sites.

At the L'Ambiance Plaza project in Bridgeport, Connecticut, the designer and contractor chose to construct a 16-story office building by the lift-slab technique. The structural engineer of record developed conceptual plans for the post-tensioned flat slab building, delegating to the contractor the details of the post-tensioning and structural details, including the lifting connections. In fact, several sub-subcontractors designed different portions of the structure. During a lift of several slabs to the top of the columns, the system of slabs collapsed to the ground, killing 28 construction workers (Hager, 1991). Forensic studies after the collapse indicated problems in the post-tension design in the areas of elevator openings as well as some possible deficiencies in the column-slab connections and column design for the construction lifting conditions. The design work was to have been performed by registered engineers engaged by the contractor and reviewed by the engineer. None of the various subcontractors' drawings contained registration stamps, but all were reviewed by the engineer of record, who expressed no objection. The only stamped drawing was the original concept plan by the engineer of record. It is unclear who, if anyone, coordinated this "committee design." The L'Ambiance Plaza site remains vacant today. The project was completely thwarted by a failure in coordination of final design between the engineer of record and the contractor and in design of the means and methods of construction, which were the sole responsibility of the construction contractor.

6.4.1 Shallow Foundations

Suitable subgrade for shallow foundations may be disturbed by groundwater pressure and flow, a criterion frequently more stringent than the dewatering performance required by the contractor to safely excavate the ground and install the foundations and slabs. Contract documents may have to prescribe limitations on the contractor's temporary dewatering methods to preserve subgrade stability.

Dewatering may be required to construct new foundations in areas where nearby structures, streets, and utilities could be affected by settlement induced by temporary dewatering. Once construction is complete, the contractor will allow groundwater to reach its preconstruction level. This is a common problem where surrounding structures are supported on friction piles through shallow, compressible soils. Temporary groundwater lowering may initiate rotting of timber piles and will add down drag loads to induce settlement of piles even of concrete or steel. The engineer must provide concise criteria in the design documents so that the contractor can plan (and price) his or her temporary means and methods appropriately to prevent such difficulties during construction.

The engineer may need to design a basement or cut-and-cover tunnel to be below the permanent groundwater level. To construct the building or tunnel, the contractor will generally lower the water table by pumping. In the permanent condition, the basement or tunnel will be sealed, and the weight of the structure plus any tie-downs must resist the buoyant forces of the groundwater, tending to lift the floor. The engineer must specify that the contractor operate the temporary dewatering system until sufficient superstructure weight is completed to resist the permanent buoyant water forces. If the contractor allows the floor to lift, it is rarely possible to restore it to the original grade, and insurers may be reluctant to pay to demolish and reconstruct the damaged work. Certainly, the schedule will suffer.

6.4.2 Driven Pile Foundations

Driven pile foundations are installed below the ground surface and can be inspected directly after construction by static load testing, an impractical option for production piles for ordinary construction. The engineer must adopt special means, often utilizing the contractor's means and methods of installation, to verify that the pile foundations satisfy the permanent support requirements.

An engineer ordinarily evaluates the capacity of driven production piles based on observation of resistance to driving under blows of the contractor's pile hammer. The correlation between driving resistance and bearing capacity is typically determined by static pile load tests, although dynamic pile testing is increasingly accepted by engineers around the world. Uncalibrated dynamic driving formulas are accepted in some cases. Before the pile is load tested, the contractor and engineer must agree on a provisional driving resistance for the piles which will be load tested.

The bearing capacity of piles driven with a given hammer to a given penetration resistance is not a unique value; rather, the bearing capacity is a random variable affected at least by the variation in output of the contractor's driving equipment and in variable soil conditions. When an engineer specifies a required pile capacity, what really is required is that a large percentage of the piles, whether 90, 95, or 99+ percent, will have at least the required driving resistance. Rarely is a percentage stated. Often, the engineer assumes that 100 percent will be above the specified minimum capacity. Theoretically, the reliability will always be less than 100 percent.

If the engineer intends that only a small percentage of the piles have a capacity below the specified lower limit, then the average (mean) bearing capacity of all the piles must be well above the required lower limit. If the contractor and engineer choose too large a final set or penetration resistance, the mean and range of capacities of the production piles will lead to a significant percentage of piles having a lower allowable capacity than that adopted by the engineer for design of the superstructure. If the selected final set or penetration is small, the average bearing capacity will increase, but the stresses on the pile materials imposed by the contractor's equipment will also escalate, leading to a higher risk of breaking the piles during installation. In trying to achieve a higher percentage of piles with capacities above the limit selected for design, the engineer may lead the contractor to break a higher percentage of piles during construction. Presumably, the contractor and engineer will be able to detect which piles are broken.

The owner and engineer must decide who will pay the contingency costs for broken piles if a high provisional penetration resistance is selected, and who will pay for the cost of a pile load test if the provisional test pile does not reach the specified capacity. In addition to replacement piles and pile cap redesign, broken production piles can wreak havoc on the construction schedule, in which the owner has an ultimate interest. For these reasons, the engineer may want to specify limits on the pile installation equipment so that the engineer will have confidence that the contractor's equipment consistently achieves the pile design capacity without unduly raising the risk of pile breakage. Since the engineer will use the installation equipment to gauge pile capacity, he or she may need to use the results of wave equation or other analyses performed by either the engineer or the contractor to determine a provisional or final pile driving resistance. Therefore, engineers commonly require contractors to submit details of the driving equipment and the contractor's proposed provisional driving resistance for review by the engineer. It is an element of the means and methods of construction in which all members of the construction trichotomy have a vested interest.

If the owner decides to pay for a pile load test for all test piles that do not support the design capacity, the engineer will want a level of confidence that the test pile is likely to achieve the required design capacity on installation. The engineer will need to evaluate the contractor's installation system with the goal of consistently achieving the required end product. There is no safe harbor for the engineer in selecting provisional driving resistance. As more issues are turned back into the contractor's responsibility, the contractor is likely to add contingency allowances to cover the risks. If driving conditions are much worse than anticipated, the contractor can always turn to a claim for differing site conditions.

6.4.3 Excavated Foundation Piers

Foundation piers drilled, bored, or excavated in the dry can be visually inspected by the engineer prior to concreting. The subgrade soil can be identified, and, in some instances, the consistency of the subgrade can be tested by field engineering tools. However, piers excavated under slurry cannot be inspected directly. The engineer is limited to prescribing certain criteria for construction means and methods that are likely to achieve a constructed work product that meets the engineer's intended design outcome with limited means of verification. Minimum slurry density limits may be specified to provide confidence that the walls of the pier or panel will resist collapse. The bottom cleaning process may be stipulated with limits on maximum sand content and slurry density to demonstrate removal of spoils and conditions that are conducive to proper tremie concrete placement. Although it is possible for an engineer to "probe in the dark" at the bottom of a slurry-filled hole, the results are uncertain. In the end, the engineer has limited tools available to achieve a quality final product unless the contractor delivers good, workmanlike construction in this type of subsurface work.

6.4.4 Temporary Earth Support Systems

Selection and design of temporary earth support systems are ordinarily left to the sole discretion of the contractor. Engineers, however, will frequently impose specific conditions on a contractor's discretion in this area to assure temporary earth support system performance consistent with a level of conservatism that the engineer and owner judge compatible with the owner's long range interests, a level of performance that otherwise might not be selected by a contractor due to cost pressures and a prime objective to get the job done quickly.

Several issues enter the allocation of responsibility and risk between contractor and engineer:

- *Government regulations*—At a minimum, a temporary earth support system must protect the contractor's workers. Earth support system failures, generally the collapse of unsupported trenches, are the number-one cause of worker deaths on construction sites and are responsible for 1 percent of worker deaths for all types of employment (Carper, 1987). State and federal regulations[3] and building codes[4] provide minimum criteria for protection of excavations and slopes where worker and public safety will be affected.

  The contractor's financial responsibility to the workers is set forth in various states' workers' compensation laws. These laws serve to limit the contractor's liabilities to his or her workers as well as provide a certain minimum medical and disability protection for injured workers.

  The engineer is not included in the contractor-worker relationship encompassed by workers' compensation statutes. If the engineer is found to have some responsibility for the design or implementation of the earth support system, he or she may, in some states, incur liability to injured workers without liability limit.[5] To limit the engineer's liability to construction workers, the engineer should attempt to preserve sole responsibility for design and construction of temporary earth support systems in the hands of the contractor. Nevertheless, the engineer cannot ignore the contractor's failure to carry out those responsibilities as set forth in regulations and contract documents. At the very least, the engineer has a legal obligation to evaluate the workplace for the safety of the engineer's own employees, who may be jeopardized by inadequate temporary measures designed and erected by the contractor. Section 6.5.4 suggests an approach to observe safety concerns that attempts to achieve a safe construction workplace without the engineer inadvertently accepting responsibility and liability for the contractor's design and construction of safe temporary earth support, scaffolding, forming, and other systems that may affect worker safety.

- *Deflection criteria*—At a minimum, the contractor will base the support system design on a factor of safety against failure or collapse. If there are nearby structures, such as buildings, utilities, tunnels, or streets, an experienced engineer engaged by the contractor will recognize that displacement of the earth support system due to earth removal may be the governing design criterion. The engineer may choose to impose displacement limit criteria for the contractor's design of the system to assure that the contractor protects adjacent structures. In order to keep deflections within the limit objectives, the engineer may additionally choose to impose certain high-quality, workmanlike construction means or details that will help achieve the specified deflection limits. These means might include maximum spacing of support levels, blocking between wales and sheeting or soldier piles, and preloading of rakers or struts prior to final attachment in order to remove slack in the system. Although these measures begin an intrusion by the engineer into the contractor's realm, they can be communicated to indicate only the level of conservatism and effort that the owner and engineer expect the contractor to spend on his or her own design of temporary means and methods of construction.
- *Primary support installation*—Tiebacks, struts, and rakers form the primary support elements of a braced-earth support system. Failure of one of these elements can lead to a cascading collapse of a portion of the entire temporary earth support structure. Such a collapse can lead to loss of life and substantial property damage, both on and off the project site. The engineer may wish to increase the reliability of the temporary system by imposing installation and testing criteria for these primary elements of the contractor's design and temporary construction.
- *Depth of embedment*—Embedment of the support system below the bottom of the cut is used to resist earth pressures outside the excavation. But embedment may also serve to reduce water pressures and flow if the support system also provides a continuous groundwater cutoff wall, such as by interlocked sheeting or concrete diaphragm wall. The support system, often a steel H-Z wall or a concrete diaphragm wall, may provide temporary or permanent vertical support for other elements of the construction. If a low-permeability layer is present some distance below the bottom of the cut, the engineer may choose to impose wall embedment that reaches to the cutoff layer to reduce groundwater flow into the excavation and attendant groundwater withdrawal outside the excavation, and to reduce long-term pumping below the floor if it is required. Foundation supports may have to bear on deeper strata than required to simply support the earth excavation. Extending the cutoff wall to the low-permeability layer is generally more reliable in maintaining groundwater levels surrounding the excavation than any recharge system developed by the contractor to restore groundwater caused by water flow beneath the bottom of the coffer dam.
- If the ultimate performance of the contractor's designed means and meth-

ods can be achieved only by a single means, then the engineer should specify the means necessary. This should result in fewer disputes between the contractor, owner, and engineer, since the contractor can allow for the specified means in his or her pricing. The engineer and owner can always agree to accept less if the contractor develops a preferred option that achieves equal performance and, possibly, mutual economy.

6.4.5 Dewatering

Groundwater pressure and flow are the greatest single factors influencing soil and rock behavior. High water pressure leads to lower soil and rock strength. Lowered water pressure will lead to settlement of compressible soils. Water allowed to collect on the surface of an exposed subgrade can lead to disturbance and softening of an otherwise stable soil.

In some cases, one element of the contractor's means and methods will serve multiple functions—intended or unintended. An earth support system can also serve to cut off groundwater flow. Dewatering below the base of a cut may serve to increase horizontal resistance to earth stresses on the embedded portion of the support system. Temporary groundwater lowering may be used to reduce heave from excavation of soil, thereby reducing recompression of the soil under the weight of the permanent structure. In order to achieve the ultimate performance of both the temporary and permanent constructions, the engineer may specify restrictions on the temporary means and methods of dewatering available to the contractor to avoid unintended and undesirable effects.

6.4.6 Tunnels and Shafts

Engineering of tunnels is intimately connected to constructability. Means and methods issues enter the design considerations from the outset. The most basic means and methods evaluation by the engineer is how a tunnel will be advanced—open-faced or closed-face shield in soil, drill-and-blast or tunnel boring machine (TBM) in rock. Decisions regarding means of excavation will dictate the type of temporary support or lining that the contractor can install, because excavation means and methods and temporary ground support are an integrated system. Consequently, the temporary ground support system will affect the design of the permanent lining that might be required.

The tunnel engineer must also assess the locations of shafts and portals in order to acquire rights and permits to use intended access locations prior to letting the construction contract. Spacing of access shafts may be limited by capacity of the type of equipment available to contractors. The access shafts are sometimes temporary, secured and abandoned after completion of construction. But they frequently must be incorporated into the overall design by the engineer in order to achieve the ultimate project goals.

Choice of tunnel construction means and methods, particularly when a TBM is considered, requires a choice of equipment that can bore and stabilize the entire range of ground conditions through which the tunnel is proposed. The engineer will spend years planning and exploring the route of a proposed tunnel, whereas the contractor will spend only months planning the details and costs of constructing the work. If the engineer identifies certain adverse subsurface conditions in critical portions of the alignment that restrict the performance criteria of the TBM necessary to complete the entire work, the engineer may choose to restrict the means and methods criteria for the contractor's choice of TBM equipment. For example, requiring the contractor to be able to install a temporary tunnel lining within a specified length behind a tunnel shield or other restrictions on the contractor's means and methods may lead the contractor to a choice of equipment that is readily able to respond to unstable ground conditions that might be expected over limited lengths of an otherwise stable tunnel alignment. On the other hand, the engineer will be encroaching on the contractor's means and methods decisions, which may shift some responsibility for performance of the equipment onto the engineer. This risk is in exchange for allowing the contractor to choose equipment that, in the engineer's opinion, will be unable to complete the entire alignment.

6.4.7 Earthwork

Earth fills are used in many areas of civil engineering construction. Compacted fills support structure foundations, pavements, and utilities. Dams, reservoirs, and levees are frequently constructed of compacted earth. Compacted silt and clay are used to retain or preclude water flow in dams and reservoir liners or as a precipitation barrier for landfill caps or seals around buried municipal or hazardous waste.

The basic engineering properties relied on for structural support are shear strength and compressibility. These properties can be measured in the field or in the laboratory, but the cost and time for such measurements makes them impractical for common evaluation of a contractor's earthwork construction product. Engineering research has correlated strength and compressibility of compacted soil fill with compacted soil dry unit weight (U.S. Naval Facilities Engineering Command, 1982). Fill unit weight has become the common acceptance criterion for compacted fill, using either the measurement of the weight of a small specimen of soil from a measured volume or an indirect measure of soil unit weight by nuclear gamma ray dispersion.

In place, dry unit weight is an unreliable indicator of soil permeability for compacted soil fluid barriers. Engineers frequently require sampling of in-place compacted material to obtain laboratory measurements of soil permeability. But research has shown that compaction water content influences permeability—higher initial water content yields lower permeability, even if there is some reduction in compacted dry unit weight. Therefore, the engineer may restrict the contractor's placement water content (means and methods) to a higher range to raise the likelihood that the compacted soil will ultimately achieve a sufficiently low permeability.

For either of these applications, the frequency of sampling is extremely small, perhaps less than 0.025 percent. Therefore, it is essential that the contractor use uniform means and methods to construct the fill so there is some confidence that the infrequent tests used by the engineer are representative of the entire fill. Soil engineers have discovered that the density of soil achieved is affected by the water content prior to compaction (U.S. Bureau of Reclamation, 1974; Hilf, 1975). There is a range of water contents near which the contractor can achieve a density above a certain lower limit. There is also potential variability in the amount of compaction effort that will be received over the entire fill. A single pass of the contractor's equipment may skip portions of the exposed surface. Therefore, the engineer may specify limits on the contractor's compaction means and methods, such as compaction water content range and minimum number of compactor passes, not necessarily to achieve the ultimate acceptance property of minimum dry unit weight, but to achieve a level of fill uniformity so that the infrequent testing can be extrapolated to represent the large volume of fill.

Thus, in the area of earthwork engineering and construction, the engineer may intrude on the contractor's means and methods of construction in order to achieve a constructed end result that is likely to and ultimately meets the engineer's design requirements.

6.5 SUGGESTED APPROACHES TO MANAGING THE RISKS

There is no single correct way to manage the interface between design and construction in civil engineering, particularly in subsurface projects. These two elements are so intertwined that it is impossible and undesirable to ignore the construction process in design. At the very least, the engineer must evaluate the design to ascertain that the project is constructable. The best that the engineer can accomplish is to manage the interface between design and construction in such a way as to avoid transferring risks of performance from the party who can directly control and supervise an aspect of the work, to one who has a lesser ability to control that aspect and has only secondary responsibility for a component influencing the outcome of the project. Even more critical for the engineer is to preclude ambiguity in the assignment, authority, and responsibility for design and execution so that allocation of responsibility does not become uncertain. In this section, the author holds an underlying philosophy that the responsibility for an outcome of an element of the project should reside with the party in control of that element of the work.

The following sections address issues and suggested approaches to manage the risks associated with engineers' involvement with contractors' means and methods of construction.

6.5.1 Qualifications of the Contractor's Design Engineer

The contractor must be able to competently design the required elements of the temporary means and methods of construction. If the engineer intends to allocate responsibility for design of construction means and methods to the contractor, he or she should evaluate the contractor's qualifications to perform the means and methods design on which the owner and engineer will rely. If no qualified design professional is assigned specific design responsibility, the engineer of record may inherit that responsibility even for temporary means and methods of construction. By requiring that a qualified engineer design critical elements of the contractor's temporary works, the engineer can minimize the risk of inadvertently accepting responsibility for the successful execution of the construction means and methods designed and implemented by the contractor.[6] Some contractors have the requisite design expertise and qualifications in-house; however, most contractors are required to obtain design services from an independent consultant.

Some portions of the temporary means and methods of construction may be of such a complexity or importance as to make it undesirable or impractical to transfer the responsibility totally to the contractor. These are situations where the most practical solution may be for the engineer to take on some of what is customarily the contractor's obligation, using suitable contract provisions to appropriately define and allocate the risks.

It is customary for the contractor to design, or to engage an independent engineer to design, the temporary means and methods of construction. In some portions of the work, it is also common for the contractor to design elements of the permanent structure as well, such as structural connections and specialty foundation elements. To unambiguously allocate design responsibility to the contractor for either temporary means and methods or permanent structures, the engineer should include in the specifications that the contractor's engineer satisfy qualification criteria comparable to those that the owner required of the engineer. These requirements include (1) current professional registration in the locality of the project and (2) financial resources to stand behind the adequacy of the design, *i.e.*, professional liability insurance. Without both registration and insurance, the assignment of responsibility for design to the contractor may be to an absent professional or to an empty financial resource, and ultimately may lead to claims against the engineer should the design prove to be deficient.[7]

Delegation of design responsibility for elements of the project does not necessarily transfer responsibility for coordination of the contractor's portion of the design into the project as a whole. The engineer of record must verify that the elements are consistent with the design concept. The 1981 collapse of the Hyatt Regency in Kansas City showed the consequences of a contractor-developed connection design that failed to incorporate a key design concept. The design engineer failed to adequately review the detail, even when it was repeatedly called to his attention by the contractor's engineer during the shop drawing preparation and structural fabrication (Rubin and Banick, 1987).

6.5.2 Contractor Submittals

Engineers include submittal requirements in construction specifications for a number of reasons; the most common reason is to verify that materials and equipment that the contractor will be providing and incorporating into the completed project satisfy the design concept and requirements set forth in the plans and specifications (contract documents). Most frequently, the contractor will submit information on products he or she intends to supply to meet specification requirements and shop drawings showing the assembly of elements of the structure to meet the design requirements. In subsurface, as well as in other portions of the work, contractor's submitted documents may contain information that relates to construction means and methods. In some instances, the engineer may have a legitimate need to review elements of the contractor's means and methods to evaluate how the engineer will use the means and methods to verify that the completed construction meets the ultimate design goals. Finally, there are areas of construction where the distinction between design and construction is indistinct. The contractor may be called upon to design a portion of the permanent structure. The engineer must verify that the contractor's design meets the overall design intent of the engineer.

The following sections present suggested approaches to deal with each of these types of submittal information.

What submittals to request—Before an engineer includes a requirement in specifications that the contractor submit a document detailing an element of the contractor's work, the engineer must ask the following questions: "Why do I need to know this information? Why do I need to know this now (before the contractor installs it)?" The engineer is ordinarily responsible for verifying only that the contractor has furnished and installed the designed elements in accordance with the plans and specifications. There is no general requirement to evaluate the equipment and work before they are installed.

In order to contain the engineer's responsibility, the purpose of each submittal should be stated in the specifications. If the engineer has no identified need for a submittal, it should not be requested. Reducing the number of unnecessary submittals will allow the engineer more time to concentrate on essential submittal review and will reduce the chances that the engineer will inherit responsibility for the contractor's work.

What items to review—The engineer must be careful to review the contractor's submittals only for those factors that satisfy the specified purposes for which the submittal was requested. In this way, the engineer can distinguish between portions of submittals that relate to requirements of the design documents and portions that may relate only to the means and methods of construction, which generally are the sole responsibility of the contractor (Clark, 1985). The engineer should clearly distinguish on the submittal documents those portions reviewed and those portions ignored as outside the scope of requested information.

Product submittals—The generally preferred method to specify a product is to list or refer to a set of performance criteria. In such circumstances, the submittal review involves checking that the submitted product meets the specified performance criteria.

Engineers will sometimes use the common shorthand practice of specifying a product or piece of equipment by one or more specific manufacturers' make and model with "or equal" substitute language. The question then arises: "Equal in what manner?" Without specifying the performance criteria important to the engineer in judging equality, the engineer is open to claims from the contractor for additional compensation for refusing to accept a substitute that was equal in the contractor's mind by criteria not stated in the contract documents.

The recommended practice for specifying products is to provide performance criteria in the specifications required by the design. Review of a submittal of a product proposed by the contractor involves checking manufacturers' specifications against the contract performance criteria. If the engineer chooses to specify a particular manufacturer's item "or equal," the engineer should still include the necessary performance criteria on which the design depends in judging "or equal" status.

A frequent pitfall in this area is that product submittals, often prepared by the equipment supplier, will also contain information regarding assembly and installation. In approving the submittal, the engineer must be careful to mark or state that the review includes only evaluating adherence of the material for criteria set forth in the contract documents and specifically excludes any review of information contained that relates to such aspects of means and methods of the construction.

Temporary means and methods—If the contractor is to be completely responsible for temporary means and methods, then no design criteria by the engineer or submittal by the contractor should be necessary. If the temporary means and methods will affect the overall project goals, the engineer may choose to provide design criteria and require a certain level of professional qualifications for the person or organization responsible for the design of the temporary means and methods. In these instances, the engineer should be careful to state the purpose or purposes for engineer-established design criteria and submittals. The stated purpose should relate to risks associated with achieving the overall project goals. The engineer must restrict his or her review of submittals solely to those areas specifically requested and required in the contract documents. If the contractor must employ a registered engineer to supervise the temporary means and methods design, then the reviewing engineer's role should be limited to verifying that a registered professional engineer prepared and stamped the design documents and shop drawings and that professional liability insurance requirements have been certified. If limiting temporary design criteria are specified to protect adjacent structures, the engineer should verify that the contractor's designer has demonstrated that the design meets the criteria. Even if temporary design criteria are provided in the contract documents, the contractor's designer has an independent professional responsibility to evaluate the design criteria to confirm that the design will achieve the intended performance requirements.

If the engineer goes beyond reviewing the specified criteria contained in the contract documents in performing his or her review, such as by commenting on the design approach or calculations of the contractor's designer, the engineer risks accepting partial responsibility for the contractor's design of the temporary means and methods. In addition to potential liabilities to the contractor and to third parties (most frequently, construction workers) in the event of a failure of the contractor's means and methods, OSHA may seek to impose citations on the engineer for the contractor's design and execution deficiency in temporary means and methods of the work.[8] (See also Section 6.6.)

Typically, the geotechnical data and design summary information will be provided to the contractor for use in temporary design. Sometimes the engineer will suggest or specify soil and rock parameters for means and methods designs involving subsurface construction. Should the contractor be permitted to adopt his or her own interpretation of the soil and rock parameters? If the contractor is required to use the specified parameters, the engineer will assume a certain degree of responsibility for temporary means and methods design. If the contractor adopts a different temporary construction concept from that anticipated by the engineer, the engineer should evaluate whether this difference results in a change in soil or rock parameters suitable for the design. Regardless, the engineer must evaluate the degree of control and responsibility for temporary means and methods design to be allocated between contractor and engineer.

Connection detail design—The contractor is frequently charged with designing the detailed connections between elements of the structure. Connections relate to the fit between different elements and are affected by the contractor's choice of connection hardware and sequence of assembly. The engineer must review the detail design to verify that the contractor's connections properly interpret the engineer's concept of the load carrying flow through the structure adopted in the overall design. It is essential that the engineer check this coordination of overall and connection design, yet responsibilities become blurred if the engineer takes on the responsibility to check the contractor's engineer's detail design. If the engineer intends to verify the contractor's detail design, this significant effort must be documented in the owner-engineer agreement and adequately budgeted. If the contractor will be solely responsible for the adequacy of the detail design, this allocation must be clearly defined in the construction specifications, must include the detailer's professional qualifications and insurance requirements, and must be reinforced by the engineer in his or her review markings on the contractor's submittals.

Contractor-designed permanent elements—This case is similar to connection details. It is typically encountered in precast (prestressed) concrete structural elements, including superstructure, permanent tunnel linings, and driven piles. The final design of the precast structural elements is influenced by the manufacturing equipment of the precaster. Often, contractors or precast concrete material suppliers employ engineers who are especially familiar with these special design practices. The engineer must allow sufficient autonomy in the specifications so the contractor has significant involvement in the final design to implement his or her own plan within the limits of the engineer's concept. The contract documents in this area should not to be so restrictive as to be interpreted as a design specification with the contractor implementing the engineer's specified design.

Drilled mini-piles and other specialty foundation work frequently call for contractor design of the permanent foundation elements, since the choice of structural elements is influenced by the equipment, means, and methods available to particular specialty contractors. The engineer's provision of only a concept design allows for maximum flexibility by the contractor to increase price competitiveness.

In reviewing a contractor-designed permanent element, the owner and engineer must decide how to allocate design responsibility for these selected elements of the overall structure. If the contractor will be solely responsible for design of these elements, the engineer must still verify that the contractor's designer satisfies the professional qualification requirements of registration and professional liability insurance. The engineer must also verify that the contractor's designer has used the design criteria set forth in the contract documents and that the design coordinates properly with the engineer's overall design concept. However, the engineer should clearly note on the submittal that he or she has not checked the contractor's design for accuracy, correctness, or compliance with codes and regulations and the ordinary standard of professional care, since the contractor is solely responsible for this element of the design.

Deep foundations construction means and methods—Deep foundations describes bearing elements either installed by displacement methods, such as driven or jacked piles, or installed by nondisplacement methods, such as machine-excavated piers or drilled piles.

Piles—Engineers typically evaluate the load-bearing capacity of driven foundation piles using measurement of pile penetration into the ground under blows of the contractor's pile-driving hammer. Sometimes, this relationship between driving resistance and load capacity is verified by static or dynamic pile load tests on a small percentage of the installed piles (Bowles, 1982). Nevertheless, driving resistance under the pile hammer is used to evaluate installed permanent capacity of the majority of the contractor's work.

The engineer sometimes evaluates the contractor's driving means and methods to assess at least a preliminary acceptable driving resistance, including the sensitivity of pile capacity to measured penetration resistance. In other words, the engineer looks for a final acceptance penetration resistance, which indicates load resistance with acceptable precision. If the contractor's selected hammer is too large, the change in driven capacity will be large for small variations in driving resistance (blows per inch). If the pile hammer is too small, the contractor may barely be able to develop the required capacity even driving the piles to practical refusal, depending on local soil conditions.

If pile load tests will be used to determine the final driving resistance, the owner—and, therefore, the engineer—have an interest in the contractor installing test piles that are likely to meet the required design capacity. The specifications may allocate to the owner the contingency cost for pile tests that do not achieve the required design load, as long as the contractor has installed the test piles and performed the test in a good, workmanlike manner according to the initial driving criteria specified or reviewed and approved by the engineer. Even if this risk is allocated to the contractor, repeated cycles of test pile installation and load testing will delay the construction process at the beginning of the work where the delay will typically flow through the entire process of construction. Finally, if the contractor becomes totally frustrated, he or she may claim that the work is impossible because the engineer's pile design is defective or because the subsoil conditions are materially different from those fairly anticipated.

Foundation Piers—Foundation piers are excavated by nondisplacement means to bear on suitable subgrade soil or rock at depth. Some drilled piles or excavated piers cannot be inspected by visual or other direct means. Geophysical methods can be incorporated in the foundation elements to verify competent concrete after the unit is poured (Brettman and Olsen, 1996). Generally, the engineer must rely on indirect techniques using the construction means and methods to assess suitability of the permanent work to meet the design intent, or the engineer may require construction means and methods that lead to the desired end result with high likelihood. For example, the engineer may require a certain excavation fluid cleaning process (slurry desanding) for foundation elements constructed through bentonite slurry. It is difficult, or sometimes impossible, to observe loose material at the bottom of the slurry-filled excavation. Since poor-quality cleaning will prevent high-quality concrete placement, the engineer may want to restrict or provide criteria for the contractor's temporary means and methods of construction. Repair of poor-quality concrete after pouring is a poor substitute for the contractor's good-quality slurry construction means and methods in the first place, regardless of who pays the repair cost.

Small-Diameter Drilled Piles—Good workmanlike construction means and methods may be the only approach available to an engineer to assess that the contractor has produced a full-diameter hole over the entire length of a small-diameter drilled pile. Where the contractor chooses to install temporary or permanent casing throughout the depth of the piles, the engineer is confident of a full-diameter hole through the entire pile length. Centering hardware or other devices and geophysical soundings can be used by the engineer to verify open holes in uncased zones. Regardless of the drilling method, the contractor must demonstrate that the drilled hole is completely filled with concrete or grout. This function is especially difficult where a hollow stem auger is used to drill and grout the piles. The engineer may specify that the contractor measure the grout materials as installed so the engineer can verify that each length of the drilled hole is completely filled with grout or concrete as the casing or auger is installed. The contractor's construction equipment is the only means available to the engineer to assess the adequacy of the finished work product without load testing each foundation unit.

Contractor construction schedules—An early submittal frequently required of a contractor is a detailed schedule of the entire project construction. Often, the schedule is developed using a critical path or Gantt-type chart, which identifies the activities that are essential to meeting the overall completion date. The owner's and engineer's underlying interest is in seeing that the contractor completes the project within the overall allotted time. But there are additional reasons that the engineer may require a detailed schedule. The engineer should verify that the contractor has, in actuality, developed a practicable, detailed plan to complete the work on time. The engineer may be able to anticipate elements in the contractor's plan that are impractical, given constraints in permits or long-lead material delivery or installation. For example, roofing and paving may be precluded in the winter, and wetlands restoration may be impractical in the rainy spring. Long-lead equipment may not be manufactured by the time the contractor anticipates installation. The schedule permits the engineer to anticipate the need for inspectors for various elements of the work. The schedule will permit the owner to anticipate when the value of the construction will be earned so the owner can finance the cash flow. The detailed schedule allows the engineer to identify, during the course of the work, if the contractor is behind schedule. If necessary, the engineer can recommend to the owner that the contractor accelerate the work to meet the contracted completion date. Finally, the submitted schedule can form an objective baseline for evaluating the legitimacy of a potential delay claim if changes force the contractor to modify the schedule.

There are three areas of submittals that require special caution on the part of the engineer:

- *Information not requested*—Sometimes, contractors will submit information not requested in the specifications. The engineer should promptly return the documents marked appropriately, such as "Returned without Review—No Submittal Required," or a similar marking. The contractor's submittals may arrive with information not requested mixed with required submittal information, such as assembly instructions marked on manufacturer's catalogue cuts. The engineer should clearly delineate those portions reviewed, and mark the extraneous information as "Returned without Review." Without this clear communication, the engineer may inherit responsibility for an area of the contractor's means and methods of construction that was totally unintended.
- *Contractor-Proposed Design Changes*—Contractors will frequently propose changes in the design that they believe will achieve economy, at least for the contractor. Sometimes the allocation of responsibility for the outcome of these contractor-proposed design changes are set forth in the contract documents. Even if the allocation of risk and responsibility are not addressed beforehand, the engineer must work with the owner to carefully consider who will take on the responsibility for (1) design of the changed element, (2) integration of the change into the overall project design and concept, (3) requirements for submittals associated with the design changes, and (4) construction observations to verify that the implementation in the field meets the design concept. The issue of contractor's designer qualifications and insurance goes without saying.

  If the engineer agrees to accept some of these responsibilities, he or she should be certain to receive adequate time and compensation to perform a complete review of the proposal, particularly to assess integration of the change into the overall design concept. The engineer will typically develop his or her design by thinking through the project in sequence from top to bottom (or reverse). A contractor-proposed change in the middle sometimes leads to the oversight of consequences that were properly handled in the original sequential design. The engineer should be suitably wary of contractor-proposed changes, not because they may be unsound, but because they may not integrate well with the engineer's overall project design concept.

- *Submittal review response*—The engineer's response to contractor's submittals requires a careful choice of words and means of communication. The communication must reinforce the allocation of risk and responsibility that are set forth in the contract documents and in the owner-engineer agreement. Some owners require that the engineer "approve" contractor submittals. Other contracts may be less specific, allowing the use of "reviewed" without indicating endorsement of the contractor's proposed means and methods of construction. Regardless, the response should reflect the purpose of the engineer's review as set forth in the contract documents. For example, paraphrasing the EJCDC language (EJCDC, 1990),

> Review and approval of Shop Drawings and Samples and other data submitted by Contractor is only to evaluate conformance with the information given in the Contract Documents and compatibility with the design concept of the completed Project as a functioning whole as indicated in the Contract Documents. Such reviews and approvals or other action will not extend to means, methods, techniques, sequences or procedures of construction or to safety precautions and programs incident thereto.

> The engineer may also add specific limitations as review comments, including notation of portions not reviewed, reviewed only for credentials of designer and provision of insurance certificate, and so on. If more lengthy notations are judged appropriate, the engineer should mark the submittal with a reference to a companion document, which can be prepared and attached to the response.

6.5.3 Engineer's Advice to the Contractor

The best-laid plans, policies, and procedures regarding the engineer's involvement in construction means and methods established and carried out in the engineer's office can be instantly undone by imprudent actions or inactions by the engineer or his or her representatives on the construction site. When pressed with the urgency of workers and equipment lying idle while waiting for technical guidance, it is tempting for the engineer with a can-do spirit to step in to provide the needed direction. If the element requiring the technical decision is the contractor's sole responsibility, however, the engineer's ready advice may convert the contractor's responsibility into the engineer's ultimate headache and liability.

A contractor may sometimes ask the engineer for advice on how to pursue a certain aspect of the means and methods of construction. Or, the engineer may observe the contractor's work and believe that the engineer can identify a cheaper, better, and faster way to get something done, or may identify a temporary requirement that the contractor has, in the engineer's opinion, overlooked or has pursued inadequately. Directing, or even suggesting, means and methods to a contractor exposes the engineer and owner to claims that the engineer interfered with the contractor's responsibility for and, therefore, the cost of completing the work. Whether the engineer's contribution does actually achieve improved and more economical construction is determined by both the quality of the engineer's advice and the effectiveness of the contractor's implementation. Reconciling this question is often an expensive dispute resolution process in itself. Determining whether certain specific temporary measures are absolutely required to construct the project is the contractor's sole burden. By treading on the contractor's territory, the engineer risks initiating a disagreement as to whether the engineer is helping the project or hindering the construction process.

The engineer's overall goal is the same as the owner's and contractor's—completing the work on time, within budget, and in accordance with the original design concept. If the engineer has expertise that can help the contractor complete the work, he or she should not have to withhold it simply to avoid transferring overall responsibility for construction means and methods from the contractor to the engineer. However, the engineer should adopt a few precautions to help clarify the nature and significance of the engineer's communication.

The engineer should consider whether such advice should be incorporated in the contract documents in some way. Always keep the ultimate responsibility for choice of method and design implementation in the hands of the contractor. Whether in the specifications or during construction, the engineer must communicate with language which clarifies that the technical guidance is just that—guidance. Reinforce that the contractor is free to choose his or her own means and methods of construction. State that the contractor's adoption of the engineer's guidance is the contractor's sole decision, not a directive of the engineer, and that the contractor retains sole responsibility for selection and implementation of the means and methods of construction.

One method the engineer can use to suggest without directing is to ask leading questions: "Have you tried this?" "Is your dewatering system achieving the required drawdown?" "Is the soil water content near the optimum density?" The questions can communicate the emphasis and concepts that the engineer may choose to impart without making a definitive statement. Questions may also point out observations affecting the contractor's means and methods without advising the contractor how to use this information to facilitate the construction process.

6.6 HEALTH AND SAFETY ISSUES

The health and safety of construction workers is directly and inextricably tied to the means and methods by which the work is performed. For that reason, in circumstances in which the contractor is exclusively responsible for construction means and methods, it is logical and natural that the contractor would likewise have the same level of responsibility for health and safety at the construction site. Nevertheless, efforts are frequently made to bring the engineer into the liability loop for construction accidents, particularly since workers' compensation statutes limit the financial recovery that may be obtained from an employer. In recent times, courts have tended to lean farther into the lap of workers, to the financial detriment of design professionals.

Except in special cases, the engineer should avoid accepting direct responsibility for worker health and safety. There are situations, such as sites where hazardous materials may be present, in which the engineer is well advised to fairly warn the contractor of these hazards and clearly specify the type of special precautions for worker protection that the contractor must meet to manage the presented hazard. However, the biggest risk management challenge for the engineer is responding to site conditions where the contractor is ignoring serious health and safety requirements.

6.6.1 Health and Safety Plans

It is good practice for a construction contractor to develop and implement a worker safety plan on all job sites. If the site contains chemical contamination, federal regulations require additional measures for worker safety protection, such as medical monitoring of personnel, personal protective gear, and specific training to recognize and adapt to potential chemical exposure on the work site.

Where such exposure exists on a construction work site, the engineer may specify the general nature of the measures required of the contractor in order that the contractor include the considerable cost of these measures in preparing his or her estimate. This is an area clearly affecting worker safety, so that the engineer should be careful to keep all responsibilities for planning and implementing health and safety measures in the contractor's arena. The engineer should define the specific nature of the hazards presented, in the same way soil, rock, and groundwater conditions are presented, but the contractor must retain sole responsibility for developing the health and safety plan to address these hazards for his workers.

To verify that the contractor has taken safety measures required by regulations, the engineer may require that the contractor submit a health and safety plan developed to protect the workers. If special conditions exist on the site, the specifications may also require that a qualified safety professional (such as a certified safety professional or a certified industrial hygienist) oversee the contractor's plan to assure that the contractor is relying on his or her own professional employees or consultants to develop these plans, and not merely trusting the work of the engineer to protect the construction workers.

If a health and safety plan is requested, the engineer should review the submittal to see that the plan was prepared and that an appropriate safety professional reviewed and signed it, if required. The engineer should also review the design hazards or conditions described in the plan to verify that the preparer has accounted for those special situations identified and reported in the engineer's design studies. The engineer must avoid commenting on the protection plan itself, since this may be an intrusion into construction means and methods, shifting responsibility for worker safety from the contractor to the engineer.

6.6.2 Response to Contractor's Health and Safety Violations

Responsibility for safety measures for construction personnel is always assigned to the contractor. The contractor is in a position to control the safety of the workers in practice, and statutes allow the contractor to insure the workers for injury through workers' compensation insurance. The engineer will sometimes observe situations in which the contractor fails to meet ordinary safety precautions. Sometimes, these situations present immediate potential for serious personal injury or death, and the engineer is faced with the dilemma of (1) taking some action to help prevent a tragedy and risk incurring potential liability for the outcome of his or her action or (2) ignoring this "sole responsibility of the contractor" and suffering both moral and possible financial liability for standing by to watch injury that the engineer could foresee. Neither option is without risk, but the author suggests that taking action appropriate to the urgency of the situation is a more defensible position than the often-preached and selected approach of observing but keeping silent about a serious safety violation.

The contractor is responsible for overall control of the site. This includes all safety precautions for his or her own employees and seeing that all subcontractors and visitors to the site, including the engineer, comply with the contractor's site safety program. The engineer is responsible for his or her own safety and that of his or her own employees and for complying with the overall site safety program implemented by the general contractor. While on site, the engineer has the opportunity to observe many aspects of the work, not just the end result of the contractor's efforts. After all, the contract documents normally require the contractor to follow reasonable safety precautions and OSHA regulations, and the engineer is charged with observing the contractor's compliance with the contract plans and specifications. The engineer also has the responsibility to observe and evaluate the safety measures implemented by the contractor on a site as they may affect the safety of the engineer's own employees.

Minor deficiencies in safety requirements that are judged not likely to lead to immediate serious injury may be left to the contractor, unless they affect the engineer's needs to function on the site. Serious deficiencies, which could readily lead to serious injury or death to personnel on the site, present situations that call for carefully planned action, not procrastination or inaction:

- *Imminent injury or damage*—If a worker is about to suffer an injury, the engineer should take whatever action is necessary to prevent the tragedy.

An example of such a situation might be an unsupported trench showing signs of imminent collapse, the odor of sulfur dioxide (rotten eggs) in a poorly ventilated area, a crane approaching electric wires, or a previously undetected gas valve cover near an excavation in progress. These are situations that could result in serious injury or death in a matter of seconds or minutes. Communicating through a normal "chain of command" will be much too slow to protect life safety.

In these types of situations, the engineer should use whatever means possible to protect those persons who may suffer injury. This may include giving direct orders to workers to leave a hazardous location as promptly as possible. Following removal of workers from the hazard, the engineer should follow a communication procedure with the contractor as described below.

- *Serious safety violations*—Occasionally, the engineer will observe conditions on a site that do not conform to generally accepted safety practices. The most common of these, and the one that results in the most frequent construction site injuries and death, is inadequately shored or sloped trenches and excavations with workers in the bottom.[9] Absence of safety barriers around trenches and elevated openings are another common safety deficiency. These safety provisions may be the responsibility of a subcontractor or the general contractor.

  In these situations, there may be time for the engineer to implement appropriate measures. The author recommends that the engineer develop a written procedure to deal with this situation, which is understood by all field personnel. The first step in such a circumstance is for the engineer's personnel to remove themselves from the exposed hazardous situation immediately. Then, the engineer's safety violation response procedure should include a form that the engineer's field representative can use "on the spot" to request that the general contractor correct an identified unsafe act or condition that the engineer's representative notes as not being in accordance with the contract documents or referenced safety regulations. The engineer must be careful in how he or she communicates the information:

1. Communicate only with the general contractor's available senior representative who has authority to direct changes in the work. This is normally the general superintendent or his or her assistant. Follow up with written confirmation of the notice to the contractor's project manager, generally by facsimile transmission. Do not communicate with subcontractors' personnel or forepersons.
2. Describe only the hazardous condition; do not provide any guidance on how to remedy the hazard.
3. Use a written (even handwritten) communication at the time the hazard is noted. Time is critical.

4. Be sure to advise the contractor that the notice of a safety violation in one instance does not mean that this is the only safety violation that may exist at this time or that the engineer will notify the contractor of any noted violations in the future. In notifying the contractor of any specific deficiency, it is critical that the engineer does not create a duty to identify and report other safety violations that the engineer may encounter now or in the future.

A sample letter an engineer might adopt to inform the general contractor of safety violations is shown in Figure 6.1. This is a powerful tool, to be used with discretion. It puts a written notice of a safety violation in the contractor's hands, where it is difficult for a responsible superintendent or project manager to ignore. At this point, it is a one-on-one communication between engineer and contractor. It does nothing to remove control of the safety situation from the contractor, but advises the contractor to evaluate the observed condition and to develop his or her own means and methods to correct the deficiency if he or she deems it appropriate. Finally, it states that the notice does not alter the contractor's sole responsibility for health and safety on the site.

Should this approach fail to result in an acceptable response from the contractor, the engineer should provide a similar notice to the project owner within 48 hours after the original contractor notice. Finally, if the contractor fails to maintain a safe work site for the engineer's access or the contractor's own personnel, the only safe response on the part of the engineer is to leave the site and terminate services, if necessary. The engineer's retreat from his or her own contract obligations to the owner results from the owner's failure to insist on a safe work site under the control of the owner and the general contractor. A call to OSHA may result in correction of safety violations, but it will probably do little to modify a casual attitude about safety held by the contractor and owner.

6.7 ENGINEERS' EXPANDED RESPONSIBILITY FOR CONSTRUCTION MEANS AND METHODS

The reasons the engineer may intentionally become responsible for particular elements of construction means and methods are described in Section 6.2.2. The engineer's control of construction means and methods may be identified and implemented at the outset of the project, may come about in final contractor negotiations when the contractor's limited knowledge and experience in designing certain portions of the temporary means are identified, or may be recognized and adopted during the course of construction. Regardless, in order to maintain the fair allocation of risk among owner, engineer, and contractor, this variance from the customary roles of engineer and contractor described in Section 6.1 must be reflected in both the engineer-owner agreement and the construction contract documents.

[ENGINEER'S LETTERHEAD]
NOTICE OF NON-COMPLIANCE WITH CONTRACT
HEALTH AND SAFETY REQUIREMENTS

Date: ____________________ Time: ______________

Project: ____________________________________

Location: ____

General Contractor: ____

Address: ____

City, State, Zip Code: ____

Attention: ____

RE: Unsafe Acts or Conditions or Non-compliance with Regulations Related to Health and Safety at the ________ ____ project (the "Site").

Gentlemen:

This is formal notification that your company, or subcontractors controlled by your company, have placed [Engineer's] field personnel and other employees at risk of accident or injury due to noncompliance with contract requirements or OSHA or state regulations related to health and safety at the above-referenced Site. In our opinion, the specific acts or conditions not in compliance with the contract terms or regulations are as follows:

1. ____
2. ____
3. ____

You are hereby requested to correct the unsafe acts or conditions cited above in a timely and appropriate manner. Unsafe acts or conditions described above as serious must be addressed immediately upon receipt of this notice. Failure to correct the unsafe acts or conditions noted above may require [Engineer] to remove our employees from the Site or take other action to reduce the likelihood of injury to our employees. Please notify us, in writing, of your intended action within 24 hours or immediately if the health and safety concern is serious.

In providing you this notice of unsafe acts or conditions, [Engineer] makes no representation that this is a report of all unsafe acts or conditions that may exist on the site presently nor do we undertake to identify unsafe acts or conditions that may exist on the site at any other time. As general contractor, you retain sole responsibility for overall health and safety compliance at the project site.

Very truly yours,
ENGINEER'S FIRM RECEIVED BY:

_______/S/_____________________ ______________________________

______________________________ Printed Name: ______________________
Printed Name

Date: ______________________

cc: Engineer-in-Charge

____

Figure 6.1 Safety violation notice to general contractor

When the engineer takes on an expanded role for construction means and methods, the owner-engineer contract should include provisions that:

- The engineer is not responsible for the design or any other aspect of any other construction means and methods other than the particular means and methods which the engineer has specifically undertaken responsibility to design.

- If any shop drawing or other submission will be required of the contractor with respect to the means and methods designed by the engineer, the engineer's contract (and the construction contract documents) should clearly define the respective duties and responsibilities of the engineer and contractor with respect to the preparation and review of such submissions. Correspondingly, the contractor's role and responsibility during construction with respect to the engineers' designed means and methods with respect to preparation of shop drawings and submittals, installation, implementation, and coordination should be specifically set forth in the contract documents.
- The engineer's role, if any, with respect to field observation, construction, installation, and implementation of the particular means and methods should be specifically defined in both the owner-engineer agreement and the construction contract documents. Since the success of the engineer's design is intimately tied to the interpretation and implementation care and skill of the contractor, the engineer may be well advised to insist on a broad role during construction of this element of means and methods for which he or she is responsible in order to limit extra construction costs should unanticipated conditions surface in the field.
- The engineer must recognize that he or she may accumulate liabilities that encompass both the design and construction activities. For this reason, the engineer may need, and fairly require from the owner, higher fee schedules for this type of work than the fees that may customarily apply to design work alone. The higher fee is to compensate the engineer for the added level of risk involved in design of construction means and methods on behalf of the owner. Without the higher fee structure, the engineer will be unable to fund the higher frequency of possible claims that may arise out of this type of work.

When these issues arise early in project planning, appropriate clauses can be included in the owner-engineer agreement at the outset or by a contract modification. The engineer can then incorporate the reallocation of responsibilities for specific means and methods designed by the engineer into the contract documents as they are developed. When these issues are identified at the time of the negotiation of the construction contract, the engineer must insist on both an amendment to his or her contract with the owner and a contract modification between the owner and contractor (at least a no-cost change order). Then, both contracts will accurately reflect both the increased responsibility accepted by the engineer and the distinct limits of that responsibility. Otherwise, the engineer, by engaging in limited activities that expand the normal role regarding means and methods of construction, may find others ascribing to the engineer broad responsibility for construction means and methods.

6.8 DESIGN OF MEANS AND METHODS ON BEHALF OF THE CONTRACTOR

An engineer may be engaged directly by the contractor to design construction means and methods. This role does not involve a conflict with the customary role of engineer of record, since an individual different from the engineer of record is performing the design. The risks involved working in this alternative role are discussed in Section 7.6 of Chapter 7.

6.9 CONCLUSION

Although the traditional roles of engineer and contractor separate them with respect to design and construction means and methods, the complexities of subsurface projects may result in contractual arrangements that blur the division or responsibility. This fusing of the roles has led to innovative subsurface designs proposed by contractors and economical construction means and methods developed by engineers. The ultimate result, in many cases, has been an overall elevation of project quality with a reduction in project cost—a double win for the owner. Both of these effects are considered beneficial as long as the party who receives the principal advantage is also the party who accepts the fair share of the risks and pays an equitable cost. All parties in the construction triangle must remain aware of the roles assigned to each party as the project progresses, so that the party in control of each element of the work is assigned the responsibility for controlling and financing the risk. The quality of the contractual arrangements that define the roles and responsibilities must emulate the quality of the design and the construction.

NOTES

1. Refer also to Chapter 10, Section 10.2.2, which addresses the legal principles that apply to engineering consulting and construction means and methods.
2. See also Chapter 10, Section 10.2.2.
3. See, for example, 29 C.F.R. Part 1926—Safety and Health Regulations for Construction, Subpart P—Excavations (1989).
4. See, for example, *The BOCA National Building Code* (1993). Country Club Hills, IL: Building Officials & Code Administrators International, Inc., Section 3308.0—Protection of Adjoining Property and Section 3310.0—Demolition and Excavation, pp. 306–07.
5. Even if the engineer is found ultimately to have no liability for the injury or death of a construction worker, or if statute protects the engineer from such claims, the cost and professional time in handling an investigation and in defending even a groundless claim can quickly dwarf the amount of the fee that the engineer might earn on a particular project. Recently, the Occupational Safety and Health Administration (OSHA) has been more aggressive in attempting to assign safety responsibility for

construction worker injuries to engineers following serious construction accidents. See Section 7.5.2 of Chapter 7, and Section 10.2.2.2(B) of Chapter 10. Professional errors and omissions insurance will not protect the engineer totally from this exposure.

6. In a complementary fashion, the engineer, either on his or her own or in consultation with the owner, should carefully consider the advisability of attempting to transfer to the contractor responsibility for the success of temporary elements of the project design. Does the contractor have the design experience and qualifications to complete the design for these elements? Similarly, the owner, with the engineer's cooperation, may attempt to transfer to the contractor the risk of discovery of subsurface conditions that are unforeseen by all the parties to the construction processes. A contractor may include a contingency allowance in his or her price to accommodate unforeseen subsurface conditions yet still render a claim for additional compensation if such unforeseen conditions arise.
7. Refer also to Chapter 11 for a general discussion regarding the use of insurance to transfer and finance risks inherent in the several aspects of the construction process.
8. See *Secretary of Labor v. Simpson, Gumpertz & Heger*, 3 F.3d 1 (1st Cir. 1993) and *Secretary of Labor v. Foit-Albert Assoc., Architects & Engineers, P.C.*, OSHARC Docket No. 92-0654, 1997 WL 197016 (O.S.H.R.C., April 21, 1997).
9. Usually, the engineer's personnel do not have to enter the excavations, so it is not the engineer's personnel who are directly subjected to the hazard.

REFERENCES

American Institute of Architects, "Standard Form of Agreement between Owner and Architect," Document No. B141, 1987a Paragraphs 2.6.6, 2.6.10 and 2.6.12.

American Institute of Architects, "General Conditions of the Construction Contract," 1987b, Document No. A201, Paragraphs 3.3.1, 4.2.3, 4.2.7, and 9.4.2.

Bowles, J. E., *Foundation Analysis and Design*, New York: McGraw-Hill Book Co., 1982, pp. 638–69.

Brettman, T. and Olsen, L., "Defect Detection," *Civil Engineering*, Vol. 66, No. 7, 1996, pp. 2A-7A.

Carper, Kenneth L., "Structural Failures During Construction," *Journal of Performance of Constructed Facilities*, Vol. 1, No. 3, August 1987, pp. 132–49.

Casagrande, A., "Role of the 'Calculated Risk' in Earthwork and Foundation Engineering" 1965, in *The Terzaghi Lectures 1963–1972*. New York: ASCE, 1974, pp. 72–109.

Clark, John R., Esq., *Commentary on Agreements for Engineering Services and Construction Related Documents*, Alexandria, VA: Professional Engineers in Private Practice. EJCDC Publication No. 1910-9, 1993.

Clark, John R., Esq., "Focus on Shop Drawings," Alexandria, VA: Professional Engineers in Private Practice, EJCDC Publication No. 1910-9-C, 1985.

Engineers Joint Contract Documents Committee, *Agreements for Engineering Services and Construction Related Documents*, Alexandria, VA: Professional Engineers in Private Practice. EJCDC Publications No. 1910-1 through 1910-30, 1993 ff.

Engineers Joint Contract Documents Committee, "Standard Form of Agreement Between Owner and Engineer for Professional Services," Alexandria, VA: Professional Engineers in Private Practice. Document No. 1910-1, ¶6.01.I, 1996, p. 4.

Engineers Joint Contract Documents Committee, "Standard General Conditions of the Construction Contract," Alexandria, VA: Professional Engineers in Private Practice. Document No. 1910-8, ¶6.01.A, 1990, p. 18.

Essex, Randall J., "Geotechnical Baseline Reports for Underground Construction—Guidelines and Practices," prepared by the Technical Committee of the Underground Technology Research Council (American Society of Civil Engineers, 1997).

Gould, James P., "Geotechnology in Dispute Resolution," *Journal of Geotechnical Engineering*, Vol. 121, No. 7, 1995, pp. 521–34.

Hager, Frank J., "Public Safety Issues in Collapse of L'Ambiance Plaza," *Journal of Performance of Constructed Facilities*, Vol. 5, No. 2, August 1991, pp. 92–112.

Hilf, J. D., "Compacted Fill," in Winterkorn, H. F. and Fang, H-Y, eds., *Foundation Engineering Handbook*, New York: Van Nostrand Reinhold Co., 1975, pp. 244–311.

Peck, R. B., "Advantages and Limitations of the Observational Method in Applied Soil Mechanics," *Geotechnique*, Vol. 19, no. 2, 1969, pp. 171–187.

Rubin, Robert A. and Banick, L. A., "The Hyatt Regency Decision: One View," *Journal of Performance of Constructed Facilities*, August 1987, Vol. 1, No. 3, pp. 161–67.

Terzaghi, K., "Consultants, Clients and Contractors," *Journal of the Boston Society of Civil Engineers*, Jan. 1958, pp. 1–41; 255–277 (closure).

U.S. Bureau of Reclamation, *Earth Manual*, Second Edition, Denver, 1974, pp. 51, 53, and 176–181.

U.S. Naval Facilities Engineering Command, "Correlations of Strength Characteristics for Granular Soils," *Soil Mechanics*, Design Manual 7.1. Alexandria, VA, 1982, Figure 7, p. 7.1–149.

7

RISK MANAGEMENT CONSIDERATIONS FOR ENGINEERS ENGAGED BY CONTRACTORS ON SUBSURFACE PROJECTS

Joseph D. Guertin, P.E.
GZA GeoEnvironmental, Inc.

7.0 INTRODUCTION

Risks faced by professional engineers working for contractors are diverse, ranging from the business risk of fixed-price contracts to litigation risk often resulting from job-site safety issues. This chapter presents descriptions of the diverse professional services commonly provided for contractors, plus a discussion of the different nature of the relationship between engineers and contractors from that between engineers and owners, both public and private. Some of the more common risks faced in this type of work are discussed, along with recommended strategies for managing them.

Subsurface Conditions: Risk Management for Design and Construction Management Professionals, edited by David J. Hatem
ISBN 0471-15607-8

7.1 OVERVIEW

Contractors often need professional engineering services to assist them with a variety of issues, beginning with prebid assistance, extending to construction issues, and concluding with construction litigation, which can last for many years. It is not uncommon for an engineer to work with a contractor on a single project from inception to completion for a period of ten or more years. Relationships between engineers and contractors are often confrontational when the engineer is working for an owner, due to the often conflicting objectives of owner and contractor on work procured by the public bidding process. When engineers work for contractors, however, the relationships can be quite satisfactory because of shared common goals, although often for different motivations.

The risks of suits against engineers working for contractor clients are probably less than when providing project design or construction management services, either directly or indirectly, for a public or private owner. The risk of claims against engineers from contractor clients can result if the work does not proceed as expected, but the risk is generally quite small. Typically, the risk is limited to the engineer's fee. However, the risk of third-party claims arising out of safety issues, and the related risk of citations by the Occupational Safety and Health Administration (OSHA), is significantly greater for an engineer working for a contractor than for an engineer working as a designer or construction manager for the owner, especially if the engineer is in substantial control of the work, and/or has directed or is in a position to direct construction means and methods. Should there be an accident or safety violation, the engineer may be subject to a third-party suit or a citation by OSHA.

When engineers and contractors work together on the same team, the work can be extremely satisfying because of the direct link between engineering service and construction results. When working for a contractor, the engineer may see the physical manifestations of his or her work being constructed within a matter of days after design completion; when working for an owner, especially a public agency, the time between design and construction is typically measured in years, not days.

In the sections to follow, risks faced by engineers when working for contractors will be examined, as will the measures that can be taken to effectively manage those risks. In some instances, the risks may arise from the contractor client; however, the risks of third-party suits and OSHA citations have increased, largely because of the proximity of the engineer to the construction process. When problems arise, the magnitude of these problems can be large, with attendant financial risk to all parties associated with the work, i.e., owner, contractor, and engineer.

On large multicontract infrastructure projects, conflict-of-interest statutes and ethics codes must also be considered. Questions arise when an engineer working for an owner on one portion of a project accepts an assignment from a contractor on some other, unrelated, project section. Is that engineer prevented from working for contractors on other portions of the project? This issue is a current topic of debate and is relevant to very large citywide infrastructure projects, such as Washington, D.C.'s rapid transit system or Boston's Central Artery/Tunnel Project. Numerous recent examples illustrate situations in which engineers have successfully provided professional engineering services to owners and contractors on the same project without conflict.

7.2 CONSULTING ENGINEERING SERVICES TO CONTRACTORS

Consulting engineers provide a wide variety of services to construction companies before, during, and after construction. Typical assignments are described below.

7.2.1 Prebid Evaluations

Prior to bid, contractors often retain engineers to review the project documents and to evaluate conditions as an aid to the contractor during bid preparation. Contractors use such evaluations to develop bid strategies, develop construction methods and costs, determine likely work progress, and consider future claims potential. Claims-potential evaluations are considered by some to be ethically questionable, but in the public bidding system, such evaluations are nothing more than sound business practice. Prebid evaluations can be very important to contractors' bids, and, as such, contractors are likely to rely heavily on the engineers' analyses.

7.2.2 Design of Temporary Structures

Often, the contractor working on a subsurface project must design and build temporary structures to facilitate construction. Designs are most commonly required for lateral support and underpinning, but other structures may require the design of facilities such as temporary buildings or bridges. Many large contractors are capable of preparing such designs with their own staff, but often, small companies, and, in some instances, even some of the larger companies, seek specialized engineering services.

7.2.3 Development of Means and Methods of Construction

The responsibility for means and methods of construction is typically left to the contractor. However, contractors may seek design or professional consulting assistance with specialty construction methods, such as soil and rock excavation support, dewatering, grouting, rock blast design, optimization of pile driving equipment and procedures, and tunneling methods.

7.2.4 Development and Conduct of Specified Test Programs

Performance testing is often required as part of construction contracts, especially large infrastructure or building projects involving deep excavations, deep foundations, or tunneling. Contractors may engage the services of consulting engineers to develop test details, make submittals for review, and perform the tests. Typical performance test requirements include static load testing of piles and caissons, dynamic testing of piles to determine behavior during driving and load-carrying performance, pump tests for design and evaluation of dewatering systems, and test blast programs conducted as part of vibration control measures.

7.2.5 Construction Monitoring

The responsibility for construction monitoring may be given to the contractor, with oversight by an owner's representative. Typical monitoring duties may include geotechnical instrumentation to measure lateral and vertical deformations, piezometric levels, and the like; structural instrumentation to measure loads and deflections; blast monitoring to measure vibration levels and air overpressures; and environmental water quality and soil quality monitoring to ensure compliance with environmental laws and regulations.

7.2.6 Environmental Support

Environmental considerations have become increasingly important to contractors as the volume of environmental laws and regulations has increased dramatically over the past 20 years. Increasingly, contractors are being required to obtain permits and to continuously monitor various soil, water, and air quality parameters during construction to demonstrate compliance with environmental specifications based on current laws and regulations. Engineers and environmental consultants are retained to provide these services, which can have a large impact on construction progress. If specified criteria—such as discharge water quality—are not being met, a project can be stopped until measures are taken to correct the environmental problems. The contractor's project progress can be controlled by the speed with which environmental testing and monitoring can be accomplished.

7.2.7 Health and Safety Plans

In addition to environmental monitoring, contractors may be required to provide project-specific health and safety plans that address both construction safety and environmental risk exposure. Most contractors are not equipped to develop these plans with respect to protection from special chemical hazards and, therefore, turn to consultants.

7.2.8 Value Engineering

Contractors often perceive technically superior and less expensive ways to achieve the owner's design intents through design modifications and/or alternate construction methods. Value engineering cost proposals (VECP) provide a contract mechanism sometimes included by owners to motivate contractors to take advantage of these project improvements. Under the VECP process, contractors can submit new concepts, which, if less expensive while meeting the owner's needs and if accepted by the owner, become part of the contract; the cost savings for an accepted proposal are shared between the contractor and owner. Value engineering cost proposals frequently require a significant design and cost estimating effort, which is provided by consulting engineers retained by the contractor. Since the design authorship now rests with both the engineer and contractor, engineers working for the owner and engineers working for the contractor must consider risk and responsibility allocation of the VECP portion of the design.

7.2.9 Claims Support

Providing professional assistance to contractors in their pursuit of claims for equitable time extensions or cost reimbursement due to differing site conditions or other changes in the work has proven to be a substantial service area for consulting engineers. This service area is so important that several small specialty engineering companies and individual consultants now provide services exclusively to assist contractors with such claims and any subsequent litigation.

7.2.10 Litigation Support

Consulting engineers often provide litigation support services. These services typically include technical input into legal strategy development, engineering analysis of construction events, review of opposing rebuttal presentations, and expert witness testimony in construction litigation.

All these service categories are crucial to the successful completion of construction contracts. There are resultant risks to those who provide such services, as litigation can be initiated by both the contractor client and third parties. These risks can be substantial, especially in suits arising out of construction accidents, regardless of whether the engineer was directly involved. Some of these risks, and ways to mitigate them, are discussed in subsequent sections of this chapter.

7.3 THE NATURE OF SERVICES FOR CONTRACTORS

Providing engineering services to contractors is a significantly different experience from working for other types of clients, especially public owners. The emphasis is on competent, quick action. Contractors make money by being decisive in an informed way. Unlike public owners, contractors may not have time to study alternatives in a detailed, highly documented fashion; they tend to zero in on key issues, always viewing cost as a paramount consideration.

The competitive nature of the construction industry leads the contractor's engineer to technically challenging assignments because contractors typically seek new, more cost- and time-efficient ways of doing things in order to be increasingly competitive. As a result, work for contractors is often fairly aggressive and "pushes the technical envelope," more so than work for other client types. Contractors are rewarded for new, innovative ideas and cannot be competitive if they just do the job the same way as before.

Bureaucracy and procedural impediments to completing the work are minimized when working for a contractor. Contractors are more interested in results than in process. Consequently, there is typically a strong sense of teamwork between contractor and client due to this focus on results.

Engineering services are often provided on a lump-sum basis for a defined scope of work. This is especially true for prebid evaluations, designs of temporary structures, instrumentation monitoring, and so forth, i.e., work that can be well defined in advance. For service areas such as litigation support, this obviously cannot be the basis for compensation, and work is typically done on a time-and-materials basis or possibly a series of fixed-price work scopes.

There is probably less risk of claims by the contractor against the engineer than with other client types. Because contractors are business people often experienced in construction-related claims, they generally realize that it is not worth the cost and effort to sue an engineer for perceived damage resulting from the allegedly negligent engineering services. They may not hire the engineer again or may denigrate the engineer's reputation, but they generally do not pursue litigation against their consulting engineers.

7.4 CONTRACTOR EXPECTATIONS OF ENGINEERS

Contractors expect engineers to understand that construction is a highly competitive business. As such, contractors expect engineers to maintain strict confidentiality on all project matters that do not have to be reported legally. When pursuing the design work for other client types, engineers, especially as subconsultants (e.g., geotechnical or environmental consultants), may pursue the design work as members of several competing teams. This is rarely the case for work with competing contractors, whether in traditional publicly-bid work or in the increasingly common design-build projects. Contractors insist on a higher commitment to confidentiality from their retained engineers, than engineering and architecture companies that retain other engineers, because the construction cost risks are much larger than the design fee risks, and because a contractor's submittal may be based on proprietary innovative concepts, which represent the heart of competitive advantage.

Contractors expect a higher tolerance for risk than typically expected to be assumed by professional engineers engaged by owners. Contractors are risk takers; they therefore expect their engineers to guide them in the assumption of calculated risks. Conservative, proven designs typically do not provide an advantage in a highly competitive fixed-price business environment. In addition to the assumption of certain technical risks, contractors expect engineers to be risk-tolerant in business terms. They expect engineers to take business risks, just as all contractors do every time they submit a bid. With this business norm, contractors usually negotiate fixed-price engineering services contracts whenever possible. Contractors expect engineers to understand that construction is a business first, and that business issues take precedence over technical issues as long as technical requirements are met.

Contractors expect high technical competence in both design and constructability evaluations. Construction knowledge at a detailed level is necessary to effectively provide services to contractors. Constructability issues may be as important, if not more so, than detailed design concepts. This detailed knowledge is expected to lead to aggressive designs, but designs that contemplate a thorough understanding of the risks involved; these risks, of course, must be effectively communicated to the contractor, who ultimately must make the decision of whether or not to assume risk and, if the risk is assumed, how to manage it.

Contractors expect their engineers to have an inclination for action. Delivery schedules are often measured in hours, not weeks or months. This business culture is not unlike a military environment, where the strong, intelligent, and bold are ultimately the ones rewarded.

7.5 ENGINEERS' EXPECTATIONS OF CONTRACTORS

Engineers who work for contractors expect contractors to understand the difference between professional practice obligations, and the business pressures and culture of the construction industry. Contractors are rewarded for completing work very quickly, whereas, in general, engineers are not. Engineers are motivated by doing a technically proficient job in a reasonable amount of time. They expect contractors to understand that the risk of making errors is increased by rushing work.

Engineers expect contractors to favor lump-sum, defined-scope contracts over open-ended, time-and-materials contracts. Provided the scope can be adequately defined, this is usually advantageous to engineers as well. Prompt payment is expected, as opposed to payment based on construction progress.

Engineers provide services under codes of ethics of professional practice and environmental reporting laws, which they expect contractors to understand. The engineer's license to practice can be put at risk. For instance, engineers are required to follow the Fundamental Canons and Professional Obligations of the Code of Ethics of the National Society of Professional Engineers (NSPE), which provide as follows:

Fundamental Canons:

- Hold paramount the safety, health, and welfare of the public.
- Perform services only in areas of one's competence.
- Issue public statements only in an objective and truthful manner.
- Act for each employer or client as a faithful agent or trustee.
- Avoid deceptive acts.
- Conduct oneself honorably, responsibly, ethically, and lawfully so as to enhance the honor, reputation, and usefulness of the profession.

Professional Obligations:

- Engineers shall be guided in all their relations by the highest standards of honesty and integrity.
- Engineers shall at all times strive to serve the public interest.
- Engineers shall avoid all conduct or practice which deceives the public.
- Engineers shall not disclose, without consent, confidential information concerning the business affairs or technical processes of any present or former client or employer, or public body on which they serve.
- Engineers shall not be influenced in their professional duties by conflicting interests.
- Engineers shall not attempt to obtain employment or advancement or professional engagements by untruthfully criticizing other engineers, or by other improper or questionable methods.
- Engineers shall not attempt to injure, maliciously or falsely, directly or indirectly, the professional reputation, prospects, practice, or employment of other engineers. Engineers who believe others are guilty of unethical or illegal practice shall present such information to the proper authority for action.
- Engineers shall accept personal responsibility for their professional activities, provided, however, that engineers may seek indemnification for services arising out of their practice for other than gross negligence, where the engineer's interests cannot otherwise be protected.
- Engineers shall give credit for engineering work to those to whom credit is due, and will recognize the proprietary interests of others.

7.6 RISKS TO ENGINEERS

The major risks faced by engineers working for contractors are not that different from the risks faced by engineers working for public agencies or private owners. The most significant area of increased exposure for an engineer working for a contractor is probably the risk associated with construction incidents from injured workers and/or from safety infraction citations by the Occupational Safety and Health Administration (OSHA). These risk areas include:

- Alleged liability to third parties as a result of construction incidents in which the engineer may or may not have been directly involved
- OSHA citations for not enforcing safety regulations or project safety procedures for contractor's personnel
- Alleged liability for damages resulting from not incorporating project-wide standards in proposed designs
- Breach of confidentiality

7.6.1 Third-Party Suits

Third-party suits are common and are difficult to avoid, regardless of whether the engineer is working for a contractor or an owner. Suits may result from construction accidents, such as cave-ins, form collapses, falls, and the like. Because, in many states, injured workers cannot sue their employers due to workers' compensation laws, they seek relief wherever it can be found. A likely target for this relief is the project design organization, which may be carrying substantial insurance coverage. Suits can also come from interests affected by, but not directly involved in, the construction. A passing pedestrian or abutting property owner may seek relief of perceived damage by suing the engineer. Engineers working for the contractor may be perceived to be in control of the work and are, therefore, probable targets for suit.

The case of *National Foundation Company v. Post, Buckley, Schuh & Jernigan, Inc.,* 465 S.E.2d 726 (Ga. Ct. App. 1995), serves as an example of a third party's unsuccessful attempt to hold a design professional liable for injuries sustained on a work site. In this case, a carpenter who was employed by the general contractor on a construction project was injured when he fell from an unguarded walkway along a shoring wall at the project. The shoring wall was designed by an engineering firm that worked on the project pursuant to an agreement with the project owner. The carpenter sued the engineering firm, among others, alleging that the firm was negligent in maintaining the work area and in designing and preparing subsequent shop drawings for the shoring wall. Specifically, the carpenter argued that the engineering firm should have placed temporary handrails or barricades on the work-site area to guard against injuries.

The court rejected the carpenter's argument, noting that the placement of handrails or barricades is a safety issue, rather than a design issue. Accordingly, because the engineering firm was not contractually obligated to exercise control or supervisory responsibility over safety measures on the job site, the engineering firm was not held liable for the carpenter's injuries.

7.6.2 OSHA Citations

Typically, engineers are not involved in construction safety, and they frequently negotiate contract terms that specifically state that the engineer is not responsible for construction safety, no matter who the client is. OSHA has cited engineers in construction accident cases that have led to the development of a substantial body of case law that is helping to define these risks.

The Secretary of Labor, through OSHA, has recently attempted to hold engineering design professionals responsible for compliance with and/or enforcement of construction safety regulations under 29 CFR §1926. In *Secretary of Labor v. Simpson, Gumpertz & Heger, Inc.*, 3 F.3d 1 (1st Cir. 1993), the Secretary of Labor attempted to hold Simpson, Gumpertz, an engineering firm, responsible for alleged violations of construction safety regulations relating to the collapse of a concrete form. Ultimately, an administrative law judge and the OSHA Review Commission dismissed the citation against Simpson, Gumpertz. The U.S. Court of Appeals for the First Circuit affirmed these dismissals, but did so on the narrow ground that the project site was not a "place of employment" for Simpson, Gumpertz, and, therefore, Simpson, Gumpertz had no duty under the OSHA construction work safety regulations to protect the project site. In so ruling, the court declined to explicitly and more broadly rule that design professionals are not engaged in "construction work," which is a jurisdictional prerequisite for application of the OSHA construction work safety regulations.

The case of *Secretary of Labor v. Foit-Albert Assoc., Architects & Engineers, P.C.*, OSHARC Docket No. 92-0654, 1997 WL 197016 (O.S.H.R.C. April 21, 1997), involved an OSHA citation alleging that Foit-Albert, an engineering firm, violated standards resulting in the collapse of concrete form work. Specifically, OSHA claimed that Foit-Albert had failed to ensure that the form work was properly designed, fabricated, erected, supported, braced, and maintained and that it had failed to properly inspect shoring equipment immediately prior to, during, and immediately after concrete placement. The form work collapsed during concrete placement, injuring three workers, including two Foit-Albert inspectors. Ultimately, the Foit-Albert citation was vacated by an administrative law judge. On appeal, the OSHA Review Commission affirmed the administrative law judge's dismissal, finding that the Foit-Albert employees had no contractual authority to direct the activities of the trade contractors, and had no contractual responsibility over safety means and measures. The Review Commission noted that although Foit-Albert performed inspection activities that may have unavoidably addressed safety issues, it did not manifest the ability to control or direct matters of safety. Accordingly, the Review Commission concluded that Foit-Albert did not exercise "substantial supervision" over the construction work that would have brought it under the OSHA construction work safety regulations and would have required it to protect the project site.

The cases described above are cases in which the engineer was working for an owner, not a contractor, but the issues involved are relevant no matter who the client is. These cases illustrate a significant risk to engineers against which they must guard when working for contractors, i.e., not to become engaged in construction or to exercise substantial supervision over the work. If engineers agree to undertake such tasks, as when providing construction management services, they must take appropriate steps to protect themselves, as will be discussed in the section on risk mitigation. (For a more detailed discussion of similar issues and other relevant case law, see Chapter 10, Section 10.2.2.2(B).)

7.6.3 Meeting Design Standards

Any professional designs provided on behalf of contractor clients must meet project-wide standards, in addition to the ordinary standard of professional care. Failure to do so can result in substantial liability on the part of the contractor and, ultimately, the engineer. Assuming design standards are met, the engineer's responsibility is then to provide services using that degree of care and skill ordinarily exercised under similar conditions by professional consultants practicing at the same time in the same or similar locality. An engineer who agrees to a higher standard of care or performance may be found liable for breach of contract, even though not negligent. (See Chapter 10, Section 10.1.1.2.)

7.6.4 Breach of Confidentiality

Construction is a highly competitive business. Engineers working for contractors must understand this fact fully and, therefore, have a high professional obligation not to disclose confidential proprietary information to anyone. The consequences of a breach of confidentiality can be substantial and may cause a contractor to seek contractual damages against an engineer who improperly discloses information that affects the contractor's competitive or negotiating position.

A related risk to engineers is that contractors may try to impose a gag order on their engineers regarding issues that must be reported by law, ethical codes, and/or regulations. This is especially true of environmental laws and regulations. In anticipation, the engineer should propose a contract clause that would authorize the engineer to make a report to relevant authorities within some reasonable time after making the report to the contractor. This issue comes to the heart of the understanding by contractors that engineers have certain legal and professional obligations that transcend their contractual obligations to any one client. Holland (1996) suggests the following language, which could be inserted in a contract between engineer and contractor:

> Engineer agrees to maintain confidentiality regarding any information obtained as a result of services provided by Engineer under this agreement. Contractor agrees, however, that Engineer's obligation to maintain confidentiality shall not apply to disclosures that Engineer must make pursuant to law, regulations, or an order or subpoena of a court. Engineer will make a reasonable effort to notify Contractor promptly after discovery of any condition that requires reporting. Engineer does not, by this agreement, assume Contractor's legal responsibility to notify appropriate agencies as required by law.

7.7 CONFLICT OF INTEREST

Engineers working for contractors must consider conflict of interest issues, particularly when working on large, multisection projects. Is an engineer ethically precluded from working for a contractor on a project section if that engineer has worked for the owner on some other, but unrelated, section of the same project? Engineers, owners, and contractors have had to examine this issue in recent years on some very large urban infrastructure projects in the United States. As long as the engineer has had no previous involvement in the design of the project section to be built by the proposed contractor-client, it has generally been determined that there is no conflict of interest, and the engineer is free to work for the contractor. If, however, the engineer was involved in the design—or even the preliminary design of the project section to be built by the contractor—he or she will likely be determined to have a conflict of interest in working for the contractor, and the owner is likely to strongly object to the engineer's providing professional services on that section of the project. In some cases, the engineer can work for a contractor on a section of a project for which he or she provided design service if the work for the contractor is not professional in nature, such as providing field or laboratory testing service without professional interpretation of test results or any other exercise of professional judgment. Conflict-of-interest issues must be examined in detail for each situation, and, in some instances, the issue may have to be decided by an impartial body, such as a city or state ethics commission or board.

7.8 RISK MANAGEMENT STRATEGIES

Risk management for engineers is essentially the same whether working for an owner or for a contractor. The basic risk management practices described elsewhere in this book should be followed. The major differences stem from the fact that, by working for the contractor, the engineer is more likely to be closer to the liability associated with construction impacts and accidents. The risk of liability resulting from affected third parties, whether from construction effects or from construction accidents, is therefore greater. The best way to manage this risk is to avoid getting so close to the construction process as to be perceived as engaging in construction work, or being in control of or substantially supervising the work. Recent case law indicates that engineers can successfully avoid these liabilities if the construction site is not their place of work, if they are not engaged in construction, and if they are not in substantial control of the work (Hatem, 1995; Holland, 1996).

A less tangible risk is that resulting from pressures to compromise design quality to achieve the least expensive construction solution possible. This is a professional risk that, if not carefully considered, can result in loss of professional reputation and, ultimately, loss of professional practice.

Conflict of interest is a complex legal issue that is usually decided not by engineers or contractors, but by administrators and attorneys working for public agencies. There is business risk to engineers who work for contractors, in that public agencies may be unwilling to employ them on future projects and, therefore, a reduction in business occurs. Management of these professional risks are, at best, subjective, and are decisions that are made by individual engineers. The best management strategy is to be completely professional and unbiased when working for contractors and dealing with owners. Engineers who do only what their contractor clients tell them to do are ultimately ineffective.

Management of the business risks associated with lump-sum contracts with contractors requires a clear definition of the scope of work and expectations for the agreed-upon fee. When determining fees, the engineer must consider the business and professional risks and take them into account in the negotiation of a mutually acceptable contract.

7.9 CONCLUSION

In the highly competitive construction industry, contractors must take calculated risks to succeed. They frequently turn to consulting engineers to help them evaluate these risks and, therefore, they value decisive, competent, and timely consulting and design services. Engineers engaged in such assignments should expect that the work will be technically challenging and will require new, innovative approaches. In this regard, working for a contractor can be satisfactory and rewarding, provided, of course, that both the contractor and the engineer enter into their relationship with a clear understanding of each others' needs and expectations.

Working for a contractor, however, carries with it certain risks. In addition to the typical liability exposure faced by engineers (and discussed in other chapters throughout this book), engineers working for contractors face an increased risk of claims arising out of worker safety issues and claims by third parties resulting from alleged damages caused by construction activities. Similarly, conflict of interest issues must be considered by an engineer who is asked to perform services for a contractor—particularly on large, multisection projects. Thus, in working for contractors, engineers must always be mindful of their potential liability exposure, as well as their ethical and legal obligations to their contractor clients and to third parties.

REFERENCES

Hatem, D. J., "OSHA Continues Dispute Against Design Professionals," *The CA/T Professional Reporter*, Vol. 1, No. 1 (Boston: Burns & Levinson LLP, December 1995), pp. 7–10.

Holland, J. K., "Risk Management for the Professional Consultant," lecture notes of presentation to GZA GeoEnvironmental, Inc., 1996.

8

ROLE OF THE PROFESSIONAL CONSULTANT IN THE EVALUATION OF DIFFERING SITE CONDITIONS CLAIMS

Steve J. Poulos, Ph.D., P.E.
GEI Consultants, Inc.

8.0 INTRODUCTION

This chapter will focus on the role of the design and construction management professional in the investigation and evaluation of claims based on "differing site conditions."[1]

Claims for differing site conditions are inherent in subsurface construction. Geotechnical investigations in advance of construction cannot realistically disclose all of the information about the subsurface that becomes available or is unveiled during construction. There is a practical limit for each project on the amount and cost of investigation that should be done to balance the benefit of further investigation against the potential of incurring a differing site conditions

Subsurface Conditions: Risk Management for Design and Construction Management Professionals, edited by David J. Hatem
ISBN 0471-15607-8

claim. There may be time constraints that limit the ability to conduct an extensive investigation. In addition, known site conditions may "behave" in a manner different from that anticipated during construction operations due to a variety of factors, including the interaction of the contractor's selected construction means and methods with the ground. Therefore, it is common to encounter subsurface conditions during construction that (1) are materially different from those indicated in the contract and/or in the data or reports derived from the time-limited investigation previously conducted; or (2) could not have been reasonably anticipated. Such circumstances lead to differing site conditions claims.

The purpose of construction contracting is to produce a quality project consistent with the contract documents, at the least cost to the owner and with appropriate profit to the contractor. Success of this process depends greatly on the design and construction management professionals. Therefore, it is vital that the owner retain the appropriate professionals to provide adequate scopes of services to accomplish the project objectives; that the services of those professionals meet the appropriate standard of care; and that the professionals continuously keep in mind design, constructability, and contractual issues that should be addressed to reduce the probability of claims (Gould, 1995). The professionals must anticipate, based on experience, the types of claims that are likely to arise for a particular site and project, and should recommend that the owner include, in the contract documents, provisions that address site conditions and that fairly compensate the contractor for valid differing site conditions claims, but which do not invite unfounded claims (see discussion in Section 4.3 of Chapter 4).

To some degree, claims for differing site conditions are encouraged by the "low-bidder" contractor procurement method, which is the preferred contracting method of many owners and typically is the required method for contracting public construction work at the local, state, and federal levels. Competition encourages (if not requires) bidders to use their ingenuity but also to take risks so that the project can be competitively bid and won. Simply put, the contractor must assume favorable conditions based on an optimistic view of anticipated site conditions or run the risk of losing the job. For the latter reason, in evaluating the specifications, drawings, and the site, the low bidder may look carefully for inconsistencies that could lead to change orders that could later improve profit, and may bid the project at a price that does not reflect the reality of anticipated site conditions. Unquestionably, this approach has a substantial propensity for generating differing site conditions claims during the construction process.

When claims arise, design and construction management professionals must expeditiously evaluate the situation and assist in facilitating the prompt resolution of these claims, preferably during construction when all the facts are readily available. For this reason, it is important that the owner retain those professionals to provide an adequate service scope, which includes services during both the design and the construction phases. In the investigation and evaluation of differing site conditions claims, the professionals have the obligation to act fairly and impartially, even in the face of pressures imposed by the owner and the contractor.

Legal issues usually are not within the expertise of design or construction management professionals. Nevertheless, it is helpful for such individuals to increase their awareness of the ways in which claims arise and are resolved. There is, as always, no substitute for care, diligence, and mutual respect among project participants. In providing services according to the applicable professional standard of care, design and construction management professionals should always take care to see that responsible and timely actions are taken when issues arise, and that these actions are documented so that the process can be reconstructed more accurately and fairly in the future, if necessary.

8.1 THE DIFFERING SITE CONDITIONS CLAUSE

One typical differing site conditions clause, presented in the Geotechnical Engineering Notebook of the U.S. DOT (1996), provides that:

> During the progress of the work, if subsurface or latent physical conditions are encountered at the site differing materially from those indicated in the contract or if unknown physical conditions of an unusual nature, differing materially from those ordinarily encountered and generally recognized as inherent in the work provided for in the contract are encountered at the site, the party discovering such conditions shall promptly notify the other party in writing of the specific differing conditions before the site is disturbed and before the affected work is performed.
>
> Upon written notification, the Engineer will investigate the conditions, and if it is determined that the conditions materially differ and cause an increase or decrease in the cost or time required for the performance of any work under the contract, an adjustment, excluding anticipated profits, will be made and the contract modified in writing accordingly. The Engineer will notify the contractor of the determination whether or not an adjustment of the contract is warranted.
>
> No contract adjustment which results in a benefit to the contractor will be allowed unless the contractor has provided the required written notice.
>
> No contract adjustment will be allowed under this clause for any effects caused on unchanged work.

The contractor bears the responsibility for notifying the owner or the owner's representative when the contractor encounters conditions that it believes constitute a differing site condition.

The purpose of a differing site conditions clause is to "decrease the contingency that bidders must otherwise include" in their price and to address unanticipated conditions (Gould, 1995). The clause transfers to the owner the time and cost impacts due to certain differing subsurface conditions, as defined in the clause. The differing site conditions clause, in general, defines the character of the site conditions (i.e., those materially different and physical in nature) that would support an equitable adjustment (additional compensation or time) and the precedent requirements to recovery (e.g., prompt notice). By defining those conditions and requirements in advance and as part of the contract documents, it is anticipated that the inclusion and proper administration of a differing site conditions clause will reduce conflict and antagonistic, protracted, and disruptive claims. In summary, the differing site conditions clause establishes an understanding between owner and contractor of risk allocation for site conditions.

Often, the owner, with the aid of legal counsel, attempts to place the risk for the actual site conditions encountered entirely on the contractor even if those conditions are materially different from those indicated in the contract documents or reasonably expected by the contractor when bidding the project. The owner does this by not including a differing site conditions clause in the contract. Perhaps the owner and/or its counsel assume that the experienced contractor is in a better position to know how to handle the subsurface than the owner or the owner's geotechnical or design professional. In most well-planned projects, however, the owner has adequate time prior to letting the contract to explore the subsurface conditions; and in virtually all projects, the owner has more time than the contractor to investigate site conditions. In contrast, a bidder usually cannot be expected to know as much about the subsurface conditions as the professionals who have had the opportunity to evaluate the site and to develop site information.

Historically, there are two types of differing site conditions (Gould, 1995):

Type I—Subsurface conditions at the site that differ materially from those indicated in the contract documents, and

Type II—Unknown or unusual physical conditions at the site, i.e., conditions that differ materially from those ordinarily encountered or that would be generally recognized as inherent in the contracted type of project work.

These two types of differing site conditions should be borne in mind when investigating and evaluating the entitlement to an equitable adjustment. Typically, differing site conditions clauses offer relief for both Type I and Type II differing site conditions. (For further discussion of the differing site conditions clause, see Section 9.1.4 of Chapter 9.)

8.2 THE PROFESSIONAL CONSULTANT'S ROLE IN THE INVESTIGATION AND EVALUATION OF DIFFERING SITE CONDITIONS CLAIMS

8.2.1 The Professional Consultant's Role During Design and Construction Phases

Subsurface projects inherently entail the possibility of differing site conditions claims. Designers cannot predict subsurface conditions with certainty, and the scope of the subsurface investigation cannot, in practice, be sufficient to precisely anticipate and/or precisely characterize or quantify all site conditions and, hence, eliminate the risk of such claims. Thus, many differing site conditions claims are valid; that is, site conditions are encountered during construction that are either Type I or Type II conditions, as defined above. Despite this inherent uncertainty and the associated risk of claims, however, steps can and should be taken during the design phase to reduce the number, scope, and potentially disruptive impact of differing site conditions claims.

To reduce the likelihood of a claim for differing site conditions, it is necessary for the owner to engage appropriately qualified professional consultants and also to authorize an adequate scope of subsurface investigation prior to design and construction. In addition, the owner and the professional consultant should work out the appropriate terms of the professional's scope of services during construction. The reader is referred to Chapter 1 for a guide on how to select consultants and other professionals, and to Chapter 2 for recommendations regarding contracting practices. The professional consultants should understand the constructability and legal issues associated with subsurface projects. They should also be able to identify potential sources of claims as early as possible, so that claims can be avoided or resolution promptly reached.

The ability to untangle events long after they have occurred is difficult and costly, and sometimes not even possible. Records may not exist, or may be scanty or lost. After construction is completed, access to the site usually is limited or not useful, and memories of activities during construction are cloudy, at best. At this late stage, an expert, a judge, or a jury must rely, to a large extent, only on what (if anything) is written about past events. Facts that are not documented are essentially lost and/or perceived as less reliable and credible.

It is imperative to identify and record differing site conditions as early as possible during construction to help reduce the consequences and contain the impacts of differing site conditions claims (irrespective of whether the claims are made by the contractor or by the owner). To this end, some suggestions follow:

1. Prior to the letting of the contract documents out to bid, the professional should read the documents carefully and, from the contractor's point of view, identify ambiguities or inconsistencies that could lead to claims. These documents should also be reviewed from the standpoint of constructability considerations. Subsurface data provided to the contractors, all reports on the subsurface conditions, the subsurface design and the reasoning for the design should all be carefully reviewed by the professional—again, from the perspective of the contractor competitively bidding on the project. Ambiguities, inconsistencies, or potential problems should be brought to the attention of the owner and addressed early.

2. The professional should discuss the specifications with the contractor during prebid conferences or at the inception of the project; verify that

the contractor's site superintendent understands the specifications, subsurface information, and the subsurface design; and resolve any differences between the designer's and the superintendent's understanding.

3. The full implications of the designer's description of subsurface conditions, with respect to the cost and schedule of construction may not be clear, and may not be taken into account by the bidder (or the contractor's superintendent during construction). The designer's wording in the specifications and the notes on the drawings may be unclear (see Section 4.3 of Chapter 4). Therefore, the professional's early discussions with the contractor are vital in helping to promote understanding and cooperation. It is also possible that the contractor who has underbid the project may be looking for opportunities to make a claim. If so, these early discussions can reveal that attitude so that early, high-level discussions can be set up between the owner and contractor to anticipate and, hopefully, avoid such issues. Partnering should be considered at this time.

   During bidding and before award, the contractor generally is cooperative. After award, the contractor must make a profit; he or she controls the site, and can take advantage of that control if he or she feels it is necessary. Thus, the prebid period provides fertile ground for productive discussions between the professional and bidders, the fruits of which can potentially reduce the possibility and/or impact of differing site conditions claims.

4. Prior to the start of construction, the owner's design professional or the construction manager should set up a recordkeeping system that permits instructions to flow through an established line of authority in written form for both standard and emergency situations.

   The responsibilities of the owner, the owner's representative, the designer, the construction management professional, the geotechnical engineer, the superintendent, and the contractor should be defined. For smooth communication, the construction management professional (engaged by the owner) should be the individual who deals directly with the superintendent (the contractor's representative).

   Memoranda recording verbal instructions or agreements should be promptly prepared. One should assure that the decision-making recipients of these memoranda review them on a timely basis so that misunderstandings can be resolved before the start of the related work, or very soon thereafter.

5. Prior to the start of construction, a defined system should be established to resolve issues as they develop.

6. When potential differing site conditions issues are suspected, the contractor should, if it is safe to do so, cease any activity that would disturb the conditions as they exist. This requirement is stated in the typical differing site conditions clause. In this way, an immediate investigation may be undertaken by the owner's and contractor's respective representatives

to understand and evaluate the situation and to determine how to proceed with construction and to ultimately resolve the issues.

The design and construction management professional must be technically competent, diligent, fair, and assertive when handling claim-related subsurface issues that will affect the potential profits of the contractor in a competitively bid contract. Quite often, even the owner may want to cooperate with the contractor so that positive relationships are encouraged and the project is carried out with high quality and on schedule; consequently, the owner may tend to lean in favor of giving the contractor significant latitude for recovery with respect to a differing site conditions claim. However, the professional should never lose sight of the possibility that the owner subsequently may attempt to pursue "cost recovery" against the professional at the conclusion of the project, for claims that the owner has paid (perhaps imprudently, or for "expediency" reasons) to the contractor. (For further discussion of cost recovery claims against the design and/or construction management professional, see Hatem, 1996.)

The way to maintain high quality while reducing the likelihood of claims is to start out with a clear understanding that the contractor must fairly meet the specifications, and that fair consideration must be given by both parties to changes that might reduce or increase the costs to the owner. The observational approach is essential to accomplishing this objective. The observational approach requires a close coordination between the design and construction processes. The use of the observational approach begins during development of the design and the contract documents. At that stage, the designer should introduce procedures to observe movements, water levels, and other parameters during construction. These measurements are selected judiciously so that they can be used to indicate, in advance, the measures that may have to be taken by the contractor to help prevent specifically selected criteria from being exceeded. These measures usually are required as part of a contractor submittal soon after construction begins. If such an observational approach was not planned during design, the construction manager should take similar steps during construction so that he or she may detect when critical aspects of the subsurface work are not proceeding as expected. For example, the construction manager could require the contractor to dig test pits or drill borings in questionable or critical zones. The cost of such exploration is likely to be small, compared to the costs of potential construction disruption, difficulties and claims. Also, the contractor can start in a zone that will have the least effect on cost or on safety so that any difficulties that arise can be resolved and applied in the more complicated or important zones of the site.

By planning and using the observational approach, all project participants become aware of potential problems in a timely fashion. That awareness will lead to a more rapid recognition and resolution of differing site conditions and related claims. The increased cost and other potential impacts that result from encountering conditions that materially differ from conditions expected may thus be reduced for both the contractor and the owner.

Too often, the owner takes the position that he or she can rely on the subsurface design as if it embodies and expresses a complete and accurate understanding of the subsurface conditions, which is an impossible standard of performance for any exploration or design program. Unfortunately, design professionals, desiring to appear optimistic to the owner, sometimes tend to give the owner the unrealistic impression that their own understanding of the subsurface is essentially complete. Professionals have a duty to discuss the range of risks about subsurface conditions with the owner and to make appropriate recommendations.

In summary, during the design phase, the professional consultant should make it one of his or her tasks to consider methods for reducing the likelihood that claims will arise, and for limiting the negative and disruptive effects of such potential claims on project costs and schedule, as well as on the relationships among the owner, contractor and professional. This task entails selection of a proper exploration program suited to the site. The documents should contain appropriate mechanisms for identifying, as soon as possible, when conditions arise in the field that materially differ from what was anticipated, and, when implementing these mechanisms, the contractor's perspective should be considered. Prior to the start of construction, the owner and the contractor should decide how differing site conditions claims are to be handled. Finally, a good system of controls during construction should be set up—controls that are fair to the owner and to the contractor.

8.2.2 Investigating and Evaluating Differing Site Conditions Claims

An outline for investigating differing site conditions claims is set forth below. In performing such an investigation, the professional consultant must be timely, and be impartial and fair in providing specific recommendations to the client. The client ultimately must make the decisions with respect to claims. It should be noted here that some owners and their attorneys make every effort to influence the direction of technical opinions. However, the professional has the obligation to exercise his or her independent judgment, to be fair and impartial in the evaluation of differing site conditions claims, and to perform these activities in a diligent manner.

The professional who investigates a differing site conditions claim must conscientiously and diligently compare the information in the contract documents with actual site conditions that are claimed to be materially different. The professional should commence his or her investigation of a differing site conditions claim by gaining a preliminary understanding of the incident that initiated the claim and the issues as perceived by the owner and the contractor.

The professional should then examine the site to gain firsthand information on the nature of the conditions that form the basis of the claim. It should also be determined whether the construction activities in the areas of the alleged differing site condition can or should be suspended while the evaluation takes place, or whether an interim plan of action is needed so that delays are minimized. The professional should begin thinking about alternative designs that could be used to mitigate or avoid the time and/or cost impact of changed conditions. A preliminary plan for any necessary explorations should be formulated. The professional also should judge whether other professionals, such as drilling experts, estimators, structural specialists, and the like, should be engaged to handle specific items for which additional expertise and/or experience is needed.

As the next step in the process, all relevant portions of the contract documents—the specifications, drawings, and referenced information—should be studied to understand their intent and application when applied to the specific claim.

Next, the professional should study relevant correspondence, field records, and the details of the construction activities that preceded the claim. The professional should interview personnel or other representatives of the owner and contractor to gain information that is not in the documents. During this process, the professional should again identify the types of technical information that must be understood and the best persons to provide that information.

During the review of documents, the professional should develop a preliminary view of the issues and finalize any exploration or measurement program that is needed. This program should be carried out expeditiously as the review continues. The professional should observe the explorations and measurements in progress, to assure that the information gathered will suit the professional's purpose.

With the exploratory and measurement data in hand, and after a thorough review of the documents, the professional should formulate his or her opinion about whether or not a differing site condition occurred, options on how it can be addressed, and whether the contractor is entitled to an adjustment due to any such differing site conditions.

This opinion of the professional, whether in writing or verbal, should then be discussed with the client and, if necessary, with the client's attorney, to establish a procedure for bringing the issue to resolution.

The role of the professional in the investigation and evaluation of differing site conditions claims requires knowledge of the contract documents; any other data or reports developed during design and construction; knowledge of the means and methods used by the contractor; a thorough understanding of the specific issue regarding the subsurface that triggered the claim; and an understanding of the nature and extent of the condition forming the basis of the claim. The professional bears a heavy responsibility in such evaluations. It is a time when the professional's competence and integrity are severely tested.

8.2.3 Role of the Professional Consultant in the Dispute Review Board Process

The dispute review board (DRB) is empaneled by the owner and contractor soon after contract award and is typically provided for in the bid and contract documents. It is composed of a group of individuals selected because of their experience in the design and construction of subsurface projects. The DRB will consider disputes between the owner and contractor and make impartial recommendations concerning their resolution. The members of the DRB typically view the project site on a regular basis to acquaint themselves on a firsthand basis with the work and site conditions that may become, or are the subject of, a differing site conditions claim. (For a more detailed discussion of the dispute review board, see Section 9.3 of Chapter 9.)

The professional who assists the owner in preparing for the DRB process should follow the procedures outlined in Section 8.2.2. For the more formal process associated with a DRB, it is vital that the professional consultant ensure, early in the process, that all of the experts have been tapped who are necessary (and affordable) to understand and present information about the critical issues of the claim. It is also vital to study the weaknesses of the information, both pro and con, so that the owner can be given an objectively reasonable and realistic understanding of the issues that relate to whether or not the claim is valid.

In essence, the role of the professional consultant is to understand whether the claim is valid, to be able to describe the strengths and weaknesses of the information to the owner, and to present the information clearly so that it can be used by the DRB.

The professional consultant should focus on carefully planning and preparing exhibits that are directed toward proving or disproving whether materially differing site conditions were actually encountered, as defined in the differing site conditions clause. Clarity and specificity should be paramount. Were the site conditions materially different from those shown in the relevant documents? Were the site conditions unusual for the particular location and time, such that the contractor could not have reasonably been expected to anticipate them? The exhibits and related documentation should support each other. Contradictory information should be explained. The backup information should be readily accessible, and the applicability of the information to the issues of the claim should be stated clearly. The professional consultant may wish to consult with his or her own legal counsel for advice in connection with the legal aspects of the claim and ways to enhance the presentation. Also, as previously noted, the professional consultant should always be mindful of the possibility that the owner may pursue cost recovery against the professional, based on the outcome of the contractor's differing site conditions claim against the owner (particularly when the claim includes allegations of defective design; see Chapter 10, Section 10.2.2.3). Therefore, the professional's consultation with its own legal counsel is advisable throughout the differing site conditions claim evaluation and resolution process, including proceedings before the DRB.

8.3 CONCLUSION

Differing site conditions claims are an inherent part of subsurface construction, especially when contracts are awarded to the low bidder. Competition requires that a contractor limit contingencies in his or her price or that he or she interpret conditions in an optimistic (low-cost) light in order to win the work. This process often results in subsurface projects being awarded to contractors who have slim profit margins. Under these circumstances, substantial tension can develop between the owner and the contractor, and, on a secondary level, between the latter and the design and construction management professional.

Despite these inherent risks and the potential tensions created by the low-bid procurement process, a fair method for evaluating differing site conditions claims can and should be established. Such procedures may include use of the observational approach, partnering, and DRBs.

Many differing site conditions claims arise because prior to the dispute—i.e., during the performance of conceptual development, design, report preparation, and contract document preparation services—insufficient attention is given to how the products of those services will be understood by the contractor during the bid and construction process. However, the professional, in performing these services, should consider the project, the geotechnical reports, the design, and the related contract documents from the low-bid contractor's perspective to help reduce the bid price and the probability of claims. With attention to this issue, it is more likely that all parties will be sensitive to each others' perspectives so that mutual trust can be developed during construction.

The information in this chapter has been aimed at assisting all parties involved in subsurface construction to perform in a manner that promotes the goals of quality construction with reasonable profit to the contractor, but with lowest cost to the owner after any (hopefully, few) differing site conditions claims have been paid. The design and construction management professional has a vital role in helping to anticipate, control, and manage the potentially disruptive and other negative effects of differing site conditions claims.

NOTE

1. Most contractual and statutory provisions that provide for an equitable adjustment for differing site conditions offer a basis for recovery for both the contractor and/or the owner in the event that the site conditions differ substantially or materially from those indicated in the contract documents or ordinarily encountered in projects of the character under construction. These provisions could result in a credit to the owner if the site conditions are more favorable than anticipated. Because owners traditionally have seldom sought credits, this chapter will focus on the evaluation of differing site conditions claims that originate with the contractor.

REFERENCES

Gould, James P., "Geotechnology in Dispute Resolution," The Twenty-Sixth Karl Terzaghi Lecture, *Journal of Geotechnical Engineering*, Vol. 121, No. 7, July 1995.

Hatem, D. J., "Errors/Omissions Cost Recovery Claims," *The CA/T Professional Liability Reporter*, Vol. 1, No. 4 (Boston: Burns & Levinson LLP, June 1996), pp. 1–10.

U.S. Department of Transportation (DOT), "Geotechnical 'Differing Site Conditions,'" *Geotechnical Engineering Notebook*, GT Guideline No. 15, Federal Highway Administration, April 1996.

9

DISPUTE RESOLUTION MECHANISMS FOR DIFFERING SITE CONDITIONS CLAIMS

Hugh P. Caspe, P.E.
Woodward-Clyde Consultants

9.0 INTRODUCTION

Chapter 8 focused on the role of the professional in the investigation and evaluation of differing site conditions claims.[1] This chapter will focus on the mechanisms available for the resolution of differing site conditions claims. Before discussing that subject, it is important to provide some general background regarding risk allocation for site conditions on subsurface projects.

9.1 RISK ALLOCATION ON SUBSURFACE PROJECTS

9.1.1 Discussion of Risk

Subsurface construction is fraught with risk due to claimed (or actually encountered) differing site conditions, the resolution of which, in many cases, is directly tied to the contractor's financial success on the project and/or the

Subsurface Conditions: Risk Management for Design and Construction Management Professionals, edited by David J. Hatem
ISBN 0471-15607-8 © 1998 John Wiley & Sons, Inc.

owner's ability to fund the project. In building a structure above ground, one is able to anticipate in a relatively effective manner where and how items are to be placed and how they relate to and interact with each other, both during and after construction. On subsurface projects, in contrast, elements are placed in or removed from a relatively unknown existing medium, which may or may not have been adequately investigated by owners or engineers, and may or may not have been accurately represented or interpreted within the contract documents or in other information made available to the contractor. In addition, even if actual conditions are accurately represented and interpreted, there could be unanticipated difficulties due to ground behavior at a specific location or area and/or the interaction of site conditions with the contractor's selected construction means and methods. Some of these factors are within the control of project participants, but many are not. Although there are other significant elements of risk associated with all types of construction, which (presumably) will be reflected in a contractor's bid for any particular project, subsurface construction is usually categorized as high risk due to the risk potential for these relatively unknown and/or unquantifiable factors and variables.

As an example of how factors unknown at the time of bid can influence final costs, consider the experience of Water Tunnel No. 3, constructed in the early 1970s in New York City. At the time of bid, the contract documents did not contain any specific geological information or data, but did indicate that approximately 750 linear feet of steel rib supports would be required within the 13.7-mile-long excavated tunnel, which was a small proportion of the total tunnel length and, therefore, a clear indication of competent, self-supporting rock. The tunnel was either 27 feet or 23 feet in diameter. The steel rib supports consisted of 8-inch-deep bent wide-flange shapes, weighing approximately 32 pounds per foot, placed at a 4-foot spacing. Additionally, timber blocking and lagging were to be placed between the ribs and rock, to transfer the rock load to the steel. The tunnel itself was excavated by the drill-and-blast process, which effectively established the following general mining production cycle:

1. Drill blast holes.
2. Load explosives and shoot.
3. Excavate spoils.
4. Scale loose rock and install rock bolts as required.

During the bidding process, the contractor anticipated that daily production per heading would be about 48 feet, based on crews working three shifts per day and completing two cycles per shift. Those assumptions were based on information made available to the contractor, including the limited requirement for steel supports called for in the bid quantities, and, as such, formed the basis of the contractor's construction schedule and anticipated costs.

In reality, the ground was significantly different than anticipated and implied in the contract, as the rock stress relieved itself shortly after excavation and, in so doing, created a situation in which there was significant probability of loose wedges or blocks of rock falling out of the tunnel roof. (In one location, a 42-foot-deep wedge fell from the arch about 2,000 feet behind the heading face.) This condition required the use of markedly more steel ribs to support the excavated tunnel. Production was drastically reduced to levels substantially below the bid rate originally anticipated.

The associated cost increases due to this unforeseen situation were disastrous for the contractor. In effect, an additional, fifth step was added to the mining production cycle: "Erect steel rib supports and place timber blocking and lagging." This additional production step, along with its associated time and cost impacts, was considered by the owner as the "risk" that the contractor had assumed in signing the contract. When the contractor expended more than $100 million in nonreimbursed costs and effectively reached the limit of its financial resources, and was not able to reach an agreement with the owner to resolve the dispute, the contractor abandoned the project and commenced litigation to recoup its losses. The owner then countersued the contractor for abandoning the work. An eight-year legal battle followed, in which an estimated tens of millions of dollars in litigation costs were expended. In the end, neither party won. For the owner, completion of the project was delayed for more than 15 years, and, as part of the negotiated settlement, the owner also paid the contractor a sum of money. For the contractor, the losses during construction were significant; some of the joint venture partners went bankrupt or were left in poor financial condition (some never to fully recover), and the settlement amount received was hardly sufficient to cover the contractor's litigation costs.

In the final analysis, while the risks associated with subsurface construction cannot be entirely avoided, with forethought and consideration, they can be anticipated, defined, and fairly allocated among the parties involved in the construction process in an effective manner, so that they can be successfully managed. Essential ingredients for a successful, proactive risk management approach include (1) a recognition that disputes and claims are to be expected on subsurface projects due to the variable and unpredictable nature of site conditions, and the interaction of the contractor's selected construction means and methods with those conditions; and (2) the need for mechanisms in the contract documents that encourage and facilitate the fair, prompt, and cost-effective resolution of disputes and claims.

9.1.2 Risk Allocation—Who Owns the Ground

Risk allocation raises the following question: "Who owns the ground that a project is to be built on or through?"

Over the years, owners have recognized that there is significant risk involved with respect to what the ground reveals during construction, as compared to what was anticipated prior to construction. Additionally, many owners have identified a need for cost "certainty" to be established in the construction contract, which typically results in one or all of the following: (1) a competitive "lowest" bid procurement process, culminating in a fixed price contract; (2) a minimum of "extras" acknowledged by the owner as valid; and (3) the assignment of as much risk as can reasonably (legally) be justified to the contractor. In the realm of public construction projects, cost certainty has been a significant consideration, as public agencies generally are legally required to award the construction contract to the lowest (not necessarily the most qualified) bidder to construct a project within fixed funding or other appropriate limits. These limits are often established many years before the actual construction, and, in many cases, are based on the anticipation of favorable ground conditions in order to maintain the budget within available funding.

Faced with such considerations, owners and their lawyers have traditionally accomplished risk allocation by preparing contract documents containing broad exculpatory clauses intended to disclaim any liability of the owner for the accuracy and/or completeness of subsurface information presented, (either data or any interpretation of data), disclosed or otherwise made available to bidders. Under this method of risk allocation, the owner may disclaim any information used by the contractor in formulating his or her bid and/or in planning and performing the work (i.e., reports, borings, specifications, geotechnical interpretive reports, and so on). This type of contracting and risk allocation philosophy purports to assign to the contractor the entire risk for any differences in site conditions between those anticipated (based on information, data, or interpretive reports furnished by the owner) and those actually encountered; as such, the risk must be covered with the funds presumably included in the contractor's bid proposal. In legal parlance, this philosophy became embodied in the so-called common law rule (see Chapter 10, Section 10.2.1). Under this common law rule, the contractor bears all the risk of unanticipated site conditions. In some situations in which the rule has applied, some contractors further compounded their eventual burdens by bidding subsurface projects on the assumption that the actual or encountered ground conditions would be consistent with the most optimistic interpretation of those conditions in the contract documents. In many instances, if the contractor were to assume a worst-case (or even a more realistic) assessment, the bid may not be competitive. These contradictory, "collision course" approaches by owners and contractors have led to significant conflict and have generated discussions on the need for improved approaches to risk allocation for site conditions on subsurface projects (Hatem, 1997).

9.1.3 Recommendations Regarding Risk Allocation Subsurface Proposal

In 1968, the federal government initiated a differing site conditions clause in its contracts, which today reads as follows:

Differing Site Conditions

(a) The Contractor shall promptly, and before such conditions are disturbed, give a written notice to the Contracting Officer of (1) subsurface or latent physical conditions at the site which differ materially from those indicated in this contract, or (2) unknown physical conditions at the site, of an unusual nature, which differ materially from those ordinarily encountered and generally recognized as inhering in work of the character provided for in this contract.

(b) The Contracting Officer shall investigate the site conditions promptly after receiving the notice. If the conditions do materially so differ and cause an increase or decrease in the Contractor's cost of, or the time required for, performing any part of the work under this contract, whether or not changed as a result of the conditions, an equitable adjustment shall be made under this clause and the contract modified in writing accordingly.

(c) No request by the Contractor for an equitable adjustment to the contract under this clause shall be allowed unless the Contractor has given the written notice required; *provided*, that the time prescribed in (a) above for giving written notice may be extended by the Contracting Officer.

(d) No request by the Contractor for an equitable adjustment to the contract for differing site conditions shall be allowed if made after final payment under this contract.

48 CFR §52.236-2 (1996).

With the adoption of this differing site conditions clause, the federal government recognized that, within certain defined parameters, the owner would assume the risk for unanticipated ground conditions, and that, under certain circumstances, a contractor may be entitled to an equitable adjustment due to differing site conditions. Likewise, with the implementation of this clause, contractors generally may rely on information made available to them at the time of bid and estimate their costs accordingly. Some of the "risk" that was formerly exclusively allocated to contractors was now reallocated to the owner with the recognition that the ground belonged to the owner, at least as far as a remedy was available under the differing site conditions clause.

In November 1974, contracting practices on subsurface projects benefited with the publication by the U.S. National Committee on Tunneling Technology (USNCTT) of "Better Contracting for Underground Construction." The report was sponsored by a number of federal agencies, foundations, private corporations, and associations, including the U.S. National Science Foundation, U.S. Department of Transportation, U.S. Urban Mass Transportation Administration, U.S. Bureau of Reclamation, and U.S. Army Corps of Engineers. The subcommittee that prepared the report was composed of leaders in the subsurface construction community, including engineers, contractors, owners, educators, and lawyers, and recognized that the then-current methods of contracting practices had serious "barriers to the application of improved technology and thus to the achievement of more economical utilization of underground space." It also recognized that the use of competitively bid, firm fixed-price contracts for subsurface construction under the common law rule had a number of serious drawbacks. Among the specific issues recognized were:

- While, ideally, contract documents should be explicit in all detail at the time of award, on subsurface projects actual subsurface conditions only reveal themselves as construction proceeds.
- Adversarial relationships develop because unforeseen and/or unpredictable site conditions may require a change in project planning, design, sequencing, or in the selected construction means and methods, and may result in work disruption with an associated time and cost impact, sometimes leading to claims, disputes, and, finally, litigation.
- Without an opportunity for contractual relief (such as that afforded by a differing site conditions clause), contractor bids on subsurface projects contain large (often undisclosed) contingencies, which could represent a major element in the overall bid price. These contingencies represent the risk that the contractors need to retain and accept, and include, among others, cost for unknown subsurface conditions, and allowances for significant amounts of change orders that disrupt the work. Should the contractor be lucky and conditions during construction be favorable, significant profit can be made at the owner's expense. Likewise, should conditions not be favorable and the contractor's contingency not be sufficient, the owner stands to benefit at the contractor's expense.
- Implementing advances in technology and innovative construction techniques are not necessarily in the contractor's best interest on subsurface projects, in circumstances in which the contracting practices impose all risk on the contractor.

The study's objective was "to develop improved contracting methods" and, by implementation of those methods, lower the cost of subsurface construction for the owners while allowing for a fair profit to be made by the contractors. To that end, the USNCTT made a number of specific recommendations covering many of the elements that affect the risks encountered on subsurface projects. The report made 17 specific recommendations which, if implemented, would significantly reduce and/or appropriately reallocate, or minimize, some of the risk associated with subsurface work. Summarized below are the specific recommendations related to dispute resolution and differing site conditions claims:

- *Disclosure of subsurface information, professional interpretations, and design considerations*—All factual subsurface data, professional interpretations thereof, and design considerations thereby raised should be made available to bidders, but with a careful distinction drawn between factual data and interpretation or opinion.

- *Disclaimers*—Contract documents should not disclaim responsibility for factual data provided.
- *Differing site conditions clauses*—A differing site conditions clause (preferably with wording similar to the federal differing site conditions clause quoted earlier in this chapter) providing, in essence, for assumption by the owner of certain risk concerning unknowns in subsurface physical conditions, should be included in all contracts for subsurface construction.
- *Extraordinary water problems*—Special contractual provisions should be made to provide for the contingency of water problems.
- *Change negotiations*—Certain bid items and specification provisions should be included in the contract documents so that effort and time now devoted to negotiation of contract price adjustments for changes can be substantially reduced.
- *Publication of engineer's estimate*—The engineer's estimate should be disclosed by public agencies with the opening of bids, and, if there is a limit on the amount available for contract award, whether for a public or private project, it should be announced in the invitation for bids.
- *Arbitration*—An industrywide system of arbitration with qualified arbitrators was proposed.

With the basis now established to move forward on the issues associated with ownership of "risk" in contracting subsurface construction work, the USNCTT, in 1978, published another report, titled "Better Management of Major Underground Construction Projects," which outlined major functions and associated responsibilities for different phases of a project. Although it did not directly discuss differing site conditions or contracting practices, it did outline 34 recommendations to be instituted in the management of complex subsurface projects.

Each of the recommendations sought to promote one of the following six objectives:

1. Establish project goals and objectives and organize the project to facilitate their accomplishments.
2. Plan the project to achieve the owner's objective.
3. Achieve effective organization, supervision, and accountability.
4. Achieve effective construction methods, procedures, and supervision.
5. Achieve sound management of the project.
6. Achieve successful startup of the project.

A number of the 34 recommendations, when implemented in conjunction with the findings of the prior USNCTT report, became the building blocks in establishing new approaches to reduce or fairly allocate the risk associated with contracting practices for subsurface projects. These include:

- An independent group of senior consultants should be utilized in an advisory role on key issues during the life of the program.
- The participating parties should encourage prompt identification of problems, and should encourage and empower people to make appropriate and timely decisions within their responsibility and authority.
- Procedures should be established for solving field or design issues as they develop.
- A review board of experts should be established to assist in settlement of claims that cannot be settled by the normal contract administrative process.
- Owners should incorporate within their contracts provisions for some form of mediation or arbitration, should parties fail to reach agreements on issues that develop during construction. This recommendation referred to a 1977 USNCTT publication titled "Recommended Procedures for Settlement of Underground Construction Disputes," which outlined mediation and arbitration methods that could be employed.
- Sound contracting procedures should be established by incorporating the recommendations of the 1974 USNCTT publication titled "Better Contracting for Underground Construction" (discussed earlier in this chapter).
- A thorough risk and liability plan should be developed, and risk should be assigned appropriately.

At a 1979 conference on construction risks and liability sharing, J. J. Casey presented a paper in which he clearly articulated two precepts as to which party should own the risk involved in subsurface construction: All construction risks belong to the owner "unless transferred to or assumed by another party for a fair compensation," and, as such, "the principal guideline in determining whether a risk should be so transferred should be whether the receiving party has both the competence to fairly assess the risk and the experience necessary to control or minimize it."

In a 1984 report published by the Underground Technical Research Council, "Guidelines for Tunnel Lining Design," specific suggestions were presented on an approach to equitably share in the risks associated with subsurface construction and resolve disputes between owners and their contractors. These suggestions were:

- Inclusion of a differing site conditions clause in the contract similar to that employed by the federal government.
- Disclosure of geotechnical information in two reports, the first being the results of the geotechnical investigation and the second, the geotechnical basis used in design. The design report, it was suggested, should be utilized for defining existing subsurface conditions, effectively establishing a baseline.

- Establishment of a dispute review board, which would be made up of three members who would hear and make recommendations regarding disputes that the participants themselves were unable to resolve. It was further suggested that the dispute review board's recommendation be non-binding, but admissible as evidence should arbitration or litigation be used in the future to resolve the issues.
- Bid documents of the low-bid contractor should be placed in escrow. These documents would be utilized in the future to establish fair adjustments for extra work, differing site conditions issues, and the like.

Since the publication of these early reports, the informed subsurface construction community of owners, engineers, and contractors have generally agreed that the then-existing, traditional contracting practices, especially as they related to risk allocation and dispute resolution for differing site conditions, did not fairly allocate risk, and that, in effect, the owners paid for that risk twice—once in the contractor's bid contingency, and second during the claim process and in the associated settlement costs of litigation. Additionally, the traditional methods of dispute resolution were recognized as being both protracted and costly, thereby resulting in yet another layer of unfairness.

This inherent unfairness and costliness were confirmed in a 1987 study conducted by the American Society of Civil Engineers (ASCE), which reported that although owners still often used contract provisions that assigned to contractors significant risk of unforeseen site conditions, in practical reality, contractors, irrespective of those contract provisions, usually did not assume that risk within their bids and expected that the owners would be ultimately responsible; this approach is "rationalized" by the contractor's need to be competitive in a lowest bidder's procurement process and, as a result, is especially prevalent in the public construction context.

The ASCE study further revealed that contracts containing differing site conditions clauses generally did not result in a net increase in owners' costs and allowed claims to be resolved earlier. On the other hand, contracts that did *not* contain differing site conditions clauses, or other specific contract provisions providing contractors some opportunity for relief in the event of a differing site condition, did not afford the owner any savings, as contractors' claims were formulated and advanced under creative legal theories designed to circumvent a contract's exculpatory language, and were often ultimately compensated by the owner typically as part of a settlement. In many of the cases in which contractors obtained relief, they did so by pursuing legal claims based on charges or allegations of fraud and misrepresentation; failure to disclose and/or concealment of material information; breach of warranty; design error; or inadequate, incomplete, and/or defective geotechnical data, interpretive reports or contract documents. In other words, the owners effectively paid for differing site conditions claims whether or not a differing site conditions clause was contained in the contract, and the report concluded that there is more potential for claims (with more unpredictable outcomes) in the absence of a differing site conditions clause. Simply put, the absence of a differing site conditions clause and/or the inclusion of broad disclaimers provided the owner (in practical effect) with only a false sense of security and an unrealistic expectation that the risk of differing site conditions had been transferred to the contractor.

Finally, the ASCE report recommended that owners include a differing site conditions clause within the contract, as such a clause facilitated the earlier and more predictable resolution of differing site conditions claims and, at the same time, effectively reduced the overall cost of the project to owners resulting from disruptive and antagonistic claims propelled by the absence of any contractual remedy. An early and more predictable resolution of claims also has the advantage of reducing the cost of litigation associated with the dispute, and the cost of compensating senior staff, whose time and attention would inevitably be diverted away from the day-to-day work of the project and devoted to the litigation.

9.1.4 Differing Site Conditions Clause

As one court has noted, the purpose of a differing site conditions clause is to "take some of the gamble on subsurface conditions out of the bidding":

> ... Bidders need not weigh the cost and ease of making their own borings against the risk of encountering an adverse subsurface condition, and they need not consider how large a contingency should be added to the bid to cover the risk. There will be no windfalls and no disasters. The Government benefits from more accurate bidding, without inflation for risks which may not eventuate. It pays for difficult subsurface work only when it is encountered and was not indicated in the logs.[2]

Over the years, different states, federal agencies, state agencies, and engineering professional organizations have suggested their own wording for differing site conditions clauses. Although each such version may have a slightly different emphasis and thus could be interpreted differently, the underlying tone and objective of all differing site conditions clauses are substantially similar in nature.

The standard differing site conditions clause covers two types of subsurface conditions that could be encountered during construction: "Type I" and "Type II" conditions. A Type I differing site condition is one in which the actual subsurface or latent physical conditions differ materially or substantially from the conditions shown or indicated in the contract documents. The term "indicated" includes both *explicit* and *implicit* references in the contract. An explicit reference is one that clearly and precisely expresses what the subsurface condition is (e.g., the water table is located at a specified elevation in the contract documents), while an implicit reference is one in which the subsurface condition is tacitly recognized but not clearly expressed (e.g., the water table is located below a clay lens and has a high piezometric head—though not stated, an indication that an artesian condition may exist).

A Type II differing site condition is one in which an unanticipated or unknown physical condition arises that would not usually be anticipated and that differs substantially and materially from conditions ordinarily encountered and generally recognized as inherent in work of the character provided for in the contract documents. The differing site condition must be one that could not be reasonably anticipated by the contractor from his or her examination of the contract documents, the site investigation, and the contractor's general experience in the area.

In the case of either a Type I or Type II differing site conditions claim, the contractor must reasonably demonstrate that there was an additional cost associated with the unanticipated physical site condition, and that the contractor reasonably relied on information given in the contract documents. It should also be understood that even though a differing site condition may exist, a request for an adjustment will be granted only if there is an impact to either cost or time associated with the differing site condition.

9.2 DISPUTE RESOLUTION PROCESSES

Agreements between owners and contractors traditionally contain clauses that address a process for handling claims arising out of differing site conditions that may be encountered during construction. While Chapter 8 of this book focused on the role of the design and construction management professional in the investigation and evaluation of differing site conditions claims during the construction of a subsurface project, this section of this chapter will focus on the *processes* or *mechanisms* available for resolution of such claims.

Since any changes to a contract typically can only be initiated by the change order process, the steps involved in analyzing a differing site conditions claim will effectively follow the change order process that is specifically set forth in the contract documents. Figure 9.1 details a typical change order process. For purposes of clarity, the term "change" or "request for change" will describe a typical contractual procedure to incorporate the item at issue into an approved change order. The term "dispute" is used in the figure to reflect a request for an equitable adjustment or a differing site conditions claim that is rejected by the owner and becomes the subject of the contractor's efforts to litigate or to obtain relief by alternative resolution methods.

As can be seen from Figure 9.1, which outlines the typical change order process, if parties fail to reach agreement, the last step is litigation, a costly and time-consuming process in which, in most cases, there are no "winners." Litigation can go on for many years, and, if it should proceed to trial, the outcome may be the result of too many factors over which neither party may have control.

Who are the decision makers for each party, and what elements should guide

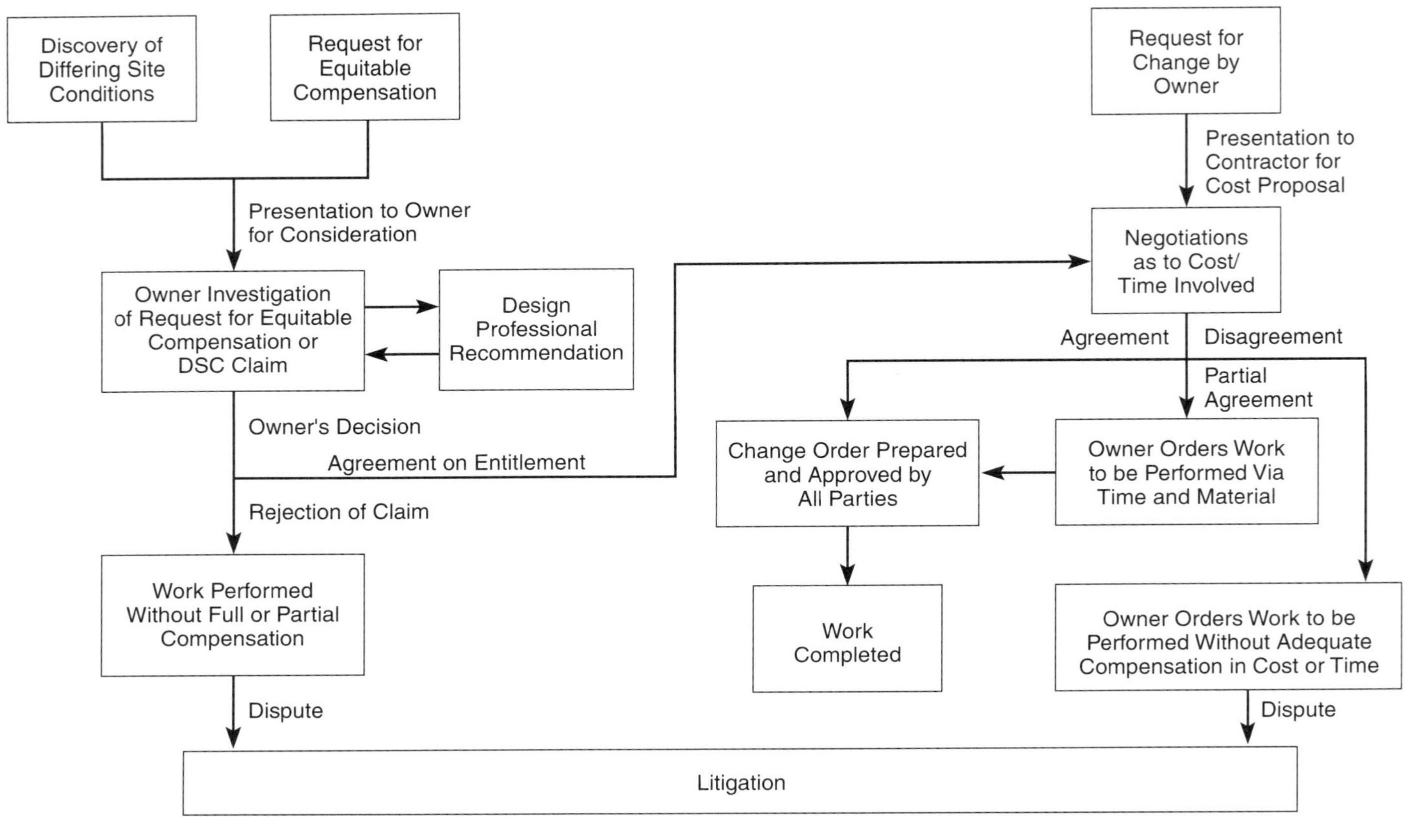

Figure 9.1 Typical change order process

them in the process of determining whether a differing site conditions claim should be made or accepted? For the contractor, that initial decision maker is usually the project manager, who typically has firsthand knowledge of the conditions encountered and is able to identify the actual or potential cost and/or time impact. The project manager makes an initial judgment on the cost impact (actually or potentially) associated with the differing site condition and any delays. The contractor's project manager then submits the claim to the owner's design professional and/or construction manager for investigation and evaluation. The design professional and/or construction manager, in turn, evaluates the claim based on an analysis of the conditions encountered, drawing on the technical expertise and opinions of the geotechnical engineer (if different from the design engineer). At that point, the design professional and/or construction manager renders a decision or makes a recommendation to the owner. Last, the owner's official representative (usually the chief engineer) reviews the contractor's claim and the design professional and/or construction manager's decision or recommendation, along with a legal interpretation of the contract documents, and makes the final determination as to whether the contractor is entitled to a time-and/or cost-equitable adjustment due to a differing site condition.

A key issue associated with the typical differing site conditions claim process is that final decisions generally are not always made at the field level, and there tends to be an intransigent attitude that develops within each group as the claim is shifted to higher levels within an organization, especially when significant costs are involved. Additionally, in many cases, proof that there is, or that there is not, a differing site condition is within the eyes of the beholder, and complex technical discussions, subjective interpretations and judgments, and arguments usually ensue. This complex process, which involves multifarious technical and subjective considerations and contested positions, does not usually lend to a direct or expedited resolution.

9.2.1 Litigation

Although litigation is the last opportunity for the contractor to resolve differences with the owner, the impact on all parties involved is usually at great cost, not only in terms of time for the disputing parties and money (i.e., for lawyers, technical experts, research, and testimony), but also with regard to the state of the parties' present and prospective relationships. For both the contractor and owner, their involvement usually demands the attention of high-level personnel, who are thus drawn away from their usually major (and "constructive") functions to concentrate on this "pathological" litigation. In addition, for the design professional, litigation is often destructive to existing professional relationships with clients.

The court system also has numerous potential drawbacks. One such drawback is the fact that judges and juries generally do not have the acumen or knowledge necessary to adequately understand or appreciate the technical subtleties that underlie the differing site conditions claim. In addition, expert witnesses may be presented to the jury on behalf of each party, expounding their respective perspectives on the technical issues, which, in many cases, only tend to confuse the jury. Because of the complex technical details involved, juries tend to make difficult decisions based on factors not necessarily relevant to the real issues involved. As such, courts tend to encourage the litigants to reach an agreement or settlement to avoid the complex, time-consuming demands, as well as the unpredictable outcomes, of the litigation process. It has often been said that the only winners in litigation are the lawyers. Because of the cost and time demanded from both parties in litigating a dispute, there has been a significant effort by the engineering and construction community, as well as owners (and some enlightened lawyers), to institute alternative dispute resolution techniques and partnering methods to effect a resolution to disputes before involving (and, hopefully, without the need to resort to) the court system.

Despite the recognized shortcomings in the litigation system, it must realistically be acknowledged that some disputes will always need to be litigated, primarily due to unreasonable positions taken by one or both parties. In those (hopefully infrequent) circumstances, there are potential advantages to litigating a dispute. The system of precedent, on which our entire common law is based, is one such advantage. Under this system, once a rule of law is established by a court in a jurisdiction, it is to be followed until the highest court in that jurisdiction identifies and articulates a good, rational reason to depart from it. Strong policy arguments favor adherence to precedent, as it allows for a degree of predictability of results, and, consequently, allows individuals to guide their business decisions and contracting practices by that prediction. Other advantages to litigation include a greater assurance of the neutrality and impartiality of the decision maker, as judges are governed by a code of ethics.

The steps involved in the litigation process are briefly outlined below.

Pleading stage—The litigation is initiated with the filing of a complaint with the court outlining the plaintiff's allegations against the defendant and defining the remedy sought for the damages or injury sustained. In response, the defendant files an answer to the complaint, responding to the allegations and asserting defenses; in many cases, the defendant may file a counterclaim seeking affirmative relief against the plaintiff.

Pretrial stage—During this stage, the discovery, or fact-finding, process occurs, which could include:

- Interrogatories—factual questions that are to be answered by the parties in writing, usually under the pains and penalties of perjury
- Depositions—questioning of parties and witnesses under oath
- Document production—demand from the opposing party to provide requested documentation relevant to the claims being litigated

Before trial, a pretrial conference may be convened by the judge assigned to the case. During this conference, the judge may establish the ground rules for the legal process that will follow. The judge can also order that the parties engage in settlement discussions and report back to the court. During the pretrial stage, each side has the opportunity to file a motion for "summary judgment," which effectively asks the court to dismiss the claim or counterclaim on the grounds that the opposing party cannot prove its case as a matter of law.

Trial stage—During this stage, a jury is chosen or a judge is assigned to hear the case and render a decision. After the jury is impaneled, the plaintiff's attorney, followed by the defendant's attorney, will make opening statements to the jury outlining their respective cases. The plaintiff then proceeds to present its case by calling witnesses and introducing other evidence in support of its claim; the plaintiff's witnesses may be cross-examined by the defense counsel. Witnesses are permitted to testify only as to facts, while expert witnesses are permitted to present opinions and conclusions. Once the plaintiff completes the presentation of its case, the defense has the opportunity to present its witnesses (both fact and expert witnesses), and to introduce other evidence in support of its defense. Of course, the defense witnesses can be cross-examined by plaintiff's counsel. Closing arguments are presented by each side and, if the case is to be decided by a jury, the judge will instruct the jury on the law that governs its decision.

Appellate stage—After the trial concludes and a verdict is rendered, either party has the opportunity to file an appeal, should it believe that legal errors of a prejudicial effect were committed during the trial. The most common appealable error is a "judicial" error (i.e., where the judge errs in ruling on the admissibility of evidence, or where the judge makes a mistake in the law that governs the issues). The appeal is usually heard by members of an appellate court, which renders a decision based on the court's majority opinion. If the appellate court finds reversible error, the appellate court may reverse, or modify the trial court's judgment, or even remand the case for a new trial. Should a party still not be satisfied with the appellate court's ruling, it may be able, in some states, to seek further appellate review, asking a higher court to review the appellate court's decision. It should be noted that at any stage during or before the trial process, the parties can settle their dispute between themselves, or with the assistance of a neutral third party.

Clearly, litigation holds significant risk to both the contractor and the owner. Litigation may extend over a significant time period, as it is not uncommon for a complex case to linger for several years within the court system. The costs for lawyers, experts, and court fees, as well as internal costs associated with the involvement of key managerial personnel, constitutes a major element of the overall impact of litigation. As such, judges often may encourage litigants to resolve their disputes by alternative dispute resolution (ADR) techniques. In reality, the vast majority of claims in litigation are settled without actually going to trial.

9.2.2 Alternative Dispute Resolution Methods

As shown in Figure 9.1, should a dispute arise over a differing site condition, the dispute resolution mechanism traditionally available is litigation. However, the owner and contractor have the ability to include in their contract, or to otherwise select, alternative methods of dispute resolution. An alternative dispute resolution process could be implemented either prior to or in conjunction with the litigation process, through the inclusion of a provision directly into the parties' contract, or by agreement at any stage of the dispute. In any event, the ADR process is consensual and may be tailored to meet the specific project or dispute requirements that are mutually desired by the parties. Given the potentially disruptive nature of disputes involving differing site conditions claims that arise during the progress of the work, the inclusion of some form of ADR in the contract documents for subsurface projects makes eminently good sense.

The manner in which the ADR process is included in the contract documents is by no means limited. A contract can be customized to permit the usage of one of a number of different approaches, each allowing for a variety of methods to be instituted in order to expedite the resolution of complex issues that may arise during construction. For example, a contract could be drafted that would require an ADR process and preclude litigation as an alternative.

The Association of Engineering Firms Practicing in the Geosciences (ASFE), in 1988, published its manual titled "ADR—Alternate Dispute Resolution for the Construction Industry," which discusses numerous ADR options that could be employed. A brief discussion on each option follows; it should be recognized, however, that these options do not constitute the full range of options available, and each may be subject to numerous variations.

Negotiation describes the process of direct discussions between parties to explore resolution of issues or disputes. These discussions should be conducted by representatives with the authority to commit their respective organizations to an agreement. Although, the negotiation process does not need to be structured, both parties must be able to "agree to disagree" on any element within their dispute in order to reach consensus on other points, and as such not get bogged down in the negotiations. For the negotiators, an attitude of compromise is essential so that a win-win settlement can be achieved. Negotiations that do not result in the resolution or settlement of a dispute are still considered successful, where issues are clarified and points of contention/agreement are isolated and limited.

Mediation is a nonbinding and confidential process, pursuant to which a neutral party assists the disputing parties in reaching a settlement, usually in the form of a compromise of disputed issues. The mediator is to be impartial in these proceedings, which are formal and follow a loosely formatted procedure. The role of the mediator is to try to understand the issues and personalities involved, and to identify the areas of flexibility within each party's positions on the major elements. With that, the mediator tries to identify alternatives, tradeoffs, and options in order to gain each party's acceptance of a middle ground and, ultimately, to reach an accord. Selection of the mediator is a critically important first step to a successful mediation. The mediator must be qualified as a mediator, impartial, knowledgeable, and experienced in the type of project and dispute issues involved. Resolution generally revolves around the financial tradeoffs that the disputing parties are willing to make, as each side knows its financial break-even point.

As in negotiations, should the process not result in settlement, the mediation process may still be considered successful if it clarifies or limits the issues in dispute.

Arbitration is a binding process in which one or more neutral individuals are chosen to decide a dispute. The process is formal and typically follows the specific procedures that are established in published manuals, such as the Construction Industry Arbitration Rules of the American Arbitration Association.[3] Unlike a trial, the strict rules of evidence do not apply, and the arbitrator has significant (and non-appealable) discretion as to what will and will not be accepted as evidence. Additionally, the discovery process is usually forsaken unless specifically agreed upon by the participants. A significant feature of arbitration is that the participants generally relinquish their rights to a trial by judge or jury, and the results of arbitration are not appealable except under the most extraordinary circumstances. It should also be noted that although an arbitrator may be versed in the technical issues associated with a complex differing site conditions claim, the arbitrator may not be well versed in the applicable law. In addition, the arbitrator is completely free to disregard the parties' contract and/or relevant legal principles in making his or her decision.

The arbitration process can be either mandatory (required by the contract documents) or voluntary, agreed to by the parties after the dispute has already arisen.

Mini-Trial—This approach is usually employed in cases where the parties have tried to resolve their dispute but have been unsuccessful. It could be either voluntary or required by the contract documents. This ADR process combines elements of adjudication and mediation, as there is an adversarial presentation of proof, followed by settlement negotiations. In a typical mini-trial, the parties to the dispute select a neutral third party. After a shortened period of discovery, an actual mini-trial is held, during which attorneys for the parties present their respective cases in an abbreviated form and expose their respective positions to the adversarial questioning of the other side. The attorneys' presentations are made before the neutral third party, and before representatives of the parties who have settlement authority. At the conclusion of the attorneys' presentations, the parties' representatives attempt to arrive at a settlement. Should an impasse in negotiations arise, the neutral third party can moderate, and, in some cases, advise the parties' representatives of the strengths and weaknesses of their positions, as well as what he or she believes the outcome of the matter would be if the case went to trial. However, the neutral's opinion is nonbinding and is inadmissible at any later trial.

Private litigation—As the term implies, private litigation is a process that allows a dispute to be presented before a retired judge (or other neutral with judicial qualifications), and can be initiated with or without court involvement. This approach has been approved in a number of states, including New York and California. The results of the process have the same "force and effect" as a trial held under a standing judge with constitutional authority. This approach may also allow for either of the parties involved to appeal any decisions, should the process go through an initial or trial phase. One significant advantage of this method is that the time frame for the litigation process is significantly reduced. Additionally, private litigation allows the parties to have input in the selection of the "judge" and allows the proceedings to be kept private and confidential.

Resolution through experts—Over the past few years, this ADR process has gained significant publicity. Essentially, experts in their field are selected to render their judgment to assist the parties in the resolution of their disputes. While these experts typically are chosen before a dispute arises, they could be court-appointed after the dispute has already progressed toward litigation. The resolution of a dispute through experts tends to be utilized to assist more in disputes involving technical issues rather than issues related to law. The experts then prepare a report outlining their findings and making recommendations, which are then used by the proponents to negotiate and structure an agreement. This approach has been expanded by the Underground Technical Research Council (UTRC) over the last number of years into a specialized ADR process utilizing a "dispute review board," which is discussed in Section 9.3.

With the basic ADR processes outlined above, a number of variations can be employed. These variations generally reflect specific requirements that may be imposed by contractual agreement, judicial oversight, or agreement between the parties themselves. Some ADR processes can be more effective than others, given the point in time in which they are utilized, as well as the "good faith" attitudes of the parties in genuinely desiring a fair resolution of the dispute.

Figure 9.2 outlines a number of ADR options available and the time frame in which they may be employed.

9.2.3 Partnering

Partnering is a process that was initiated in the construction industry by the U.S. Army Corps of Engineers (COE) in the late 1980s to assist in the resolution of disputes between the COE and its contractors before they evolve into formal claims. The partnering approach has worked successfully in the manufacturing field, where cooperative resolution of problems is used to everybody's benefit—the end result promoting the production of a superior product. The partnering process has met with great success through the years, and, as a result, a number of federal and state agencies have followed the COE's lead in implementing it into the construction industry in an attempt to reduce, to the greatest extent possible, the escalation of litigation on construction projects. The partnering process, as with all other ADR processes, does not necessarily have to be limited to differing site conditions claims, but can be applied to any issue in which a difference of opinion exists and the parties have the common goal of expediently resolving their disputes.

The basic premise of the partnering process is twofold: initially, to understand the other side's position and to appreciate the issues which are important

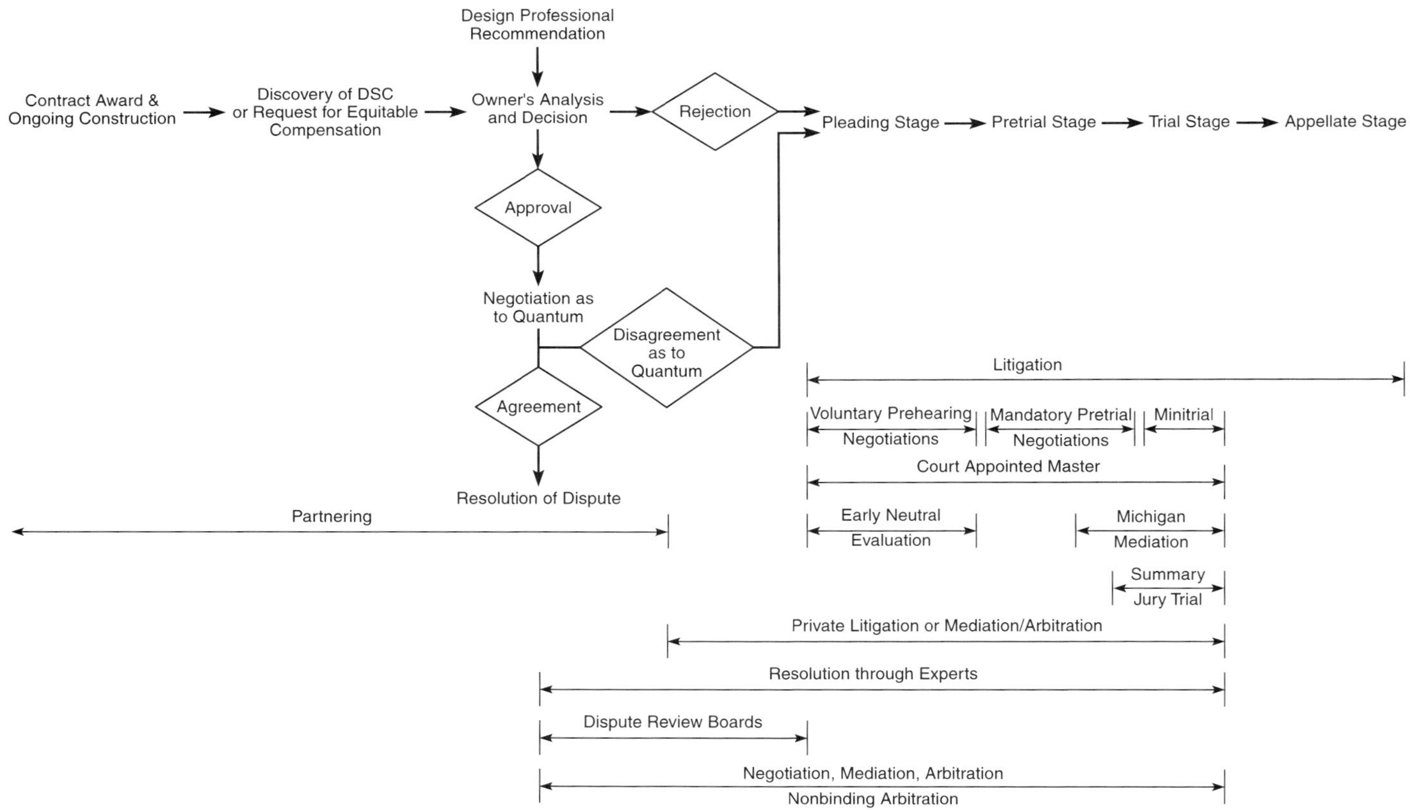

Figure 9.2 Time frames for ADR processes

to each, to create an atmosphere of teamwork among all parties so that common objectives can be achieved; and secondly, to empower individuals within each organization to make decisions that are within their level of responsibility, and not to pass that decision off to the next level of authority. With these basic teamwork structures in place, and with common goals being established among the parties, the overall objective is to resolve issues that may arise during construction utilizing the basic premises of the partnering process. In addition to the direct involvement of the parties, their lawyers may also be involved in the partnering process (Lyon, 1995). Lawyers who may not be familiar with the process may not understand its purpose and/or may try to undermine it. Conversely, lawyers who are knowledgeable and have, at some point, participated in the partnering process tend to enhance the process in a constructive manner.

There are a number of stages involved in instituting the partnering process. Initially, it must have the support of the top management of an organization (whether it be the owner, contractor, or engineer) in order to be effective. The organization must commit itself to the process and have full faith that its employees will perform and make decisions that are fair and appropriate. Senior management can not second-guess the empowered employees' decisions, and employees must be confident that they will not be criticized for their positions or placed in jeopardy for taking the initiative within their level of responsibility and authority. Effectively, the organization must establish clear lines of responsibility and document, in written procedures, the authority associated with that responsibility. The organization should perform audits, and make the necessary changes to its decision-making processes and procedures so that they will be tailored to and conform with the responsibilities intended to be given to the appropriate construction professionals. The audits should also examine the organization's procedures for resolving claims and disputes that the partnering process has not resolved.

The contract documents should be explicit in stating that the owner is committed to the partnering process and specifying the particular objectives to be achieved. The specifications should require an initial partnering meeting, as well as describe the proposed dispute resolution procedure intended to be used.

During this initial meeting among the contractor, owner, and his or her representatives (construction manager and design engineer), an intermediary is usually chosen whose role is to assist in creating confidence among parties and to facilitate the partnering process. The purpose of this initial partnering meeting, which can go on for days, is to bring together all decision-making individuals from the owner's and contractor's respective teams, and to encourage open communication and develop an understanding of each other's goals and perspectives as they relate to the upcoming project. Counterparts from each organization are encouraged to develop a team approach and to utilize their authority to solve problems and resolve differences. Project mission statements and partnering agreements are often prepared to outline common objectives and the process to be employed in the resolution of issues. The common objectives outlined in the agreement can range from a simple list to more complex criteria, against which a party's performance may be evaluated. Additional shorter workshops or meetings are also encouraged to keep the teamwork process on track during the rigors and strains that may develop during construction.

The partnering process is not all-inclusive and does allow for the implementation of dispute resolution techniques that are to be employed should it leave issues unresolved. Given the complexity of subsurface construction and the likelihood of conflicting technical interpretations relating to differing site conditions disputes, it may be advantageous to have a separate resolution process in place within the contract to resolve or recommend a solution in situations where negotiations between individuals reach a stalemate.

9.3 DISPUTE REVIEW BOARDS

Dispute Review Boards (DRBs)[4] have been used with great success in avoiding and resolving problems and disputes that would have had an adverse affect on the progress of the work and the ultimate cost. The DRB process is distinct from other ADR processes in that, unlike other processes, the DRB is established pursuant to the contract documents, convenes at the inception of the project, and meets regularly to inform itself of relevant project events, including site conditions that are actually encountered, regardless of whether there are any specific disputes. The DRB members are sent progress reports on the job and are apprised of the development of actual conditions encountered during construction. The continuity of involvement of DRB members on the project provides a balanced and informed context within which to evaluate any particular dispute presented to it—especially differing site conditions claims, the resolution of which is facilitated by firsthand knowledge of the actual site conditions encountered and presented to the DRB. Generally, the DRB is not intended to be used for routine claims or differences of opinion, but rather is reserved for resolution of major problems. DRB members are to be impartial and experienced in underground construction. While the recommendations of the DRB are not legally binding, their recommendations and conclusions, when based on a thorough knowledge of the technical and contractual aspects of the project and viewed from the perspective of prior experience on underground construction, are not likely to be reversed in litigation. By far, the DRB is the most publicized and innovative ADR process for the resolution of disputes on subsurface projects. For that reason, this section has been devoted entirely to the DRB process.

Much has been written about the advantages of the DRB process, particularly as a dispute resolution mechanism for use on subsurface projects. Although, in general, the author shares in the view that the DRB process has many advantages over arbitration and litigation, there are a number of important aspects of the DRB process that require improvement. At a minimum, it is important that design and construction management professionals understand the perceived shortcomings of the DRB process, as well as its advantages, in order to be in a position to adequately advise their clients about whether to use it and, if so, how it should be implemented through appropriate contractual provisions.

The benefits associated with the implementation of a DRB early in the construction process are numerous for all parties involved. As an ADR technique, the DRB employs the major benefits of voluntary, nonbinding mediation, utilizing neutrals knowledgeable in the specific project and dispute area at issue. The DRB concept was first employed in 1975 on the second bore of the Eisenhower Tunnel in Colorado. Its success was notable in that it heard only three disputes on this complex construction project, all of which were resolved, while the first bore of the tunnel proved to be fraught with numerous major complexities and associated claims. Since that time, DRBs have been employed on many other major construction projects. Its utilization over the years has also been expanded to include infrastructure-related, aboveground construction projects.

One very appealing aspect of the DRB process is the early selection of DRB members. Members of a DRB are typically selected shortly after the contract between owner and contractor is signed. A DRB consists of three members. The first two DRB members are selected by the owner and contractor pursuant to a nomination process, whereby the owner and contractor each nominate one prospective DRB member, and present the names of their nominees to each other for approval and acceptance. The third DRB member is then selected by the first two DRB members, who have been chosen by the owner and contractor. Typically, the third member acts as chairperson. The intention is that the three board members have the respect and confidence of both sides, and that the DRB remain impartial throughout the life of the contract. In general, DRB members are chosen because of their knowledge and background in the technical issues involved with the particular type of construction project. Although DRB members should not be representatives or advocates of the parties that appoint them, it is possible that members chosen by one party may have a bias in their approach toward the issues and assigning responsibility. As an example, what one individual may consider as an owner's design issue, another may consider to be a contractor's coordination responsibility. Therefore each party tries to choose a board member who, by experience, qualification, or general disposition, will be inclined toward that party's perspective.

Owners or agencies that have had experience with DRBs have learned the importance of the DRB member selection process, as DRBs have the potential to hear disputes that could involve significant amounts of money. Thus, an owner's interests may best be served by appointing an individual who has the design engineer's perspective in mind. Similarly, contractors, in selecting their appointments, seek individuals that may have a bias in their favor as to how they may approach issues. Accordingly, it is suggested that owners consider selecting individuals for DRBs who have design experience as well as knowledge of construction methods and techniques, rather than individuals with *only* contractor or construction-related experience. Additionally, it is suggested that owners consider interviewing their prospective choices to ascertain their views on key issues before making their actual nomination. Some owners, in order to avoid this problem, have slightly modified the DRB selection process to allow each party to prepare a list of potential DRB members. The other party then chooses a member from that list as the first party's selection. The two selected members then select the third member, as originally outlined.

With the DRB in place early in the project, it has the opportunity to observe construction from the onset of the work through completion. Its members observe, firsthand, the progress that is being made. In addition, through site visits and through their knowledge gained from other, similar projects, the DRB members can compare the contractor's construction operations to what may be experienced as the norm. In the same manner, their review of job meeting minutes, project schedules, and the like allows DRB members to develop a clearer picture of the major issues that develop, and to gain insight into the impact of these issues on construction and the associated potential differing site conditions that may arise.

Clearly, the DRB's role is to make recommendations related to disputes presented, but it also serves the role of fostering cooperation among the parties and encouraging the parties to resolve issues, through negotiations, prior to the time issues are referred to the DRB. Cooperative negotiation is fostered because both parties understand that there are knowledgeable, informed, and respected DRB members who will make fair and equitable recommendations. This tends to reduce the unproductive, win-lose attitude that owners and contractors sometimes adopt and, therefore, the negotiations that ensue tend to result in the early resolution of disputes with a cooperative win-win approach. Likewise, because each side must be prepared for a hearing before the DRB if negotiations are unsuccessful, each side needs to clearly consider its position, assessing both the strengths and weaknesses. This evaluation tends to aid in any prior negotiations, with each side making a realistic appraisal of its overall position early in the process. Effectively, the process itself tends to encourage and effectuate early resolution of disputes.

DRB hearings are usually informal in nature, with the DRB establishing its own procedures. In a typical dispute, each side prepares position papers, which are presented to the DRB in advance so it may become familiar with the issues involved prior to the hearing. At the hearing itself, each side has an opportunity to emphasize the key points in its position with the use of displays and additional documentation that it feels may be appropriate. In many hearings, after initial presentations are made, each side is given the opportunity to refute or to present counter-arguments to the position presented by the opposing side. With that process completed, DRB members are then allowed to pose questions to both sides to explore additional details that they may deem appropriate in order to arrive at a recommendation. In general, hearings take less than one day, but complex issues could last somewhat longer. Upon considering the positions presented and evidence submitted by the respective parties, along with the contents of the contract documents, the DRB determines entitlement and makes its recommendations. Negotiations between the parties for an appropriate contract adjustment, be it cost or time, are intended to occur based on the DRB's recommendations on the subject of entitlement.

Many of the DRB's recommendations center around the technical portions of the contract documents and the DRB's interpretation of them. As such, when technical issues are in question, it is recommended that the geotechnical engineer and design professional of record be present and be allowed the opportunity to discuss their design intent, as well as their opinion on the claim being considered.[5]

Another advantage of a DRB is that members can be made available on relatively short notice to hear a dispute and render a recommendation. This has proven to be invaluable in a number of cases in which adversarial positions are taken by the owner and contractor, which could affect ongoing work, and an early resolution to the dispute must be made. By utilizing the DRB's recommendations in these situations, the natural tendency of forming adversarial relationships is often avoided, all to the project's advantage.

DRBs are also considered an appealing mechanism for resolving disputes because DRB recommendations are generally made within a short time frame—two to three weeks, if not sooner. It is not unreasonable to expect that a DRB could present its preliminary findings shortly after a hearing has concluded.

A key point of the UTRC's guidelines is that a DRB's recommendations, though nonbinding, would be admissible evidence in any future litigation or alternative dispute resolution process. This approach has both benefits and disadvantages. Clearly, the opinion of trustworthy neutrals, who, for all intended purposes, are considered technical experts and are chosen by both sides to make fair and equitable recommendations, cannot be readily dismissed by either party, and, in most cases, their recommendations will foster a resolution to any dispute. Yet, at the same time, should either party decide that the DRB's recommendations are inappropriate, reject the DRB's findings for whatever reason, and wish to proceed toward litigation or another ADR process, the results of the DRB may have "pragmatic finality." In other words, given the highly technical nature of the dispute presented to the DRB, there is a great likelihood that a court would not independently hear the case on a "clean slate" and would simply adhere or defer to the DRB's findings (even though not legally required to do so). This "pragmatic finality" highlights the prominence and significance of the DRB process and the importance of a board's findings and associated recommendations. As a result of this pragmatic finality, some owners have placed limitations on the DRB's recommendations, such that in addition to being nonbinding, they are also *not* admissible as evidence, should future litigation be initiated.

Not all disputes involving differing site conditions claims between owners and contractors concern only technical issues; many of these disputes arise out of the interpretation of the contract documents and/or other contractual obligations, or other "legal" issues that can and do develop into disputes and/or claims. Legal precedent and principles relevant to the interpretation of statutes or contract provisions play a prominent role in the evolution and resolution of such "legal" issues, and may vary from state to state. DRBs usually consist of members who are technically knowledgeable but relatively inexperienced with the legal precedent established within a jurisdiction and are generally not qualified in the application of legal principles. In some instances, it has been observed that DRB members may not give adequate attention to the legal and contractual aspects of a dispute, or may completely disregard them, all in an attempt to achieve "fairness" or "justice." What the DRB may perceive as just or fair may, in reality, be in conflict with the reasonable assumptions and other bases upon which the project was bid, the contract documents, the law of the particular state, and settled legal principles and precedents regarding differing site conditions claims that govern the expectations and agreed-upon risk allocations of owners and contractors. This approach of some DRB members could create a situation in which an owner, especially a public owner, will not accept the DRB's recommendation. As such, the DRB must realize its limitations relating to these issues and recognize that the contract documents—and not any broad, undefined concept of "fairness"—should be the central basis from which its recommendations should be derived.

As an example, in a recently appealed case, a contractor sued a public agency to recover overhead costs associated with a large overrun on an allowance item, even though the contract specifically excluded payment for overhead. The contractor was paid for all contract-approved allowance costs that he had expended. The appellate court upheld the lower court's finding in favor of the agency in enforcing the contract and its exclusion on payment of overhead. Although the court recognized the substantial economic hardship to the contractor, it ruled that it was constrained to limit recovery to that authorized by the contract documents on which the bids for the project were competitively based, and that the enforcement of the contract requirements was paramount. Yet, if such an issue were presented to a DRB, it is unclear whether the result would have been similar, as many DRB members may try to base their recommendations on "fairness," rather than the requirements of the contract documents themselves.

Typically, the involvement of lawyers in either the preparation of position papers or in the actual hearing presentation is discouraged. In this respect, even when a particular dispute involves a legal issue appropriate or necessary for resolution, the parties often are not afforded a reasonable, meaningful, and effective opportunity to formulate, research, and present that issue through legal counsel, and the DRB is consequently deprived of the benefit of the parties' respective positions on the legal issue. Under such circumstances, a DRB's recommendation that is based, either in whole or in part, on legal issues may well be (1) misinformed and neglect important considerations that the DRB is not qualified, by either training or experience, to appreciate or understand, and/or (2) premised on a less-than-adequate opportunity for the parties to effectively and competently express their positions on the legal issues.

DRBs, as outlined in the 1989 UTRC publication, do have the flexibility to call in experts for consultation on legal issues and interpretation of state laws and legal precedents. As such, DRBs should not be shy in requesting support in areas in which they do not have expertise. The need for legal support from counsel for the respective parties should be looked upon as positive and aiding the process. Indeed, legal support is implicitly required under the contract documents, as it is the DRB's responsibility to make its decisions and recommendations based on the complete contract document. Arguments have been presented by some that an experienced DRB is just as qualified as a construction industry arbitrator to resolve issues relating to contractual legal issues. This position, however, is not persuasive, as a DRB composed of technical experts in underground construction, for instance, does not have the technical knowledge of the law to judge the full impacts of the contractual issues involved. As such, DRBs should allow the limited involvement of legal counsel when legal issues may be involved in the dispute. Additionally, the parties should have access to legal counsel, and counsel should have a role in the DRB process when issues of contractual interpretation are involved. In many instances, the claim or dispute before the DRB ultimately may lead to an independent claim by the contractor against the professional, and/or an owner cost recovery claim against the professional (see Chapter 10, Section 10.1). Thus, the professional should consult with legal counsel before and during the DRB process.

Some owners may be frustrated in their inability to convince a DRB of the centrality of the contract documents as the basis for the DRB's recommendation. As a result, owners have tended to limit the DRB's authority to making findings on purely technical issues, rather than allowing the DRB the broad discretion that was intended under the UTRC's 1989 guidelines.

The role of the DRB is to evaluate disputes on the basis of the complete contract documents. DRB members generally recognize that they do have "jurisdictional" limitations, and those limitations are defined by the contract agreement establishing the DRB process. As such, that agreement usually states, "The DRB's recommendations shall be based on pertinent contract provisions, and the facts and circumstances involved in the decision." It is of concern to some that this language is inadequate, as it does not require the DRB to consider the contract in its entirety, including all of its legal provisions. Yet, others have interpreted the above quotation as giving the DRB the discretion to interpret the contract broadly, and to allow its recommendations to be based solely on broadly defined notions of fairness. Clearly, the latter approach of overshadowing the original contractual intent and bargain (including risk allocation) of the parties is inappropriate.

Another issue that has recently developed is the DRB's resolution of a specific dispute on a particular project based on recommendations or findings of a different DRB in an unrelated project, effectively establishing a system of "DRB case law." Although this concept may sound appealing, one should not lose sight of the fact that DRBs are established on a project-specific basis for the purpose of resolving issues under particular contract document provisions and appropriate state or federal laws, and that the resolution of each dispute or claim typically involves a detailed consideration of the specific issues, facts, and circumstances unique to the project under consideration. In addition, the DRB members' knowledge and familiarity with the project should also play an important role in the evaluation process. Given these factors, recommendations and/or findings of another DRB on an unrelated project should not be given any consideration or weight, much less deference.

Use of DRBs has proven to be an effective tool in the subsurface construction industry and is gaining additional popularity among other construction industry groups. As indicated in the recently published *Construction Dispute Review Board Manual* (1996), through 1993, 166 projects (having a cumulative construction value of $9.7 billion) have utilized the DRB process, with another 172 contracts anticipating the use of the process in the near future. Of the 166 contracts completed or under construction, none has entered litigation. These are impressive statistics and a vote of confidence by the construction industry that it has found a method to effectively address and resolve disputes that arise during construction.

Despite its apparent success as an ADR process specifically applicable to differing site conditions disputes on subsurface projects, there is room for improvement to the DRB process. One suggested improvement is to consider a limit on the minimum value of claims that may be submitted to the DRB. This limit could be relatively conservative in value. The logic behind this approach is that there is a cost associated with any DRB hearing and the preparation of associated findings. For instance, should the value of the differing site conditions claim be less than 50 percent of the DRB cost, then it may not be considered cost-productive to present the claim to the DRB. In a similar vein, it may be advantageous to establish a penalty for parties who file frivolous claims. This could be accomplished by allowing the DRB to state in its findings whether it thinks that the rejected claim was frivolous or not. If that should be the case, the cost of the hearing, and associated costs incurred by the opposing side, would be paid by the party filing the claim.

In 1996, with the publication of the *Construction Dispute Review Board Manual*, arguments were presented lauding both the successes and operational benefits of the DRB process. The general acceptance of the process by the industry as a whole paid tribute to the individuals involved in nurturing and developing the DRB process into one of the most widely accepted ADR methods used on subsurface projects. The success and acceptance of the DRB approach depends on owners and contractors having a genuine, well-founded sense of trust and confidence in the ethical standards and impartiality of DRB members. Achieving this goal can be a significant challenge, in that the "pool" of candidates who typically serve as DRB members is a relatively small and "closed" group, with the same individuals often serving on numerous boards throughout the United States. In addition, since DRB members typically are chosen based on their knowledge, experience, and qualifications in the design and/or construction of subsurface projects, it necessarily should be expected that DRB members will have had and will continue to have business, professional, or other relationships with the contractor, design professional, and/or construction manager (or, albeit less likely, the owner) involved in the project that is the subject of the dispute presented to the DRB to which they have been appointed. These realities of the DRB selection process pose significant ethical issues and challenges. Any successful challenges to the impartiality or ethics of a DRB member may result in a court excluding any recommendations in which the member participated, thereby rendering useless a primary value of the DRB process.

As an example of the ethical issues that may develop through use of the DRB process, in a recent state superior court decision, a project owner was permitted to terminate a DRB member, whom he had appointed, due to the member's bias and lack of objectivity. The case evolved out of several disputes and claims between the owner and the contractor on a project that entailed the construction of a portion of the Los Angeles Metro tunnel system.[6]

As background, in July 1995, the owner notified the contractor of its decision to terminate its work. The course of events leading up to that termination, (which are still the subject of pending litigation between the two parties), will not be reviewed here. Yet, as part of the contract, a DRB was established in accordance with a standard three-party agreement. The DRB was to act as an alternative dispute mechanism and "fairly and impartially consider the disputes placed before it and to provide written recommendations to both [parties] for the resolution of these disputes." The three-party agreement also stated that "all [DRB] members are to act impartially and independently in the considerations of the facts and conditions surrounding any dispute presented." It also provided that recommendations would be advisory and nonbinding, although they would be admissible as evidence in subsequent litigation between the disputing parties. The agreement further stipulated that the DRB and its individual members were not to give any advice to either disputing party on the "conduct of the work or resolution of problems other than disputes referred to the [DRB]."

In October 1996, the owner terminated the member that it had selected to serve on the DRB, and subsequently requested an adjudication from the court confirming that it (the owner) was justified in terminating the DRB member on the grounds of bias and lack of objectivity. The court concluded that the owner was justified in its actions. The court relied on the following evidence in reaching its decision:

1. The terminated DRB member initiated an improper ex-parte (i.e., one-sided) communication with the owner during a DRB hearing to the effect that the owner "would lose and would not like the DRB's recommendation on this dispute." The court found that the terminated member had "violated his obligation by the ex-parte communication and by prejudging the issues."

2. In finding that the terminated DRB member was biased, the court referred to a specific hearing in which the DRB "narrowed the issues to be consid-

ered during the hearing and restricted the evidence that would be heard." The court found it was improper for the DRB to unilaterally exclude evidence at this hearing and to not respond to the owner's letter which objected to this process; the DRB's subsequent act of conducting a hearing without the owner's presence was also deemed improper. As such, the court concluded that the actions on the part of the terminated DRB member, (if not on the part of the entire DRB), indicated "lack of objectivity," and evidenced that the terminated DRB member "exhibited a predisposition in favor of [the contractor] and not an objective attitude as is required by the Three-Party Agreement."

3. The court also indicated that the terminated DRB member's ex-parte telephone call to the contractor after his termination—the purpose of which was to ascertain if the contractor had also lost confidence in him and to determine the contractor's position with respect to his termination—was further proof of partiality toward the contractor.
4. Additionally, the court found that the terminated DRB member had provided improper advice, both directly and indirectly, during meetings, hearings, and site visits. This included the member's giving advice to the contractor regarding a method of payment, and also relaying his personal knowledge and past experience "when [the contractor's] representatives would solicit the terminated DRB member's views." By giving improper advice, the terminated DRB member clearly violated industry standards, as well as the three-party agreement.

In order to avoid potentially damaging situations similar to the case discussed above, in 1994, the American Consulting Engineers Council of New England (ACEC/NE) formed a subcommittee to prepare a code of ethics for DRB members that would be associated with major subsurface construction in the Boston area, yet which could also be applied nationally. The subcommittee reviewed a number of codes applicable to judges and arbitrators, as well as codes published by other organizations involved in underground construction. In 1996, ACEC/NE adopted the recommendations of the subcommittee for a "Code of Ethics for Dispute Review Board Members." In so doing, ACEC/NE also recognized a number of special aspects of the DRB process that emphasize the need for an ethical code of conduct for its members. These include:

- In the typical methods in which disputes are settled, individuals who decide the dispute are not involved in the project until they are brought into the process as a result of the dispute. In the DRB approach, members are brought on board early in the process and are kept abreast of most issues that occur during the construction process. This approach is considered an advantage in the dispute resolution process, yet the DRB members may be exposed unequally and unconventionally to project information, potentially affecting the decision-making process. The process may, therefore, be compromised by the disparity of information held to a greater or lesser degree by various DRB members.

- Based on the uniqueness of the DRB process and its need to engage highly qualified and experienced individuals who have specific expertise in the particular type of construction, there tends to be a limited cadre of personnel available from within the design/construction community to sit on DRBs. As a result, a risk of conflicts or partiality may arise.
- DRB recommendations are generally nonbinding but admissible in court. In practical effect, however, if either party rejects the findings and litigation commences, the court would, in most cases, defer to the findings of the DRB. This "pragmatic finality" underscores the DRB's influence and thus accentuates the need to ensure the highest ethical standards of its members.
- Last, while most members of DRBs are chosen for their technical knowledge and experience that can be utilized in resolving differing site conditions claims, their knowledge of contract law may be severely limited. This is especially true when one considers that the contract documents are the central basis on which a DRB should form its recommendations. Because DRBs, as well as the UTRC guidelines, have discouraged the use of lawyers in the DRB process, the potential is increased that decisions might be made that conflict with legal principles.

Based on the above, the ACEC/NE subcommittee recommended the following six canons be included as part of the requirements for DRB members:

Canon One—Financial Separation

The members of a DRB should perform their duties while maintaining financial separation from both parties in accordance with the following framework:

- A DRB member should refrain from any financial or business dealings that may reflect adversely on his/her impartiality or involve him/her in business transaction with persons or businesses likely to be involved in disputes presented to the DRB.
- DRB members shall be compensated for the time and expenses incurred in the performance of their duties. DRB members should, however, avoid engaging in communications concerning compensation that create an appearance of impropriety. Likewise, DRB members should terminate their involvement if compensation is used in an attempt to compromise judgment.
- If a DRB member becomes aware of questionable financial conduct by any party involved in the DRB process, then he/she should report this information to both the owner and the contractor.

Canon Two—Impartiality

A DRB member should be impartial, fair, and independent in accordance with the following guidelines:

- DRB members must disclose to all parties the existence of interests or relationships that are likely to affect their impartiality or that might create an appearance that they are biased against either party. Both parties have the freedom, however, to agree on whomever they choose as a DRB member. When the contracting parties, after full disclosure of a person's interests and relationships, nevertheless desire that individual to serve as a DRB member, then that person may properly serve.

- A DRB member should disqualify him/herself from serving if he/she has:

  (A) A personal bias or prejudice concerning a party.
  (B) Personal knowledge of disputed evidentiary facts concerning the proceedings.
  (C) A financial or property interest that could be affected by the outcome of the proceedings.

- A DRB member shall not exhibit, in the performance of his/her duties, bias or prejudice including but not limited to that based upon race, sex, religion, national origin, disability, age, sexual orientation, or socioeconomic status.

Canon Three—Confidentiality/Communications

A DRB member is in a relationship of trust to the contracting parties and should not use confidential information acquired during DRB proceedings for personal gain, to adversely affect the interest of another, or to damage the reputational interests of the parties.

- The DRB should accord all persons the right to be heard in full accordance with the established DRB agreement. At no time, however, should a DRB member participate in private interviews or communications outside of an officially sanctioned DRB hearing. Any attempt by either party to institute such discussions by individuals associated with either contracting party should be immediately reported to and discussed with other members of the DRB.

- DRB members should refrain from public discussion or publication of opinions and views as they may relate to specific issues, claims, or disputes pending before a DRB.

- A DRB member should not discuss the decision-making process of the Board with others or inform anyone of a recommendation in advance of the contracting parties. Once a recommendation is made, then no member of a DRB should try to influence any action of the contracting parties unless specifically required to do so in conformance with established DRB procedures.

- The ethical obligations of a DRB member begin upon acceptance of an appointment and continue even after a recommendation has been given to the parties.

Canon Four—Adherence to the Contract Documents

DRB members should act within the confines of their jurisdiction and issue recommendations based upon and consistent with the requirements of the contract document.

- The DRB's source of authority to hear disputes and to issue recommendations derives from an agreement between the Owner and the Contractor. The contract document typically defines the process by which the respective contracting parties may assert claims or disputes and the procedures by which those claims or disputes shall be evaluated and adjudicated. DRB members should be mindful of the limitations placed upon their jurisdiction as expressed in the agreement empowering them. This admonition has two important dimensions: (1) the DRB should decide only those issues actually presented to it in accordance with the requirements of the contract document; and (2) the DRB should not supplant or otherwise interfere with the respective rights, authorities, duties and obligations of the Owner, the Designer, and the Contractor as defined in the contract document.
- In making recommendations, the DRB must acknowledge the centrality of the contract document and should never propose a resolution which ignores, disregards, or significantly undermines the intention, requirements, economic allocation of risk, or work specified in the contract document.

Canon Five—Conduct of Hearing

DRB members should attend to the conduct of a hearing as expeditiously as possible

- The DRB should facilitate the prompt accumulation and distribution of appropriate project records and should assemble to hear a dispute as quickly as possible.
- DRB members should make certain that all hearings are conducted in an evenhanded and dignified manner. A DRB should make all reasonable efforts to prevent delaying tactics, harassment of participants, or other abuse or disruption of the DRB process.
- To the greatest possible degree, the DRB should decide a claim on the basis of factual data and direct observation. DRB members are free to request additional information when it is felt that a dispute presented to the Board is incomplete. DRB members should not, however, attempt to assemble documents or conduct investigations relative to a dispute independently of the contracting parties.

Canon Six—Recommendations

DRB recommendations should be expressed in writing and in a manner which respects and acknowledges the DRB jurisdiction, acknowledges the respective positions of the parties and demonstrates fair and impartial consideration of the pertinent evidence.

- A DRB recommendation should be clearly stated and succinct. As a minimum, the recommendation should include the following: a statement of the issue(s) presented for resolution; a brief statement of the progression of the dispute so as to demonstrate that all conditions precedent to DRB consideration of the dispute have been satisfied; a recitation of the DRB's understanding of the respective parties' positions; an identification and description of the relevant evidence or information considered by the DRB; an evaluation of that evidence or information in the context of the specific dispute presented; and a clear and logical discussion of the DRB's reasoning supporting its recommendation.

- In making a recommendation, the DRB should only consider information presented by the parties in the common reference documents, the parties' respective position papers, and/or observations or other information jointly obtained by and shared with all DRB members. Information, perceptions, or observations of a DRB member which are personal to that member and not disclosed to other members should not be used to form the basis of a recommendation.

- Requests for reconsideration of a recommendation should be sparingly entertained and granted only in those circumstances in which the petitioning party demonstrates that the existing recommendation disregarded relevant evidence; is inconsistent with the contract document; is based upon fraud or other misconduct of a party; and/or exceeds the jurisdiction of the DRB. In addition, requests for reconsideration may be entertained in circumstances in which new evidence is discovered which could not have been discovered by a party through the exercise of reasonable diligence prior to the issuance of a recommendation.

It is recommended that these six canons, or portions thereof, be considered as the ethical standards against which DRB members are judged when selections are being made, and that they be incorporated into the DRB members' contract. The inclusion of this code in the DRB agreement and/or the contract documents, either explicitly or by reference, while not entirely eliminating the risk of improper conduct on the part of DRB members, should serve to more clearly define DRB members' permissible and appropriate conduct with respect to proceedings in which they participate, as well as outline the conduct and communications of DRB members relating to issues which have been or may be presented to them as disputes for resolution.

9.4 CONCLUSION

Over the last few decades, the subject of risk allocation associated with subsurface construction has been the subject of intense debate among owners, design professionals, and contractors. With the progressive recommendations made during the 1970s and 1980s and the significant implementation of differing site conditions clauses within construction contracts, the onerous financial burden associated with the unforeseen risk of differing site conditions has been more fairly allocated among parties. In part as a result of the differing site conditions clause, many owners have also included provisions for alternative dispute resolution mechanisms within their contractual agreements. As discussed above, the alternative dispute resolution mechanisms most often employed include mandatory negotiations, arbitration, mediation, mini-trials, private litigation, and resolution through experts.

DRBs, which utilize experts in the resolution of disputes, have advantages and benefits when used on complex subsurface projects, as they are established pursuant to the contract documents and convene at regular intervals to keep informed of relevant project events. At the same time, however, there are a number of actual and perceived concerns associated with the DRB process. Likewise, there is a recognized need to implement a code of ethics for DRB members similar to the ACEC/NE model, which would serve to clearly define the permissible and appropriate conduct of DRB members and, at the same time, minimize the potential of bias or lack of objectivity, and subsequent challenges to the DRB's recommendations on those grounds.

NOTES

1. Chapter 10, in turn, addresses the potential professional liability risks associated with the design and construction management professional's role in the investigation and evaluation of differing site conditions claims. In general, the professional faces relatively minor risk exposure in that capacity if his or her investigative and evaluative services are performed in a fair, impartial, prompt, and professional manner.
2. *Foster Construction C. A. & Williams Brothers Co. v. United States*, 435 F.2d 873, 887 (Ct. Cl. 1970).
3. Copies of the "Construction Industry Arbitration Rules" of the American Arbitration Association can be obtained by writing to the American Arbitration Association, 140 W. 51 Street, New York, NY 10020.
4. In 1989, the Underground Technical Research Council's (UTRC's) Committee on Contracting Practices published its report, "Avoiding and Resolving Disputes in Underground Construction," (which was subsequently revised and updated in 1991). This publication was the culmination of years of work by members of the committee to try to stem the tide of "adversary relationships and litigation with added costs and construction delays" within the subsurface construction industry. The ADR process recommendations within the report were, in many cases, the natural outgrowth of earlier works by the USNCTT and UTRC (which are discussed earlier in this chapter), as well as developments and modifications that were being recommended to expedite resolution of claims and litigation. In short, these earlier works advocated that there should be equitable risk sharing between owners and contractors, and that this risk sharing would be beneficial to both parties in the form of lower bids and, therefore, lower overall costs to the owner. This specialized ADR process is composed of three elements: the preparation of a geotechnical design summary report;

the preparation of escrow bid documents; and the establishment of a dispute review board, which is discussed in this section. See "Avoiding and Resolving Disputes in Underground Construction—Successful Practices and Guidelines," prepared by the Technical Committee on Contracting Practices, Underground Technology Research Council, published by American Society of Civil Engineers, 1989 (revised in 1991).

5. Clearly, design and construction management professionals have a stake in the outcome of the DRB process, as they could potentially be the focus of future recovery actions by the owner as well as the contractor. See, generally, Chapter 10.

6. *Los Angeles County Metropolitan Transportation Authority v. Shea-Kiewit-Kenny*, No. BC 136559 (Sup. Ct. Dec. 6, 1996). For a more detailed discussion of the LACMTA case and its implications, see Hatem, 1996.

REFERENCES

American Consulting Engineers Council of New England, "Code of Ethics for DRB Members," 1996.

American Society of Civil Engineers, "Managing Unforeseen Site Conditions," *Journal of Engineering and Management*, Vol. 113, No. 2, June 1987.

Association of Engineering Firms Practicing in the Geosciences, "Alternate Dispute Resolution for the Construction Industry," 1988.

J. J. Casey, Proceedings of the American Society of Civil Engineers Conference on Construction Risk and Liability Sharing, "Identification and Nature of Risks in Construction Projects, a Contractor's Perspective," January 1979.

D. J. Hatem, "Risk Allocation and Dispute Resolution on Construction Projects: Roles and Challenges for Legal Counsel," *The CA/T Professional Liability Reporter*, Vol. 2-No. 4, (Boston: Burns & Levinson LLP, June 1997).

D. J. Hatem, "Should There Be Published Standards of Conduct for DRB Members?" *The CA/T Professional Liability Reporter*, Vol. 2-No. 2, (Boston: Burns & Levinson LLP, December 1996), at 23-25.

V. Frederick Lyon, "Your Lawyer as a Project Team Member," in K. A. Godfrey, Jr., *Partnering in Design and Construction* (New York: McGraw-Hill, 1995).

R. M. Matyas, A. A. Mathews, R. J. Smith, and P. E. Sperry, *Construction Dispute Review Board Manual*, (New York: McGraw-Hill, 1996).

Underground Technology Research Council, "Guidelines for Tunnel Lining Design," prepared by the Technical Committee on Tunnel Lining Design, (New York: American Society of Civil Engineers, 1984).

U.S. National Committee on Tunneling Technology, "Better Contracting for Underground Construction," prepared by USNCTT Standing Subcommittee No. 4, Contracting Practices, National Research Council, (Washington D.C.: National Academy of Science-National Academy of Engineering, 1974).

U.S. National Committee on Tunneling Technology, "Better Management of Major Underground Construction Projects," Assembly of Engineering of the National Research Council, (Washington D.C.: National Academy of Science, 1978).

10

PROFESSIONAL LIABILITY AND RISK ALLOCATION/ MANAGEMENT CONSIDERATIONS FOR DESIGN AND CONSTRUCTION MANAGEMENT PROFESSIONALS INVOLVED IN SUBSURFACE PROJECTS

David J. Hatem, Esq.
Burns & Levinson LLP

10.0 INTRODUCTION

This chapter will address the professional liability and risk allocation management considerations applicable to design and construction management professionals involved in subsurface projects. While some of the subjects addressed have been discussed previously in this book, this chapter will specifically focus on those and other subjects from the standpoint of professional liability and the

Subsurface Conditions: Risk Management for Design and Construction Management Professionals, edited by David J. Hatem
ISBN 0471-15607-8

legal considerations applicable to risk allocation and management for design and construction management professionals.

Design and construction management professionals[1] involved in subsurface projects face potential professional liability exposure to both their clients and third parties (with whom the professional has no contractual relationship) directly or indirectly involved in those projects (such as contractors, speciality trade contractors, construction workers, adjacent property owners, members of the public, surety companies, banks, and other construction funding sources). This potential professional liability exposure may arise out of acts or omissions of the professional in connection with a broad range of services: recommendations regarding site selection; pre-construction site investigation; preparation of data, reports, opinions, and/or studies regarding actual or anticipated subsurface conditions; preparation of design and construction documents and/or "reference" documents to be utilized in connection with the construction of subsurface projects; the administration and/or management of construction of subsurface projects, including the review and evaluation of contractor submittals, proposed value engineering or other contractor-proposed substitutions, proposed construction means and methods, and payment requisitions; the observation of construction work for conformance with the contract documents; and the investigation, review and evaluation of contractor (or owner) differing site conditions claims.

Simply put, design and construction management professionals involved in subsurface projects face a substantial risk of professional liability exposure for claims asserted by their clients, as well as by a myriad of third parties, which arise out of virtually all services typically performed by those professionals during each primary project phase activity, i.e., planning, site investigation, preparation of reports and data, preparation of design and contract documents, and the administration or management of the construction contract.

Section 10.1 will discuss the various types of professional liability claims that may be asserted by clients and/or third parties against design and construction management professionals involved in subsurface projects, and the legal defenses generally applicable to those claims.[2] Section 10.2 will discuss the professional liability and risk allocation/management concerns of design and construction management professionals in the specific context of subsurface projects (including a detailed discussion of applicable risk allocation principles in connection with subsurface conditions). Finally, Section 10.3 will provide recommendations to design and construction management professionals for managing the risk of professional liability on subsurface projects.

10.1 PROFESSIONAL LIABILITY OF DESIGN AND CONSTRUCTION MANAGEMENT PROFESSIONALS: BASIC PRINCIPLES

Design and construction management professionals involved in subsurface projects are exposed to the risk of professional liability claims from both clients and third parties—i.e., parties with whom the professional has no contractual relationship. While design and construction management professionals certainly face the most significant professional liability exposure to their clients, they are generally in a better position to mitigate and manage that risk of professional liability exposure, as compared to the related risk of third-party liability. Specifically, the risk of professional liability to a client can be more directly and effectively managed and controlled primarily through contractual and associated risk allocation mechanisms. Despite this important distinction between client and third-party liability risk, there is substantial overlap and commonality between the types of acts and errors and omissions that form the basis of both client and third-party professional liability claims. In addition, the terms of the professional's agreement with the client are relevant to the defense of both client and third-party claims (see Section 10.1.2.1, Note 20, and corresponding text).

This section will introduce and discuss the basic principles associated with professional liability to clients and third parties predicated on the legal theories of breach of contract, negligence, and negligent misrepresentation. Section 10.2 (and its subsections) will discuss in detail the application of those legal principles in the specific context of professional liability claims on subsurface projects.

10.1.1 Liability to Client

10.1.1.1 Contract-Based Liability By definition, a client-professional relationship involves an express or implied contract between the client and the professional. In Chapter 2, the general nature of contract law and many of the provisions typically contained in the owner-professional agreement were discussed. For obvious reasons, in many instances, the terms and conditions of the contract between the professional and the client have direct relevance in the context of contract-based claims asserted by the client against design and construction management professionals.

In general, the elements of proof required to establish contractual liability in this context do not differ in any significant or material respect from those generally required to prove contractual liability in other contexts. Generally, in order to establish contractual liability, the claimant (i.e., the client) must prove the existence of a contract; a breach of a material term of that contract; and that some damage was sustained by the claimant as a result of that breach.

In the specific context of contracts between clients and professionals with respect to subsurface projects, the following types of provisions are the most likely candidates on which contract-based professional liability claims are predicated:

1. Provisions that incompletely, ambiguously, or otherwise inadequately define the scope of the professional's services in a manner that results in disappointed expectations, scope and/or compensation disputes, and/or the professional's inability to adequately perform the services required in connection with the specific project undertaking.

2. Provisions precluding or inadequately limiting the involvement of the professional during the construction phase of the subsurface project.
3. Provisions that obligate the professional to perform its services in accordance with a standard that (a) exceeds the negligence-based "professional standard of care" (see Section 10.1.1.2) or that (b) warrants or guarantees perfection (or even near perfection) in the performance of those services.
4. Provisions that obligate the professional to defend, indemnify, and/or hold harmless the client (or others) for liability and/or claims beyond those based on and proximately caused by the professional's departure from the professional standard of care.

These types of provisions frequently generate contractual disputes and claims between the client and the professional.

One primary category of contractual liability disputes frequently arising on subsurface projects pertains to the scope of professional services. Many of these "scope of services" disputes involve the following factors, or combinations of factors:

1. Unclear or otherwise inadequate definition or articulation of the scope of services expected (required) of the professional.
2. The professional's acceptance of a service scope that is unduly restrictive or inadequate in relation to the specific nature of the subsurface project (i.e., the project's complexity, its anticipated constructability or scheduling constraints, the site history, anticipated behavioral or other reactive responses of site conditions during the construction process, and/or the limited information available concerning the site).
3. The professional's alleged or actual failure to advise the client about the scope of services reasonably required to accomplish the adequate planning, design, and/or construction of the subsurface project.
4. The client's failure (and/or refusal) to accept the professional's advice and recommendations regarding the scope of services which should be performed in connection with the specific subsurface project, particularly relating to (a) the scope of preconstruction site investigation and (b) the nature and extent of the professional's involvement during the construction phase.

A second category of contractual liability disputes and claims relates to contract provisions that obligate the professional to perform services in accordance with a standard that exceeds the negligence-based professional standard of care. As discussed in Section 10.1.1.2, under the professional standard of care, a design and construction management professional is obligated to perform his or her services in accordance with the standard reasonably required of other professionals in the same or similar circumstances at the time that those services are performed. In some instances, however, a client may require that the professional agree to perform his or her services in accordance with a standard that *exceeds* the professional standard of care. In addition to increasing the risk of professional liability (not only to the client, but also potentially to third parties), a "heightened" standard-of-care provision also poses a substantial risk of uninsured professional liability.[3] Courts have consistently recognized the enhanced liability risk associated with a contractually heightened standard of care and/or breach of express warranty standard, both of which impose a performance standard that exceeds the professional standard of care (such as a provision stating that the design shall be free from all errors or defects, or that the professional shall perform its services in accordance with the "highest" standards).[4]

For example, in a recent Massachusetts Appeals Court case, *Digital Equipment Corp. v. Wallace, Floyd, Associates*, 651 N.E2d 417 (Mass. App. Ct. 1995) (unpublished Memorandum and Order), the owner made a claim against a design professional for additional cost incurred in connection with the owner's purchase, during construction, of a larger-sized transformer than the one specified by the design professional in the contract documents. There was no dispute that the larger-sized transformer was required. The design professional defended the claim on the following bases: (1) the specification of the incorrect transformer size was due to inaccurate information supplied to the design professional by a power company (through the owner); (2) the additional cost for the larger-sized transformer represented value-added or betterment (enhancement) to the project and, therefore, the cost should be the responsibility of the owner; and (3) the owner failed to identify any expert who would or could offer any opinion that, under these particular circumstances, the design professional departed from the professional standard of care in the preparation of the contract documents.

The court rejected all of the design professional's defenses based on the contract between the design professional and the owner, which contained a "standard of care" provision requiring the design professional to be "fully responsible for [the professional's] acts, errors and omissions." The court reasoned that the design professional, by that contract provision, agreed to a standard higher than the professional standard of care (i.e., more than negligence), thereby rendering irrelevant whether the design professional departed from the professional standard of care, as well as dispensing with the need for any expert opinions. Specifically, the court held that "[b]y voluntarily assuming full responsibility for any acts, errors . . . , the [design professional] assumed a higher duty than [the professional standard of care] which would otherwise be imposed on it by law." Specifically, the contract did not require that the acts, errors, or omissions be negligent in nature as a precondition to liability. The court further ruled that the standard of care provision created, in effect, a warranty or guaranty of perfection in the performance of services. Hence, the court deemed irrelevant the fact that inaccurate information supplied through the owner led to the incorrect specification of a wrong transformer size, or that any benefit to the owner (in the form of betterment or value added) resulted from obtaining the larger-sized transformer.

A third category of frequent contractual liability disputes and claims between clients and design and construction management professionals arises out of broad-form indemnification provisions sometimes contained in client-proposed agreements. In general, under a broad-form indemnification provision, a professional agrees to indemnify the client for claims, liabilities, costs, or expenses regardless of whether they are caused, in whole or in part, by the professional's negligence (i.e., a departure from the professional standard of care).[5] In legal and practical effect, these broad-form indemnification provisions result in the professional's obligation to pay (or indemnify) for project cost or other risk that may not be within the control or responsibility of the professional and/or that was not (either wholly or partially) the consequence of the professional's negligence. In addition, contract provisions that impose indemnification obligations based on a standard that exceeds the negligence-based professional standard of care not only substantially increases the risk of professional liability, but may also result in liability that is not insured (or insurable) under most professional liability insurance policies.[6]

There are several prominent defenses to contract-based liability claims asserted by clients against design and construction management professionals. A number of these defenses are predicated on the rationale that the claim lacks merit because the professional adhered to and/or performed (or substantially performed) his or her obligations under the contract. Other defenses are predicated on the basis that the alleged breach of the contract did not cause any damage to the client. The appropriateness of any given defense to a contractual claim based on the professional's failure to adhere to the standard of care or performance specified (or implied) in the agreement depends on whether the professional agreed to perform in accordance with the professional standard of care, or in accordance with a standard that exceeds that standard of care. Assuming that the contractually specified standard is consistent with the professional standard of care, or negligence standard, then all of the defenses applicable to a negligence-based claim would be available (see Section 10.1.1.2). However, if the contractually specified or required standard exceeds the professional standard of care, then the defense of such a claim may be more problematic, as exemplified by the *Wallace, Floyd* decision discussed above.

As previously noted, contract provisions dealing with the scope of services to be provided by the professional are a frequent source of dispute. A well-drafted, precise, and clear scope-of-services provision will provide an effective first-line defense to most professional liability claims, regardless of whether the claims are based on contract or some other legal theory, or whether the claims are asserted by the client or a third party. Simply put, the succinct and direct answer to a number of client and third-party claims against professionals is that the problem at issue did not result from something within the scope of services required to be provided by the design or construction management professional. Again, however, the ability to effectively assert such a defense depends on the clarity, precision, and completeness of the scope definition in the agreement.[7]

The Indiana Appellate Court's decision in *Strauss Veal Feeds, Inc. v. Mead and Hunt, Inc.*, 538 N.E.2d 299, 300 (Ind. Ct. App. 1989), demonstrates how scope-of-services limitations may be successfully utilized in the defense of a client's contractual claim against a design professional. In *Strauss*, an architectural/engineering firm (A/E) contracted to provide engineering services in connection with the construction of a feed processing plant that utilized a unique process. The plant owner sued the A/E for breach of contract and negligence after a state pollution control board issued a cease-and-desist order shutting down the completed plant because the waste produced by the plant was environmentally hazardous. The plant owner claimed that the A/E breached its contract and the professional standard of care by failing to independently investigate and advise the owner of potential waste disposal problems.

Although the trial court found that the agreement between the owner and the A/E was ambiguous as to the contractual responsibilities of the A/E, based on extrinsic evidence (i.e., evidence beyond the actual terms of the agreement), the court concluded that the A/E had no contractual obligation to provide sanitation engineering services or consultation. The evidence supporting that conclusion included the fact that the plant owner did not request such services, and that the plant owner, in answering the A/E's inquiries on that subject, assured the A/E that the plant would pose no waste or sewage problems. The trial court, however, did find that the A/E had a duty to advise the plant owner that it may have a waste problem, but ultimately found that the A/E was not negligent in the discharge of that duty. The appellate court affirmed the trial court's finding in favor of the A/E, but found erroneous the latter court's determination that the A/E had a duty to advise the plant owner of potential waste problems, because the contractual scope did not include sanitation engineering services and, especially, in light of the owner's superior knowledge of its unique and secret process and its assurances to the A/E that the plant would be clean.

The Iowa Supreme Court's decision in *Shepherd Components, Inc. v. Brice Petrides-Donohue & Associates, Inc.*, 473 N.W.2d 612 (Iowa 1991), demonstrates the successful use of a scope-of-services limitation by a design professional in the defense of a third-party claim. The plaintiff in *Shepherd* owned a cinder block building located adjacent to a construction site. After a wall of its building collapsed during excavation work performed by an excavation contractor on adjacent property for a city sewer project, the plaintiff sued the engineering firm that had prepared the plans and specifications for the project. The plaintiff's action against the engineering firm alleged negligence in both the design of the sewer project and in the administration of excavation work.

The court ruled in favor of the engineering firm, concluding that the latter's duty of care did not extend to an adjacent project owner. In reaching that decision, the court relied heavily on limitations and disclaimers regarding the engineering firm's scope of services contained in its agreement with the project owner. Specifically, the agreement provided that the contractor had the sole responsibility for avoiding damage to the public. Also, under the terms of the agreement, the engineering firm had no authority or ability to control the construction work performed at the site, nor did the engineering firm have authority to stop the use of improper or unsafe construction practices of the contractor. Accordingly, the court held that the engineering firm owed the plaintiff no duty of care.[8]

The decision in *In re Reinforced Earth Corp.*, 925 F. Supp. 906 (D. Puerto Rico 1996), provides another example of the importance of carefully defining and limiting the design professional's scope of services. In *Reinforced Earth*, an engineering firm, Geo, entered into a contract with a general contractor, Reinforced Earth Corp. (RECO), on a project involving the design, manufacture, and installation of an earth retention wall. Pursuant to its agreement with RECO, Geo contracted for the limited purpose of reviewing, sealing, and signing the design calculations and construction plans prepared by RECO for the design of the earth retention wall.

The earth retention wall failed and, consequently, the owner of the project filed an action against RECO, seeking damages for alleged defects in the design, manufacture, installation, and/or construction of the earth retention wall. RECO, in turn, filed a third-party action against Geo, alleging that Geo was negligent in the exercise of its duties, in failing to "undertake the proper studies of the conditions of the soil and subsoil at the Project, as required by the professional standards pertaining to design architects, engineers, and soil examiners."

The court rejected RECO's claim against Geo, finding that, if a failure of the earth retention wall occurred, it was for reasons beyond the limited scope of Geo's contracted services. Based on its review of Geo's contract with RECO, the court concluded that "Geo was not contracted to prepare plans, install or assemble any part of the project, but to review, seal and sign documents submitted by RECO in an advisory manner, which concluded in the determination that the submitted documents were correct.... Once Geo reviewed the documents and found them to be correct, its participation in the project ceased." By implication, the court found that Geo's limited service scope did not include any services with respect to the investigation of soil conditions and, hence, Geo had no obligations in that regard. Accordingly, judgment was entered in Geo's favor based on a finding that there was no evidence suggesting that Geo was negligent in the performance or in the quality of the limited services that it was contractually obligated to perform.

Although contractual limitations on a professional's scope of services generally serve as an effective defense to certain claims against professionals, there may well be circumstances in which such a defense will not necessarily preclude those claims. Specifically, a court may reject a defense based on scope limitations on the ground that the professional had a duty to advise the client that the limited scope was not adequate for the performance of the services required by the client and/or the project undertaking.[9] In other situations, a court may decline to accept a defense based on limited scope if it determines that the scope was unclear or ambiguous, especially if the professional drafted the scope provision.[10]

There are other defenses to contract-based claims against the professional, including the client's failure to pay; the client's lack of cooperation; the client's failure to provide certain pre-existing information (available or known to it) regarding the site, or reports, opinions, studies, data, analyses, or other work product of other consultants independently retained by the client; the client's knowing assumption of a risk through rejection of the professional's advice or recommendations; and/or the client's failure to make timely decisions. There may also be more "legalistic" defenses, such as expiration of the applicable statute of limitations and the value-added or betterment defense (i.e., even assuming breach of contract by the professional, the client did not suffer any actual loss or damage since the cost claimed would have been incurred by the client in any event).

In addition, the inclusion of a limitation-of-liability provision in the professional's agreement with the client may also serve as an effective risk allocation mechanism, as well as a defense to contract-based claims. A limitation-of-liability provision represents an agreement to limit (not release) the professional's potential liability to the client to a specified maximum dollar amount. Limitation-of-liability provisions generally are appropriate when the risk assumed by the professional in connection with the undertaking materially exceeds the professional's fee and/or when an enhanced liability risk arises from events or circumstances beyond the reasonable control or responsibility of the professional. In the specific context of subsurface projects, given the significant and enhanced professional liability risk (see Section 10.2), a limitation-of-liability provision often is an appropriate and important mechanism to fairly allocate and manage that risk. While a limitation-of-liability provision may effectively limit liability exposure of the professional to its client, it does not serve to limit liability to third parties.[11]

10.1.1.2 Negligence-Based Liability In addition to contract-based liability claims, clients, as well as third parties (see Section 10.1.2.1), may assert professional negligence or malpractice claims against design and construction management professionals involved in subsurface projects. Generally, these claims are based on the theory that the professional departed from the professional standard of care, which proximately resulted in damages or other adverse consequences to the client. Although the discussion in this section will focus specifically on negligence-based claims by clients against professionals, the discussion is equally applicable in the context of third-party negligence claims against professionals (see Section 10.1.2.1).

The "professional standard of care" may be described as follows: In performing his or her services, a design and/or construction management professional is required to exercise the degree of care, skill, and diligence ordinarily exercised by other members of the profession performing under the same or similar circumstances (including the relevant terms of the professional's agreement with the client), as existing at the time the services are performed by the professional.[12] The Supreme Judicial Court of Massachusetts expressed the professional standard of care as follows:

> As a general rule, an architect's efficiency in preparing plans and specifications is tested by the rule of ordinary and reasonable skill usually exercised by one of that profession. In the absence of a special agreement he does not imply or guarantee a perfect plan or satisfactory result. Architects, doctors, engineers, attorneys, and others deal in somewhat inexact sciences and are continually called upon to exercise their skilled judgment in order to anticipate and provide for random factors which are incapable of precise measurement. The indeterminable nature of these factors makes it impossible for professional service people to gauge them with complete accuracy in every instance. Because of the inescapable possibility of error which inheres in the services, the law has traditionally required, not perfect results, but rather the exercise of that skill and judgment which can be reasonably expected from similarly situated professionals. [Citations and internal quotation marks omitted.][13]

Generally, in the context of most professional negligence claims, the claimant (client or otherwise) will be required to prove, through expert testimony (i.e., the opinion of another similarly qualified and experienced professional), that the professional did not satisfy the professional standard of care. Typically, proof of a negligence claim against a professional will require expert testimony that establishes (1) the applicable standard of care required under the circumstances, (2) a departure from that standard, and (3) a causal connection between the departure and the claimant's damages or other injuries.[14] As a general rule, the expert's opinion should be based on a consideration of all the relevant contractual terms and circumstances, including scope of services and limitations thereon (e.g., time for performance, compensation amount, and limited opportunity to investigate the site or to interpret and evaluate data), and other relevant factors (including cooperation of the client and the client's fulfillment of his or her obligations, such as to furnish data or other relevant information).[15]

The application of the professional standard of care in the context of any particular negligence claim against a design and/or construction management professional should take into account the *totality* of the design and/or construction management professional's performance (contractual and/or actual) in judging compliance with that standard. For example, even though a design may have been deficient and resulted in additional cost to the owner, the owner reasonably should anticipate errors or omissions in the performance of the professional's services to a certain level or degree, whether expressed in terms of a percentage of construction cost or otherwise. Simply put, the professional standard of care does not require perfection. This anticipation and recognition rationalizes the professional standard of care and acknowledges that professionals are not, in reality, or reasonably and legally expected to be, perfect. The professional standard of care implicitly recognizes and tolerates some degree of imperfection in the professional's performance, and acknowledges that professional services typically involve the exercise of judgment (as distinct from scientific or empirical precision). In most instances, judgments about the adequacy of the professional's performance in the context of specific claims should be deferred until such time as the particular performance at issue may be judged in relation to the totality of the professional's project performance.

Although client claims based on contract and negligence principles have distinct and independent legal foundations, there is substantial overlap between the two claim types. The provisions in the professional's agreement with the client are important in defining whether a duty of care is owed to a client and whether the professional standard of care has been breached in any particular circumstances.[16] For example, the agreement may contain a provision defining the standard of care required of the professional. Ideally, any such standard of care provision in the agreement should be based on the professional standard of care. In some circumstances, however, the client may request that the professional agree to perform his or her services in accordance with a standard that *exceeds* the professional standard of care. For example, the client may request that the professional's services be performed "in accordance with the highest standards," "without, or free of, any deficiency," and/or "to the complete satisfaction of the client." Alternatively, the client may request that the professional be "fully and legally responsible in all respects for the technical accuracy and completeness of all design" and/or be "responsible for all acts, errors, or omissions." These requests, if accepted by the professional, would be tantamount to self-imposition of an obligation to perform perfectly and to be legally responsible for any imperfection whatsoever in the performance of services, regardless of whether the professional standard of care had been breached (i.e., negligence) and absent the necessity of any expert testimony.

Typical defenses to negligence claims against professionals include the comparative or contributory negligence of the claimant, which may serve to either reduce or entirely preclude recovery; adherence to the required professional standard of care; the value-added or betterment principle; expiration of the statute of limitation and/or statute of repose (depending on the law of the particular jurisdiction); and lack of causation between any departure from the professional standard of care and any damages being claimed.[17] Since the application of these defenses to particular claim contexts generally tends to be fact- and circumstance-intensive, a more detailed discussion here would not be especially helpful.

10.1.2 Liability to Non-Clients

In addition to being subject to professional liability claims from clients, design and construction management professionals are also subject to professional liability claims from non-clients—i.e., third-party liability. This section will discuss the two most prevalent theories of third-party liability in the context of claims against professionals involved in subsurface projects: negligence and negligent misrepresentation.[18]

10.1.2.1 Negligence-Based Liability The basic principles stated in Section 10.1.1.2 with respect to negligence claims *by clients* against professionals apply equally in the context of third-party negligence claims against professionals. Similarly, even though the third-party claimant, by definition, is not a party to the professional's agreement with the client, the terms of the latter agreement may be relevant in determining not only the contract and negligence liability of the professional to the client, but also whether the professional owed and/or breached a professional duty of care (i.e., was negligent) to third parties.[19]

In the context of third-party negligence claims, the professional may have available (depending on the law of the particular jurisdiction) an additional defense known as the economic loss doctrine.[20] In substance, the economic loss doctrine is based on the principle that one who claims purely economic loss (i.e., loss other than personal injury or property damage, such as additional construction costs, disruption, delay or other consequential damages, loss of use, extended overhead cost, or increased or extended construction financing costs) may not legally recover on a theory of negligence against a party with whom it has no contract. The basic rationale underlying the economic loss doctrine is that the third party's remedy (if any) in such a situation should be defined by and confined to whatever rights and remedies it has under its agreement with the contracting party (e.g., a general contractor's agreement with the owner, a subcontractor's agreement with the general contractor), and that to allow the third party to pursue, under the theory of negligence, a recovery against the professional (with whom the third party has no contract) in such circumstances would circumvent the contractual exchange of economic benefits and burdens embodied in the agreement between the third-party claimant and its contracting party and would subject the professional to imposition of an unfair risk of liability to persons with whom the professional has no contractual relationship.

One may fairly ask: Why would a third party pursue a negligence claim against a professional with whom it has no contractual relationship? There are several possible answers (as distinguished from legitimate justifications) to the question: The party with whom the third-party claimant has a contract may be insolvent or otherwise financially incapable of compensating the claimant; the rights and remedies of the third-party claimant under its agreement with its contracting partner may be limited, or entirely precluded by (for example) a "no damage for delay" clause, disclaimer-type provisions, or the absence of a differing site conditions clause (see Section 10.2.1), or they may have expired through (for example) the lack of timely notice or failure to comply with so-called procedural requirements under a contractual remedy provision such as a differing site condition clause (see Section 10.2.1); the third-party claimant may desire to maintain a positive relationship with its contracting partner, which would be negatively affected by the pursuit of a direct claim against that party; the third party may "need" compensation beyond that offered or paid by its contracting party and view the professional as an alternate or additional source of payment;[21] and, last, the third party may have some ulterior purpose or objective, such as to impair or undermine the relationship between the professional and the client.

The construction process is fraught with the possibility of such third-party negligence claims against professionals.[22] The largest category of these claims in the specific context of subsurface projects involves attempts by contractors or subcontractors (neither of whom typically has a contract with the design and/or construction management professional) to recover increased performance costs, delay damages, or other consequential or economic losses allegedly due to the negligence of the professional in the performance of planning or design services (for example, the preparation of reports, studies, and design and/or contract documents) and/or contract administration services (for example, the review of submittals, rejection of proposed substitutions or value engineering proposals, or rejection of work).[23]

One court applied the economic loss doctrine in refusing to impose negligence liability on a design professional in favor of a third party, stating:

> The parties involved in a construction project resort to contracts and contract law to protect their economic expectations. Their respective rights and duties are defined by the various contracts they enter. Protection against economic losses caused by another's failure to properly perform is but one provision the contractor may require in striking his bargain. Any duty on the architect in this regard is purely a creature of contract.[24]

One commentator has recently articulated a "contractual approach" rationale supporting the economic loss doctrine in the specific context of third-party claims in the construction process as follows:

> Every construction project in which a claim against a third party may arise is characterized by an extensive set of entwined relationships. These relationships originate in the parties' contracts; the owner contracts with the design professional and the general contractor, the general contractor engages subcontractors, and the subcontractors contract with suppliers, for example. Each contract is likely to contain complex terms about performance obligations that contemplate the performances of other participants, whether party to the particular contract or not. For example, the owner's agreement with the architect specifies the extent of the architect's duty to coordinate or supervise the work of contractors and subcontractors ... The contract may also contain risk allocation terms, defining the scope of liability to contracting partners and other participants, either in detail or by a general provision or limitation.
>
> The complex relationships in a construction project extend beyond the terms of the written contracts. Each contract in each project takes place within a network of customs and expectations concerning the typical roles and responsibilities of the different members of the construction industry. The participants in an individual project contract and perform against the background of those customs and expectations.
>
> ...
>
> The law's role is to support the parties' private ordering by using contract law to interpret and enforce their contracts. In defining obligations and allocating risks

> and benefits, the parties create expectations. The function of the law is to fulfill those expectations by enforcing the contracts as the parties have made them. It is inappropriate for the courts to impose liability outside the contracts, as through tort [i.e., negligence] law. . . . Imposing liability would upset the parties' own allocation of rights and duties, diminish their ability to regulate their own affairs, introduce inefficiencies into the process, and raise the threat of indeterminate liability.
>
> The contractual approach to third-party liability . . . is particularly well suited to construction industry cases. . . .
>
> The economic loss rule . . . [is] the obvious doctrine . . . to be used to implement the contractual approach. The economic loss rule bars parties in a contractual setting from bringing actions in tort [i.e., negligence]. Only when there is a contractual assumption of responsibility does a contracting party owe a duty to a third party.
>
> . . .
>
> [T]he contractual approach emphasizes the [design professional's] role as the agent of the owner. Because the [design professional] bears primary responsibility for advancing the owner's interest, it should not be subject to the conflict of interest that could arise if it is subject to suit by the contractor. If the [design professional's] error injures the contractor, the contractor may have an action against the owner on the contract between them. If it does have such an action, the contractor's interest is fully protected. If the contractor does not have an action against the owner, the reason it does not is that the owner and contractor agreed in their contract that such claims would be excluded, either directly or indirectly, . . . so the contractor has no legitimate expectation to be protected.[25]

The contractual approach rationale for the economic loss doctrine is particularly appropriate in the context of subsurface projects, in which, typically, owners and contractors specifically allocate risk for subsurface conditions (see Section 10.2.1). In these circumstances, a contractor should not be permitted to avoid or circumvent contractually allocated and accepted risk through the vehicle of asserting third-party negligence claims against the design and construction management professional. In this regard, to the extent risk is fairly allocated between the owner and contractor, the risk of such third-party claims against the professional is significantly diminished (see Section 10.2.4).

Despite the general effectiveness of the economic loss doctrine (in jurisdictions that have adopted it) in precluding third-party negligence claims against professionals, the courts in those jurisdictions have engrafted an important and expansive exception to the preclusive effect of the doctrine when claims are based on the legal theory of negligent misrepresentation (which is discussed in the next section). Unfortunately, a significant number of third-party claims against professionals on subsurface projects are based primarily, or alternatively, on the theory of negligent misrepresentation, thereby substantially undercutting the utility of the economic loss doctrine defense in that context.

10.1.2.2 Negligent Misrepresentation Third parties frequently assert claims against professionals on the basis that the latter negligently made representations on which those parties relied to their financial detriment. For example, a contractor may allege that, in bidding a project, he or she relied to his or her financial detriment on reports, studies, plans, or designs that were negligently prepared by the design professional; or a construction lender may claim that a construction manager made negligent reports as to the status or value of construction completed and in place, on which the lender relied to its detriment in advancing funds or making financial commitments to a project.[26]

Subject to certain limitations discussed below (such as the third party's right to rely on the performance, work product, or opinions of the professional and the reasonableness of any such reliance), courts have generally allowed third-party recovery against the professional on the theory of negligent misrepresentation. There are several ways in which design and construction management professionals are exposed to the risk of liability to third parties who may claim to have relied on their services or work product (e.g., opinions, reports, data, surveys, designs, plans, and specifications). The following third-party claims are illustrative: contractor or subcontractor claims for additional cost and/or delays due to errors or omissions in the plans and specifications or interpretive data reports prepared by the professional; surety company or construction lender claims based on the theory that decisions were made with respect to project control, financing, and/or funding based on information, status reports, or advice inaccurately prepared or given by the professional; claims against geotechnical engineers by third parties who relied on inaccurate statements or opinions made in the engineers' reports regarding soils or subsurface conditions; and claims by purchasers of properties who relied on the accuracy or completeness of information contained in reports or other work product prepared by professionals (for others) in deciding to purchase.

The basic issues in these types of claims are whether the third party was entitled to rely on the performance, opinions, or other work product of the professional; whether the professional was aware of the possibility of third-party reliance; whether the third-party claim involved issues within the scope of the professional's contracted services with the client or other professional obligations; and whether the third party's alleged reliance on the professional's services was reasonable. In many of these negligent misrepresentation claims, it is determined that a contractor's economic losses are due to the contractor's own negligence or deficiencies in bidding the project, in having overly optimistic and/or unrealistic assumptions and expectations, in inefficiently or inadequately performing the work, and/or in selecting inappropriate construction means, methods, or equipment for the project.

Typically, in order to recover on a theory of negligent misrepresentation, the claimant must establish that its reliance upon the representation of the professional was "reasonable."[27] For example, in *Texas Tunneling Co. v. City of Chattanooga*, 204 F. Supp. 821, 831-832 (Bankr. E.D. Tenn. 1962), a subcontractor on a tunnel project filed suit against, among others, the engineering firm involved in the project, alleging that the engineering firm "wrongfully omitted, withheld and misrepresented geological data," causing the subcontractor to sustain substantial financial losses. The engineering firm admitted that the data at issue were inadvertently omitted from the information supplied to the contractor. Based on evidence establishing that the engineering firm could have reasonably foreseen that the incomplete data would be furnished to and relied on by the subcontractor, and, further, based on evidence demonstrating that the subcontractor *reasonably relied* on those data, the court held that an action for negligent misrepresentation could be maintained by the subcontractor against the engineering firm.

A recent Florida decision involving a negligent misrepresentation claim by a contractor against an engineering firm further illustrates some of these principles. In *H&S Corp. v. United States Fidelity & Guaranty Co.*, 667 So. 2d 393 (Fla. Dist. Ct. App. 1995), the contractor (H&S) that was hired by a surety company to complete construction of a wastewater collection system project when the original contractor defaulted, asserted a variety of negligent misrepresentation claims against the engineering firm engaged by the owner of that project. Specifically, H&S alleged that the engineering firm misrepresented details concerning the project site and the nature of the work to be performed in completion of the project, and also misrepresented the nature and extent of the engineering firm's authority to require testing of certain portions of H&S's work prior to authorizing payment to H&S.

The appellate court rejected all of these claims, finding that the engineering firm was not liable for H&S's misrepresentation and "changed site claim." In making that determination, the court observed that the evidence established that representatives of H&S had viewed the project site prior to re-bidding and were able to observe high groundwater and subsurface rock that had been partially excavated, and that H&S's project manager had informed the surety company (for the original contractor) at the time it re-bid the project that it "did not believe the original report regarding subsurface rock formation and water table levels (to which H&S had access) accurately depicted the pre-bidding inspection condition of the job site." Commenting on that evidence, the court noted that "the earlier report could not reasonably have been relied upon and, in fact, was not relied upon [by H&S] in making its bid." The court also noted that the record established that the earlier soil/rock/water table testing report was not false or misleading concerning the conditions present at the project site in the area where the test borings had been taken. In addition, the evidence established that H&S had been cautioned not to rely on the earlier test data.

Finally, the court rejected H&S's contention that the engineering firm misrepresented its role and authority regarding testing of the contractor's work as a condition of payment. In rejecting that claim, the court noted that the engineering firm's authority in that regard had been specified in the contractor's agreement regarding completion of the project and, therefore, no misrepresentation had been made.

In the landmark decision of *Craig v. Everett M. Brooks Co.*, 222 N.E.2d 752

(Mass. 1967), the Supreme Judicial Court of Massachusetts held that a design professional may be liable to a third party (i.e., a non-client with no contractual relationship to the design professional) who reasonably relies on erroneous information furnished by the professional in circumstances in which the professional actually knew that the third party would rely on the professional's performance. In *Craig*, the court held that an engineer hired by a developer to survey property and establish stakes for use by a road contractor (which had a contract with the developer, but not with the engineer) could be sued by the contractor on the legal theory of negligent misrepresentation if the engineer was negligent in the preparation of information furnished to the contractor. The court, however, recognized that the professional's potential liability in such circumstances should not be unlimited and acknowledged the concern of "liability in an indeterminate amount for an indeterminate time to indeterminate class." Despite its acknowledgment of that concern, the court held that liability may be imposed in appropriate circumstances in which the professional is actually aware that certain third parties (specifically or generically, by class, including the claimant) may rely on the information furnished by the professional and is actually aware of the purpose for which any such third party may so rely. The court also noted that the scope of services provided by the engineer for the client is relevant in determining whether the professional may be liable to third parties.

In *Aliberti, LaRochelle & Hodson Engineering Corp., Inc. v. Federal Deposit Insurance Corp.*, 844 F. Supp. 832 (D. Me. 1994), the federal district court ruled that an engineer and a construction manager were liable to a bank (and, hence, the FDIC, which had succeeded to the bank's interest in the project) that loaned and advanced monies to a developer based on representations made by the engineer and the construction manager. In the *Aliberti* case, the developer retained an engineer and a construction manager, under separate agreements, to provide various professional services in connection with the developer's proposed construction of a condominium/motel. The developer contacted a bank to obtain financing in connection with the project. In order to reduce project time and costs, the developer proposed to the bank that the project be designed and constructed in phases under the so-called fast-track method, in which it would not be necessary to have a complete set of detailed drawings prior to construction start. In addition, the developer reasoned that construction in phases would facilitate the ability to obtain financing within the lending limits of the bank.

In making its decision to commit financing for the project, the bank hired its own construction consultant, who had experience in the design and construction of similar projects, to verify financial and technical construction information provided to the bank by the developer and/or by the engineer and the construction manager. After some preliminary due diligence investigation, the bank issued a commitment letter to the developer that contained certain conditions. A budget prepared by the developer was appended to the commitment letter. After reviewing the budget, the engineer and construction manager notified the developer that the budget contained insufficient amounts for certain hard-cost line items and that there were scope of work items not included in the budget. In substance, the developer told the engineer and construction manager "that this was just an interim budget and not to worry about the missing budget items." At a subsequent meeting, the engineer and construction manager again expressed their concerns to the developer concerning the inadequacy and incompleteness of the budget and, again, were told "not to worry." Neither the engineer nor construction manager ever directly expressed their concerns about the budget to the bank. At a meeting with the bank, both the engineer and the construction manager were told that if, during construction, the project costs changed from the budget, then any such change in cost should be noted on the requisition form prior to submission to the bank and its construction consultant for review.

The loan closed and construction of the first phase of the project commenced. After the first month of construction, the bank's construction consultant became concerned that the project "was in trouble, but requisitions continued to display the [budget] number which was attested to by representatives of the engineer and construction manager." Shortly thereafter, it became apparent that the actual project cost would be nearly double the amount in the budget, and work on the project was suspended. The bank later purchased the project on a foreclosure bid.

After a trial, the district court found that the engineer and construction manager were liable to the FDIC (as successor to the bank's interest) based on four separate grounds: (1) inaccurate and false statements that the project could be completed for the amount stated in the budget; (2) failure to disclose knowledge of the missing line items contained in the budget; (3) failure to disclose the inadequacy of certain amounts for line items included in the budget; and (4) false statements of the total project cost on the requisition submitted to the bank. In making this determination, the district court relied on several factors, including (1) the fact that the developer's respective contracts with the engineer and construction manager obligated the latter firms to provide cost estimating services and to evaluate and update construction budget and anticipated costs; (2) the engineer and construction manager knew that the budget was inadequate and incomplete but failed to so advise the bank, despite their knowledge that the bank was relying on the information contained in the budget; (3) when the engineer and construction manager signed and submitted requisitions to the bank, they knew that the stated project cost was inadequate and substantially less than the amount needed to complete the first phase; (4) the bank justifiably relied on the representations made by the engineer and construction manager, which led the bank to wrongfully lend money and disburse loan proceeds to the developer; and (5) uncontradicted expert testimony established that the engineer and construction manager, as construction professionals, had responsibilities "to correct any misinformation, to disclose the absence of any familiarity with information, and to inform the Bank, if necessary, that estimating services which a bank would reasonably expect had not been undertaken," "that it is incumbent on the design professional to speak out," and "that construction professionals have an absolute obligation to clarify their involvement or lack of involvement in the project." After considering all this evidence, the district court ruled that the engineer and construction manager owed a legal duty to the bank to be honest and candid in communications with the bank despite the absence of a contractual relationship.

Another federal court, in California, held that, under the circumstances presented, an environmental professional hired by a purchaser of certain property to perform a subsurface soil and groundwater investigation was not liable to the seller of that property. In *Lincoln Alameda Creek v. Cooper Industries, Inc.*, 829 F. Supp. 325 (N.D. Cal. 1992), the court considered whether the seller of a parcel of land could maintain a legal action against an environmental professional retained by the purchaser of that parcel. The court held that an environmental professional's duty of care does not extend to non-clients who are not the intended beneficiaries of the environmental professional's services and/or otherwise not reasonably expected by the environmental professional to rely on the environmental professional's services, opinions, reports, or other work product produced for the client. This decision limits the exposure of the professionals to third parties, and is support for the proposition that, absent an intent that a third party be entitled to rely on a professional's services or work product, the third party is not owed a duty of care by the professional with respect to the adequacy of either the services performed or the opinions rendered.

Finally, *Bronstein v. GZA GeoEnvironmental, Inc.*, 665 A.2d 369 (N.H. 1995), demonstrates the successful use of contract provisions to negate reliance by a third party—which is an essential prerequisite to recovery under a theory of negligent misrepresentation. In that case, Stephen Bronstein hired an environmental consultant to conduct an environmental assessment of property that was the subject of a purchase-and-sale agreement between Bronstein and the owners of the property. The environmental consultant performed the assessment. Subsequently, Bronstein assigned his rights to the purchase-and-sale agreement for the property to three individuals, who purchased the property. Bronstein gave the environmental assessment report to the purchasers, and the purchasers contended that they relied on the report in purchasing the property. After buying the property, the purchasers discovered hazardous waste contamination on the site. Consequently, the plaintiffs filed suit against the environmental consultant, asserting, among other things, claims for negligence and negligent misrepresentation. In support of their claim, the purchasers argued that it was reasonable that they would rely on the environmental consultant's environmental assessment of their property, and therefore owed them a duty of care.

The court in *Bronstein* rejected the purchasers' argument, concluding that the environmental consultant owed them no duty of care. In reaching its decision, the court relied on the terms of the environmental consultant's agreement with Bronstein, which expressly provided that the environmental consultant was preparing the environmental survey report "for the exclusive use of [Bronstein]." The agreement between Bronstein and the environmental consultant further provided that Bronstein could convey the report to its lender and title insurer, but that further dissemination to other parties was forbidden "except for the specific purpose, and to the specific parties alluded to above, without the prior written consent of [the environmental consultant]." Based on these contract provisions, the court in *Bronstein* concluded that the environmental consultant only intended that its services benefit Bronstein, and, therefore, the environmental consultant owed no duty of care to the purchasers. Accordingly, judgment was entered in the environmental consultant's favor on the purchasers' negligence and negligent misrepresentation claims.

10.2 PROFESSIONAL LIABILITY OF DESIGN AND CONSTRUCTION MANAGEMENT PROFESSIONALS: SPECIFIC CONTEXT OF SUBSURFACE PROJECTS

This section will discuss some of the reasons that generally account for why the professional liability exposure risk on subsurface projects is more substantial than on other types of construction projects. In general, there are two predominant reasons for this increased professional liability exposure.

First, subsurface conditions cannot be precisely defined or predicted—in terms of either quantity or character—and the resultant variances and uncertainties typically pose substantial risk potentials for both owners and contractors. The manner in which those risks are allocated and managed between owner and contractor will materially influence the nature and extent of the professional liability risk for the design and construction management professional. Section 10.2.1 will discuss various principles regarding risk allocation for subsurface conditions between owner and contractor, and Section 10.2.4 will discuss the relationship between (1) how risk is allocated for subsurface conditions among owners and contractors and (2) the nature and extent of professional liability risk for design and construction management professionals.

In addition, subsurface projects involve a high degree of interaction and interrelatedness between the risks associated with (1) the design of the completed project and (2) the construction means and methods designed and implemented in the process of constructing the project. This interaction and interrelatedness present numerous opportunities for yet further variations between anticipated and encountered conditions, as well as for the nontraditional reallocation of risk for design adequacy and construction means and methods among the various project participants, both of which may increase the professional liability risk for the design and construction management professional. Section 10.2.2 will discuss risk allocation principles for design adequacy and construction means and methods between owners, contractors, and professionals on subsurface projects utilizing the traditional design-bid-build and design-build delivery methods.

Section 10.2.5 will provide examples of professional liability claims in the specific context of subsurface projects.

10.2.1 Risk Allocation Principles for Subsurface Conditions

Why is the professional liability exposure risk on subsurface projects more substantial than on other types of construction projects? Certainly, to a significant degree, some of the enhanced risk associated with subsurface conditions is attributable to the concealed, and generally unpredictable, nature of those conditions and the variable behavior of those conditions as a result of construction means and methods utilized in the construction process. To an equally significant degree, however, the enhanced risk of professional liability on subsurface projects derives from the way in which risk for subsurface conditions is allocated between the owner and contractor. For the latter reason, it is important to discuss in some detail the background and principles relating to risk allocation for subsurface conditions.

By definition, genuine differing site conditions—physical site conditions that are either materially different from those indicated in the contract documents and/or materially different from those normally encountered on a project of the character under construction—represent physical site conditions that, when encountered, are materially different from the conditions reasonably anticipated by one or more participants involved in the subsurface project. Typically, when differing site conditions are encountered, additional (i.e., unanticipated) time and cost normally will be added to the project schedule and budget. On most subsurface projects, conscientious efforts are made long before the commencement of construction to identify and fairly allocate by contract the risk of both known and unknown site conditions, in an effort to minimize the risk of disputes and adversarial claims (and the consequent project disruption, delay, and/or additional cost) that may result absent such a contractual allocation. On other subsurface projects, however, no effort is made on the "front end" to identify and/or contractually allocate that risk prior to the commencement of construction, which, inevitably, leads to conflicts, disputes, litigation, and project delay, disruption, and additional cost. In yet other situations, risk is identified but unfairly allocated to the contractor, and/or inadequately identified or ambiguously defined or allocated, either of which leads to similar problems, including the increased risk of professional liability claims (see Section 10.2.4).

Even in circumstances in which risk has been adequately identified and clearly and fairly allocated, and in which a contractor is entitled to additional compensation due to a valid differing site condition, owners sometimes have a difficult time accepting an obligation to grant a requested equitable adjustment because, in their perception, payment of a differing site conditions claim will not enhance or add value to the completed project.[28] In other circumstances in which a contractor has contractually accepted the risk of all or certain site conditions, the encountering of conditions *within* the contractually assumed risk may prove too financially onerous for the contractor to bear, and may result in either (1) the assertion of extracontractual claims based on, for example, fraud, misrepresentation, concealment of "superior knowledge," and/or impossibility of performance, or (2) default or some other major, traumatic, and/or "pathological" event on the subsurface project.

The primary risk allocation issue raised by differing site conditions is who—the owner or the contractor—bears the economic and temporal performance risks associated with the encountering of those conditions. For the reasons stated elsewhere (see Chapter 3), the allocation of these risks should be defined and confirmed as an initial and primary matter in the contract documents, and thereby constitute a basis of the agreement between the owner and contractor; that is, the risk allocation process should not be deferred until after a differing site condition has been encountered (or alleged to have been encountered).

While it is accurate to state that the owner "owns the site," it is not generally accurate to state that the owner "owns" all of the financial or performance risks associated with differing site conditions. In fact, absent specific contractual provisions, the contractor generally "owns" those risks.

Under the so-called common law rule, the contractor bears the risk that performance of the contractor's obligations may be more time-consuming or costly than had been (even reasonably) anticipated when the construction project was bid. In *United States v. Spearin*, 248 U.S. 132 (1918), the U.S. Supreme Court expressed the common law rule as follows:

> Where one agrees to do, for a fixed sum, a thing possible to be performed, he will not be excused or become entitled to additional compensation, because unforeseen difficulties are encountered. Thus one who undertakes to erect a structure upon a particular site, assumes ordinarily the risk of subsidence of the soil. But if the contractor is bound to build according to plans and specifications prepared by the owner, the contractor will not be responsible for the consequences of defects in the plans and specifications. This responsibility of the owner is not overcome by the usual clauses requiring builders to visit the site, to check the plans, and to inform themselves of the requirements of the work.... [T]he contractor should be relieved, if he was misled by erroneous statements in the specifications.

Similarly, in *Eastern Tunneling Corp. v. Southgate Sanitation Dist.*, 487 F. Supp. 109, 113 (D. Colo. 1979), a federal district court stated the common law rule as follows:

> It is conceded that, under the common law rule, a contractor who undertakes an entire contract for erecting a building is presumed, in absence of an express provision to the contrary, to have assumed the risk of unforeseen contingencies arising during the course of the work, unless performance is rendered impossible by an act of God, the law or the other party. [Citations and internal quotation marks omitted.]

Thus, in the absence of a specific provision in the contract documents providing the contractor with some opportunity for an equitable adjustment (either a time extension or additional compensation), under the common law rule, the contractor must bear the time and cost impacts associated with the discovery of differing site conditions.[29] Although a number of legal theories have been utilized by contractors in attempts to avoid the additional time and cost burdens that are sometimes the direct or indirect consequence of encountering differing site conditions (in the absence of contractual relief),[30] generally, contractors should not count on prevailing under such theories, and the prudent contractor in such circumstances will carry some (typically undisclosed) contingency in the bid designed to anticipate and "cover" the risk of encountering differing site conditions.

10.2.1.1 Contractual Risk Allocation—Subsurface Conditions Beyond any doubt, the most effective and generally recognized exception to the common law rule is accomplished through the inclusion in the contract documents of a differing site conditions clause, which recognizes a right of the contractor under certain prescribed and limited conditions, and subject to various requirements, to additional time or compensation when differing site conditions are encountered.[31] In fact, to a greater or lesser degree, most of the standard contract forms in use in the construction industry contain clauses that provide (upon satisfaction of varying requirements) for some degree of time- and cost-equitable adjustment to a contractor when conditions actually encountered differ substantially or materially from those indicated in the contract documents or those normally encountered on a project of the type under construction.[32]

The inclusion of a differing site conditions clause in contract documents generally results in a limited and qualified shift or reallocation to the owner of the time and economic risks (due to certain unanticipated site conditions) otherwise imposed on the contractor under the common law rule.[33] Specifically, in circumstances in which the contractor assumes the risk of increased time and cost impacts due to the encountering of differing site conditions (i.e., under the common law rule), a prudent and cautious contractor will typically include in the bid a contingency to ensure (or at a minimum provide some level of economic comfort) that the work can be profitably performed if differing site conditions are actually encountered. Since, by definition, a contractor cannot precisely predict whether any such conditions will be encountered and, if so, the nature and extent of the conditions (and any associated increased time/cost implications), the contractor may naturally err on the "high" side in establishing the contingency. Worse yet, from the standpoint of the owner, the contractor may include in the bid a contingency that is never realized or utilized due to the absence of any such conditions actually being encountered during the construction of the project. Also, a contractor may choose to disregard the risk and carry no contingency (or unrealistically or over-optimistically assess and carry an unreasonably low contingency) in order to attain the lowest bid in a price-competitive procurement process.

In order to avoid unnecessary contingencies in bids that are intended by the contractor to "insure" against the risk of differing site conditions, many owners have included differing site conditions clauses in the contract documents, which, to some limited and controlled degree, shift to the owner the risk of additional time and cost impacts due to the discovery of genuinely unanticipated site conditions. One commentator has stated:

> Large institutional owners who did vast amounts of construction (particularly the United States Government) began to realize that they were paying inflated prices on nearly every contract to insure against the possibility of encountering an unknown site condition, when these conditions arose on only a relatively few projects. As a result, they discovered that they could reduce the overall cost of construction by simply agreeing to pay more money for a differing site condition when it did occur, rather than paying extra for it in every contract whether a differing site condition arose or not. In other words, the owners realized that they could save money in the long run by assuming the risk of unforeseen site conditions themselves, which allowed the contractors to remove the contingencies placed in their bids and lower their prices.[34]

While the inclusion of a differing site conditions clause in the owner-contractor agreement is an excellent start in the process of fairly allocating risk, it does not always ensure that the contractor will obtain an equitable adjustment when a differing site condition—even a valid differing site condition—is encountered. Even when a differing site conditions clause is included in the contract documents, a contractor may not be entitled to an equitable adjustment for a number of reasons, including the contractor's failure to provide the owner (or the owner's designated agent) with timely or adequate notice regarding the conditions encountered;[35] the contractor's failure to comply with the pre-bid site investigation requirements of the contract documents;[36] the contractor's failure to prove any reliance—reasonable or otherwise—on any of the relevant "indications" in the contract documents or other information furnished or made available to the contractor;[37] the contractor's failure to prove that the time and/or cost incurred was actually and proximately caused by the alleged differing site conditions; and/or the contractor's failure to establish that the encountered conditions differed substantially or materially from those indicated in the contract documents (or otherwise described in information furnished or made available to the contractor).

Despite the inclusion of a differing site conditions clause, risk allocation for site conditions may be materially affected by the inclusion of provisions in the bid and contract documents disclaiming the accuracy or completeness of certain information, data, or interpretations furnished or made available to the contractor regarding site conditions. Similarly, bid and contract documents may include disclaimers with respect to estimates (quantity or otherwise) for unit price, allowance, or other specific components of the project work; these disclaimers may also significantly affect risk allocations.[38] Although the enforceability of these types of disclaimer provisions varies from jurisdiction to jurisdiction,[39] there is a substantial body of law (particularly from state courts) that provides strong support for the proposition that a contractor is not entitled to rely on the accuracy or completeness of disclaimed information. Put another way, disclaimed information contains no implied or other warranty from the owner as to its accuracy or completeness; in fact, the disclaimer may legally negate and contradict the implication of any such warranty, and the contractor's reliance on that disclaimed information (especially when no effort is made to independently verify the accuracy or completeness thereof) may be considered unreasonable and unjustified as a matter of law.[40]

Ample precedent exists regarding the risk allocation implications of the use of disclaimers in the related context of quantity estimates developed by a design and/or construction management professional and furnished by the owner to bidders and contractors. This legal precedent involves issues pertaining to both the liability of owners to contractors, as well as the potential liability of geotechnical and design engineers and construction management professionals to contractors.

In many instances, design and/or construction management professionals develop quantity estimates for items or components of work to be performed. The owner, in turn, includes these estimates in the bid and contract documents. In some instances, the owner may request that the contractor bid the estimated quantities of work on a unit-price basis. If there are significant overruns and/or underruns in the estimated quantities, the contractor may assert claims for adjustments in the unit price and/or otherwise seek to renegotiate the contract price based upon the theory of "cardinal change." (See J. Sweet, *On Construction Industry Contracts*, 3rd ed., Sections 11.1 and 11.6 (New York: John Wiley & Sons, Inc., 1996).) In this type of situation, the contractor typically alleges that the owner made misrepresentations regarding the actual quantities, miscalculated the basis of the estimates and/or concealed information which, if disclosed to the contractor, would have clearly demonstrated that the estimates were inaccurate. In some situations, contractors will assert negligence and/or negligent misrepresentation claims against the geotechnical engineer, design engineer, and/or the construction management professional who developed the quantity estimates. The result, in many of these types of claims, depends on whether the accuracy of the estimates is limited or disclaimed and/or whether the contractor was admonished in the bid and contract documents that the estimates were solely *approximations* rather than *absolute representations of fact* as to the physical conditions that would be encountered in the performance of the work.[41]

Finally, even with the inclusion of a differing site conditions clause, complications in the allocation of risk for differing site conditions may arise due to the action or inaction of the owner. Specifically, the accuracy and completeness—and, hence, reliability—of the information indicated in the contract documents or otherwise furnished or made available to the contractor depends on the scope of subsurface (or other appropriate) investigation authorized prior to construction by the owner. For the differing site conditions clause to function in a fair and effective manner in allocating risk, as intended, it is necessary that the scope of investigation underlying the preparation of information or data furnished or made available to the contractor (which will establish the "baseline" for purposes of applying the differing site conditions clause) be adequate. In addition, owners may not carry a sufficient contingency to fund the risk of differing site conditions, sometimes leading to owner avoidance or procrastination in addressing or compensating a valid differing site conditions claim, arbitrary rejection of a valid differing site conditions claim, and eventual litigation. Finally, the owner may (consciously or not) fail to disclose relevant information to the contractor, leading to fraud, misrepresentation, impossibility of performance, and/or concealment of "superior knowledge" claims against the owner.[42]

Simply put, the inclusion of a differing site conditions clause—while an important first step in the process of fairly allocating risk for site conditions—does not necessarily assure that risk will be fairly allocated consistent with the intent of that clause.

10.2.2 Design-Bid-Build Subsurface Projects: Risk Allocation Principles for Design Adequacy and Construction Means and Methods

At present, most subsurface projects are undertaken utilizing the traditional design-bid-build delivery method, pursuant to which the owner retains a design professional to provide design and construction administration services and independently retains a contractor to construct the project. This section will discuss the risk allocation principles for design adequacy and construction means and methods associated with that traditional delivery method. Section 10.2.3 will discuss the same risk allocation principles in the context of subsurface projects utilizing the design-build delivery method, under which the owner retains a single entity to both design and construct the project, and will also discuss the specialized professional liability issues associated with utilization of that delivery method.

Differing site conditions claims typically and frequently involve assertions that the owner breached implied warranty obligations by furnishing incomplete and/or inaccurate information regarding site conditions and/or "defective" design.[43] In some instances, owners deny these claims (in addition to asserting defenses such as inadequate or late notice, unreasonable or no reliance, and lack of causation; see Section 10.2.1), on the grounds that the contractor's chosen construction means, methods, techniques, sequences, or procedures of construction, as distinct from any material variation in physical site conditions, caused the claimed additional cost or time.[44] In addition, contractor allegations of inadequacies or omissions (intentional or inadvertent) in the preparation of site information (i.e., data or interpretive opinions), "defective" design and/or improper directions with respect to construction means and methods, asserted in support of a differing site conditions claim, may (and frequently do) trigger contract, negligence, or cost recovery claims by the owner against the design and/or construction management professional.[45] In other situations, contractors assert differing site conditions claims and defective design claims against the owner, while simultaneously asserting negligence and/or negligent misrepresentation claims against the design and/or construction management professional. For these reasons, it is important to address the risk allocation principles between owners, contractors, and professionals applicable to design adequacy and construction means and methods as they relate to subsurface projects.

Section 10.2.2.1 will discuss these risk allocation principles in the context of the traditional design-bid-build delivery method, and Section 10.2.2.2 will discuss some of the circumstances that may result in a reallocation of risk for design adequacy and construction means and methods in that delivery method. Section 10.2.3 will discuss risk allocation and professional liability issues in the context of the design-build delivery method, with a particular emphasis on subsurface projects (Section 10.2.3.5).

10.2.2.1 Traditional Risk Allocation for Design Adequacy and Construction Means and Methods

(A) Design Adequacy—Risk Allocation

Between owner and contractor—In general, an owner who furnishes a design (typically developed into a set of construction documents) to a construction contractor impliedly warrants the accuracy, completeness, and constructability of that design. This now-established risk allocation principle for "defective" design, which was first recognized by the U.S. Supreme Court in *United States v. Spearin*, 248 U.S. 132 (1918),[46] imposes on the owner who furnishes a design to a contractor an implied warranty or promise that the design is free of material defects and is constructable.[47]

The rationale for allocating the risk of defective design to the owner under the implied warranty obligation is based on the principle that since the owner (typically, through a design professional consultant) undertakes to prepare and furnish the design to the contractor, the owner should be responsible for the accuracy, completeness, and constructability of the design. As a consequence, the contractor generally is under no obligation to check and verify the adequacy or constructability of the design and documents furnished by the owner (and/or the owner's design professional consultant) because the owner assumed responsibility for the preparation of that design. In other words, if the design furnished by the owner is "defective," the owner who furnished that design is responsible to the contractor for the resultant time and financial risk.[48]

Conversely, the contractor has the obligation to exercise due diligence in bidding the project and to construct the project in accordance with the design and construction documents.[49] If the contractor fails to satisfy these requirements, he or she may not be entitled to rely on the owner's implied warranty obligation as a basis for a "defective" design claim. In addition, as discussed in Subsection 10.2.2.1(B), typically, the contractor has exclusive responsibility for the selection, design, and implementation of the construction means and methods to be utilized in the construction of the project.

Between owner and design professional—As stated above, the owner is responsible to the contractor for a "defective" design that the owner furnishes to the contractor. The owner who furnishes a defective design is potentially liable to the contractor for the resultant time and cost impacts.

As noted in Section 10.1.1.2, in general, and absent contractual language imposing a higher obligation, design and construction management professionals are obligated to perform their services, including preparation and development of design, in accordance with the degree of care ordinarily exercised by other members of the profession possessing similar skill and performing services under the same or similar circumstances (i.e., the professional standard of care).

Although the owner—under its implied warranty obligation—bears the risk associated with design defects, it is critically important to emphasize that the owner's implied warranty obligation risk is not necessarily or always the equivalent of the design professional's risk associated with a departure from the professional standard of care. Unlike the design professional's more limited negligence-based professional standard of care obligation, the owner's implied warranty risk implies a warranty that the design is virtually perfect and adequate in all material respects. Specifically, an owner may be liable to a contractor for design defects, but may be unable to obtain indemnification, cost recovery, or other recourse against the design professional who prepared that design absent satisfactory proof (typically required in the form of expert testimony) that the design professional failed to adhere to the professional standard of care (see Section 10.1.1.2).

Thus, while the project owner customarily relies on a design professional to prepare the contract documents that the owner furnishes to the contractor (and which form the basis of the owner's implied warranty obligation), a design professional's professional standard of care liability risk to the owner for defective plans and specifications is *not coterminous* with the design defect or implied warranty liability risk exposure of the owner to the contractor. The liability standard applicable to the owner for design defect/implied warranty is a no-fault, strict liability standard, whereas the liability of the design professional to the owner is measured under the negligence-based professional standard of care. Thus, the owner may be liable to the contractor if the contract documents are unsuitable, inadequate, incomplete, or otherwise defective in any respect, but the design professional will not be liable to the owner for those defects unless the professional failed to prepare the design and contract documents in accordance with the professional standard of care.

For example, in a situation in which the design professional prepares the design and the contract documents in accordance with the professional standard of care, and the contractor adheres to the requirements of the contract documents, but there are design deficiencies (although the deficiencies are short of any departure from the professional standard of care), the owner may be potentially liable to the contractor for the cost or time impacts resulting from those deficiencies, but will have no recourse against the design professional. The case of *Pittman Construction Co., Inc. v. City of New Orleans*, 178 So. 2d 312 (La. Ct. App. 1965), demonstrates this point.

In *Pittman*, a contractor sued the city of New Orleans for the balance of a contract. The project involved the design and construction of a refuse incinerator on a site selected by the city (as owner of the project). The entire project was designed by the city's consulting engineer. The contract required the contractor to perform the work in strict accordance with the detailed plans and specifications.

During the early stages of construction, the contractor encountered unexpected, unstable subsurface soil conditions. These conditions required the work to be suspended, and the engineer who had designed the project had to substantially redesign the piles for the foundation system. The contractor completed the remedial pile work, but advised the city and the engineer (to no avail) that proceeding with a floating concrete slab design would result in considerable damage to the building due to the unstable soil conditions.

The engineer inspected the work for final acceptance, found it to be satisfactory, and certified to the city that the contractor had completed the project in accordance with plans and specifications. The city, however, noting the damage done by the sunken slabs, refused to accept the contractor's work as complete and withheld the contract balance. The contractor sued the city for monies owed on the basis that it had completed all work required under its contract and had strictly complied with the "defective" design furnished by the city. The city, in turn, asserted a third-party claim based on defective design against the engineer to recover costs for the remedial pile work, as well as the amounts of any award to the contractor against the city.

Relying on expert testimony proffered by the city and by the engineer that the design was normal and standard, and "in accordance with first-class engineering practice," the Louisiana Appellate Court reversed an award in favor of the city against the engineer, stating:

> There is no charge here that the plans and specifications were such that the structure was either wholly or partially unfit for the purpose for which it was built. The only effect of the unknown soil conditions was to increase the cost of construction. The City selected the site; and in spite of [the contractor's] repeated entreaties and recommendations that changes be made in the plans and specifications, the City's representatives first refused but finally gave their consent to such changes but refused to pay the additional cost because of the limited budget the City had for the structure. The City's representatives dealt at arm's length with [the contractor] and [the engineer] and were kept fully informed at all times, during the preliminary preparations leading to the preparing of the plans and specifications and the contract, and all during the course of construction. When the City finally decided to have [the engineer] revise the plans and specifications, [the contractor] agreed to do the work with the understanding that the City was responsible for the extra cost. Since the only result from the unknown soil conditions was the additional cost of the construction, we cannot see how [the engineer] can be held liable on the ground of professional incompetency or negligence.
>
> In determining the liability of [the engineer], the same standard of care applied

> in the case of architects, physicians, attorneys, and others engaged in professions requiring the exercise of technical skill should be applied. The test is whether he performed his service in accordance with the skill usually exercised by others of his profession in the same general area; and the burden of proving he did not, is upon the party making the charge, in this case the City. . . . The City has not discharged this burden here because all expert testimony vindicates the professional skill and judgment exercised by [the engineer] in this case.

Pittman, 178 So. 2d at 320-21. In other words, even though the engineer's design was rendered inappropriate due to the unstable soil conditions, the uncontroverted evidence established that the design (while, strictly speaking and in reality, "defective") was in conformance with the professional standard of care, leaving the owner with no legal recourse against the engineer and with liability to the contractor.

(B) Construction Means and Methods—Risk Allocation Traditionally, and in accordance with construction industry practice and standard contract document provisions, the construction contractor generally is exclusively responsible for construction means and methods. The standard design professional agreements and general conditions of the construction contract published by the American Institute of Architects (AIA) and the Engineers Joint Contract Documents Committee (EJCDC), respectively, confirm the general and traditional non-involvement of the design professional of record (DPR) in construction means and methods.[50] For example, Paragraph 1.6.23 of the EJCDC "Standard Form of Agreement Between Owner and Engineer for Professional Services," Doc. No. 1910-1 (1996) provides:

> ENGINEER shall not, during such visits or as a result of such observations of Contractor(s)' work in progress, supervise, direct or have control over Contractor(s)' work nor shall ENGINEER have authority over or responsibility for the means, methods, techniques, sequences or procedures of construction selected by Contractor(s) or safety precautions and programs incident to the work of Contractor(s) or for any failure of Contractor(s) to comply with laws, rules, regulations, ordinances, codes or orders applicable to Contractor(s) furnishing and performing their work.

This provision is restated in Paragraph 9.13.2 of the EJCDC "Standard General Conditions of the Construction Contract," Doc. No. 1910-8 (1990), which states:

> ENGINEER will not supervise, direct, control or have authority over or be responsible for CONTRACTOR'S means, methods, techniques, sequences or procedures of construction, or the safety precautions and programs incident thereto, or for any failure of CONTRACTOR to comply with Laws and Regulations applicable to the furnishing or performance of the Work.

Early in their formal education and professional practice training and experience, design professionals learn about the traditional and customary distinction between (1) their roles and responsibilities in the design and administration of the construction of the completed project work as defined in the contract documents; and (2) the construction contractor's exclusive responsibility for construction means and methods.[51] This distinction is grounded upon, and traditionally has been rationalized and explained in terms of, the relative qualifications and experience of design professionals and contractors and in the notion that responsibilities between design professional and contractor should be tied and allocated according to their respective abilities to control certain activities within their areas of qualification, experience, and expertise. Hence, design professionals are charged with responsibility for the design of the *completed* project, while contractors are charged with responsibility for the *means and methods* by which the completed project is accomplished. Design professionals not only learn this distinction, but also are admonished as a matter of professional practice (by management, peer reviewers, their legal counsel, and their professional liability insurer and risk management advisers) of the critical importance of observing and maintaining that distinction in their contractual arrangements and in the actual conduct of their practice, thereby generally striving to avoid any involvement in the design and/or implementation of construction means and methods.

In general, this role distinction and responsibility demarcation/allocation is grounded in an acknowledgment of the relative qualifications, training, and experience of design professionals and construction contractors, respectively, with regard to construction means and methods, and an acknowledgment that contractual and legal responsibility (and the associated economic/liability risk) should be allocated according to those relative qualifications, training, and experience, as well as the ability to control risks associated with the activity in question—construction means and methods.[52] This reasoning and rationale lead to the conclusion that construction contractors generally should have exclusive responsibility for construction means and methods, barring unusual circumstances or some demonstrable or appreciable involvement of the DPR in the design, specification, monitoring, or direction of construction means and methods.

As a general matter, this role distinction and the resulting demarcation/allocation of responsibility generally translates itself, in standard contract documents, into a basic principle that the role of the DPR is to develop design concepts with respect to the proposed completed or permanent aspects of the project; develop the design; prepare construction documents to define and communicate that design; and, during construction, provide various administrative, advisory, and consulting services as representative of the owner for purpose of determining whether, in general, the completed work (i.e., the end result of the project) will conform to the design concept and the information contained in the construction documents. In contrast, the construction contractor's role is to execute and implement the design concept, as demonstrated in the contract documents, by constructing the work in accordance with those documents; that is, the construction contractor's efforts are directed to the execution and implementation of the construction process—i.e., the construction means and methods. (See J. Clark, "Commentary on Agreements for Engineering Services and Construction Related Documents," EJCDC No. 1910-9 (Alexandria, VA: EJCDC 1993), p. 2.)

Standard contract documents are intended to describe, in clear terms, the role of the three major participants in the construction process—owners, design professionals, and construction contractors. John R. Clark has stated:

> The aim of the documents is to set forth clearly and concisely those services for which the Engineer has been specifically trained and services they are professionally qualified and obligated to perform and should expect to perform, and also to recognize the responsibilities imposed upon Engineers by virtue of their licensure. At the same time, one must recognize that others, particularly Contractors, have their own expertise and skills which Engineers do not possess, but which are equally important for the successful completion of any project ...
>
> The Contractor has his own knowledge and training and brings to the construction skills and expertise that are quite different from the Engineer's, but are essential to successful completion of the work. The Engineer's efforts during construction are directed toward the end result, the completed project as a functioning whole, whereas the Contractor is involved in implementing the Engineer's design as presented in the Drawings and Specifications (which the Engineer may clarify and interpret for him) and concerned with the construction processes and its means and methods.
>
> The EJCDC's Construction Related Documents carefully delineate between the functions of the Engineer and those of the Contractor and adhere to the concept that each should 'do his own thing' or 'stick to his last,' as the old adage teaches us. Contractors are as concerned as Engineers (and Architects) that each party should perform his assigned functions without interference from the other. Accordingly, in describing the Contractor's functions the words supervise, control and direct are used because he must have authority over the construction activities, be in charge of the work as it is performed, and assume responsibility for how the work is done. In recognition of the Contractor's function of the supervision and control over the work as it is performed, it is stated (and the words are repeated frequently) that it is the Contractor and not the Engineer who is to have control over and be responsible for the 'means, methods, techniques, sequences and procedures of construction and safety precautions and programs incident thereto.' This is now standard language and represents a nationally accepted division of authority and responsibility between the design professional and the contractor. It has come to be recognized by the courts as having its intended meaning.

J. Clark, "Commentary on Agreements for Engineering Services and Construction Related Documents," EJCDC No. 1910-9 (Alexandria, VA: EJCDC, 1993), pp. 2, 7.

Thus, the standard contractual professional practice in the traditional project delivery context is to recognize the demarcation between the role of DPRs and construction means and methods. The standard contract documents utilized by design professionals and construction contractors acknowledge and clearly confirm that demarcation and the resulting allocation of responsibility (typically exclusive in nature) to the construction contractor for construction means and methods. These demarcation and risk allocation principles as between DPRs and construction contractors relating to construction means and methods—from both the professional practice and traditional contractual standpoints—appear to be quite clear, generally accepted, and established.

Typically, contract documents (i.e., primarily the drawings and specifications), despite their volume and detail, are not intended to address (and do not address) the construction means and methods to be utilized by the construction contractor to achieve the finished and completed project work. Thus, the design and installation of shoring, scaffolding, sheeting, bracing, dewatering, and other construction means and methods typically are the exclusive responsibility of the construction contractor and are not part of the construction documents developed and produced by the DPR. Because construction means and methods are generally not part of the DPR's design concept and/or the construction documents prepared by the DPR detailing and describing that concept, the DPR's construction administration duties (i.e., construction observation, shop drawing/submittal review and approval, and so on) do not generally involve any obligation on the part of the DPR with respect to the means and methods of the construction contractor.

A substantial body of appellate case law from several jurisdictions recognizes the basic principles that DPRs are *generally* not involved in, nor responsible for the design or implementation of construction means and methods, especially if their contract and/or the general conditions of the construction contract confirm that the contractor is exclusively responsible for construction means and methods.[53]

The case of *Shepherd Components, Inc. v. Brice Petrides-Donohue & Assocs., Inc.*, 473 N.W.2d 612 (Iowa 1991), is illustrative. In *Shepherd Components*, the owner of a building adjacent to an ongoing sewer construction project suffered property damage (in the form of a collapsed wall of its building) as a result of excavation work being performed in the project. Consequently, the adjacent property owner (the plaintiff) commenced an action naming as defendants the contractor who performed the excavation work and the engineering firm that had been retained by the project owner to design the sewer system and to provide construction administration services. After a trial, a jury found the contractor and engineering firm liable, and the engineering firm appealed. The Supreme Court of Iowa reversed the judgment against the engineering firm.

In reaching its decision, the court noted that the plans and specifications prepared by the engineering firm did not detail any protective methods for the contractor to utilize in performing its excavation work. In addition, the contractor hired its own engineer to prepare "shop drawings" for a system to retain the earth along the excavation route adjacent to the plaintiff's building. The drawings were reviewed and approved by the engineering firm and required that sheet piling (sheeting) be driven along the excavation route with the use of accompanying backfilling as the sewer line was being laid. The drawings did not provide for the excavation of the earth between the plaintiff's building and the sheeting. During the excavation work, the contractor departed from the shop drawings by removing earth immediately adjacent to the plaintiff's building before actually installing the sheeting. Shortly thereafter, cracking and deterioration were observed in the plaintiff's building.

The plaintiff claimed that the engineering firm was negligent in the administration of the construction of the project. The engineering firm's agreement with the project owner contained language that made clear that the engineer was not responsible for construction means and methods. In addition, similar language was contained in the general conditions of the construction contract and other portions of the contract documents, which made clear that the contractor was required to shore up, secure, and protect all foundations adjacent to the areas of excavation. In reversing the judgment against the engineering firm, the supreme court stated:

> The contracts between [the project owner, the engineering firm and the contractor] were prepared using forms supplied by [the engineering firm]. We have previously held that a similar professional, an architect, cannot avoid a duty by making his own rules establishing when he can be held responsible in a contract with a third party. Consequently, [the engineering firm] cannot avoid responsibility simply because the [project owner's] contract with [the engineering firm] excuses [the engineering firm] from responsibility for the contractor's acts or because the [project owner's] contract with [the contractor] places full responsibility on [the contractor] for the construction phase of the contract. We must determine whether the contract provisions gave [the engineering firm] any control over the work site and the activities that caused plaintiff's harm. Our examination of the contract provisions reveals that [the contractor] was given the power to dictate or restrict the construction methods used at the work site. [The engineering firm] was given responsibility for quality control of [the contractor's] completed work but had no power to control the work performed at the construction site. Under these circumstances, we conclude that [the engineering firm] did not owe a duty of care to others and, consequently, cannot be responsible for [the contractor's] negligence.
>
> In summary, our review of the responsibilities placed upon [the engineering firm] in the contracts leads to our conclusion that [the engineering firm] had neither responsibility for, nor control over, the construction procedures that were employed adjacent to plaintiff's building and caused the damages. We conclude that [the engineering firm] owed plaintiff no duty of care.

Shepherd Components, 473 N.W.2d at 616-17 (citations omitted).

The involvement of the DPR in the submittal and shop drawing process serves as yet further illustration of the traditional demarcation between the

DPR's project role and construction means and methods. In general, shop drawings are drawings, diagrams, schedules, and other data specifically prepared for the project work by a contractor, specialty trade contractor, manufacturer, supplier, fabricator, or distributor, which show or illustrate some portion of the work.[54] Typically, shop drawings are significantly more detailed than the DPR's project drawings. The shop drawing submittal, at least so far as the DPR is concerned, is intended to communicate to the DPR detailed information from the contractor (and/or its specialty contractors or suppliers) about specific components, systems, assemblies, or other portions of the work that are to be incorporated into the completed project.

In reviewing shop drawings, as in its observational role during construction, the DPR's focus is confined to whether the work, when completed, will conform to the contract documents; the contractor determines *how* the work will be executed and is solely responsible for construction means and methods. Specifically, in reviewing shop drawings, the DPR does so only for "the limited purpose of checking for conformance with information given and the design concept."[55] The DPR's review is not for the purpose of reviewing the construction means and methods that may be indicated or shown in the shop drawings by which the contractor intends to execute the design, and the DPR's review "shall not constitute approval of safety precautions or, unless otherwise specifically stated [by the DPR] of construction means, methods, techniques, sequences or procedures."[56] Clearly, this limited role of the DPR in the shop drawing review process is entirely consistent with the traditional noninvolvement of DPRs in, and the exclusive responsibility of the contractor for, construction means and methods.[57]

The limited focus of the DPR in the review of shop drawing submissions, therefore, is grounded in the construction means and methods demarcation which, in turn, is predicated on the distinct and respective roles, qualifications, experiences, and contractual responsibilities of the DPR and the construction contractor. As John Clark has noted:

> What are the submittals to be reviewed for? Once again it is necessary to remember the separate roles of Contractors and Engineers. The Engineer's functions are design-related, and their observations of the work of Contractors are end-result oriented. Contractors are involved in the process that produces the result called for in the Engineer's documentation. Those aspects of a submittal that pertain to the construction process (the means, methods, techniques, sequences and procedures of construction and safety precautions and programs incidental thereto) are by accepted custom and contract the responsibility of Contractors and should not be the concern of the design professionals, they should not be reviewed by the Engineer and they should not be within the scope of his responsibility. This is primarily because the Engineer does not and should not have control over them. Unfortunately, Shop Drawing submittals which usually originate with a Subcontractor or supplier often contain information that pertains to the means and methods of construction or to the fabrication, assembly, installation or erection process about which the Engineer has limited training or knowledge and as to which

> he should have no concern. The Contractor should review and pass upon these aspects ... The Engineer's review is for a different purpose—to see that what is being furnished conforms to the design concept of the project and complies with the requirements of the Contract Documents; the Engineer's review is *only* for these purposes ... It is necessary that Owners, Engineers and Contractors ... recognize and accept the separate areas of responsibility of Engineers and Contractors in this review process."

J. Clark, *Focus on Shop Drawings*, EJCDC No. 1910-9 C (Alexandria, VA: EJCDC, 1985), pp. 5–6.

Despite the traditionally articulated limited role of the DPR in the shop drawing submittal process "[t]he submittals do not ... lend themselves to a neat breakdown between design and its execution. As a result, information may be submitted (or omitted) that relates to execution of the completed project or the temporary work needed to accomplish the completed project." J. Sweet, *Legal Aspects of Architecture, Engineering and the Construction Process*, 5th ed., (St. Paul, MN: West Publishing Co., 1994), p. 221; S. Stein, *Construction Law*, ¶5.04, at 5-167 (New York: Matthew Bender, 1997).

Waggoner v. W & W Steel Co., 657 P.2d 147 (Okla. 1983), illustrates the potential for blurring the traditionally clear and limited role of the DPR in the shop drawing submittal/review process. In *Waggoner*, a construction site accident resulted from inadequate, temporary connections while the structural steel was being assembled and erected. The plaintiffs contended that the architect was negligent in approving shop drawings that did not provide for temporary bracing or connections on the expansion joints during erection of the structural steel. In holding that the architect was not liable for the accident, the Supreme Court of Oklahoma pointed to the provisions of the standard AIA documents, which allocated exclusive responsibility to the contractor for construction means and methods and safety-related issues. Commenting on the court's decision in *Waggoner*, Justin Sweet has appropriately observed:

> The court stated that it was the duty of the contractor and not the architect to see that the shop drawings included how the temporary connections were to be made. It would be more correct to state that there was no requirement that the contractor *communicate* its plan as to temporary connection to the architect through submittals. The architect could assume that a failure to include provisions in the shop drawings for temporary connections did not indicate that the contractor would use unsafe connections but indicated only that this was not information that was required by the submittal process. ...

J. Sweet, *Legal Aspects of Architecture, Engineering and the Construction Process*, 5th ed. (West Publishing Co., 1994), p. 222 (Reprinted by permission; copyright 1994 by West Publishing Company, St. Paul, MN. All other rights reserved).

Related liability concerns for the DPR are presented in the submittal process in circumstances in which a shop drawing submitted to the DPR for review contains information about proposed construction means/methods. Commenting on this subject, Justin Sweet has stated:

> A submittal may be a request by the contractor for technical advice from the design professional. The contractor may be stating that this is what it proposes to do and asking the design professional, "What do you think?" or "Can you help me out?" If this request relates to construction means, methods, or techniques, the design professional need not respond. But if the design professional knows that the contractor proposes to proceed in a way that will lead to project failure or accidents, the design professional must draw attention to this danger. This duty to act is similar to her obligation when she visits the site. During site visits, the design professional need not search for potential trouble in operations under the contractor's control, but she cannot close her eyes to obviously dangerous conditions or to information that clearly indicates likely project failure. Similarly, she should not be able to close her eyes to such information revealed during the submittal process even though she has no obligation to review that information. For example, had the submittal in the *Waggoner* case ... been a clear signal to the architect that unsafe temporary connections were to be used or if industry custom required this information be shown in the submittals and it was not, the architect should have directed attention to this problem.
>
> Whether such a duty to speak out exists depends on the balance of expertise. For example, if the subject of the submittals relates to steel detailing largely within the expertise of the structural engineer, there is a clear duty to take some action. However, if the submittal relates to an activity in which the balance of expertise lies on the contractor's side, such as vertical transportation, approval by the design professional should not relieve the contractor from responsibility. Similarly, the design professional need not answer any questions on such matters that the submittals may ask.

J. Sweet, *Legal Aspects of Architecture, Engineering and the Construction Process*, 5th ed., (St. Paul, MN: West Publishing Co., 1994), pp. 223–24 (Reprinted by permission; copyright 1994 by West Publishing Company, St. Paul, MN. All other rights reserved).

Another commentator has suggested that if a shop drawing submittal contains information relating to construction means and methods, the DPR should disclaim responsibility for them, consistent with the contractual limitations on the DPR's role in the shop drawing review process. S. Stein, *Construction Law* ¶3.01 [2], p. 3-53 n. 59 (New York: Matthew Bender, 1997).

The *Quality in the Constructed Project Manual* ¶16.11 (New York: ASCE, 1990), in discussing the subject of shop drawings for temporary construction, states:

> Temporary construction shop drawings depict components that will not be part of a completed structure, but are to be used temporarily during construction, such as sidewalk slabs, temporary lifts, temporary buildings, shoring, reshoring, form work, bracing, scaffolding, dewatering, or temporary power. The constructor has

> full authority and responsibility for these shop drawings, including design, preparation, review, and approval, since he or she develops the construction plan and has control of the construction process of which the temporary construction is a part. Construction or erection procedures, shoring, bracing for excavations, or other temporary construction requiring engineering analysis or design requires the seal of a qualified professional engineer affixed to the drawings and specifications.
>
> The role of the design professional or engineer of record is to evaluate the effect of temporary structures on the integrity of the completed structure. From this viewpoint, he or she may indicate in the construction contract documents how long such items as temporary bracing, shore-up, reshoring, and similar items need to remain in place. If a non-self-supporting frame is involved, and needs special treatment during construction, or if the design concept of the structure limits the construction sequence, he or she will also include this information. The design professional usually does not review temporary construction shop drawings except when necessary to determine compatibility with the design of the completed structure.

(Reproduced by permission of ASCE; Copyright 1990).[58]

10.2.2.2 Reallocation of Risk for (a) Design Adequacy and (b) Construction Means and Methods The traditional principles for risk allocation between owner, design professional, and construction contractor for design adequacy and construction means and methods (which are discussed in detail in Section 10.2.2.1) may be generally summarized as follows: (1) the owner impliedly warrants the accuracy, completeness, and constructability of the design and contract documents that it furnishes to the contractor; (2) the design professional is obligated to prepare the design and contract documents in accordance with the negligence-based professional standard of care, and is not responsible for the design and/or implementation of construction means and methods; and (3) the contractor is exclusively responsible for the design and implementation of construction means and methods, and for constructing the project in accordance with the contract documents.

For several reasons, these traditional risk allocation principles are more likely to vary on subsurface projects, with a corresponding risk reallocation among the project participants. Specifically, subsurface projects frequently involve and require a high degree of interaction and interdependence between the integrity and quality of the design and the construction of the completed project, and the means and methods incidental to and utilized in the process of constructing the project.

Several circumstances may result in a contractor's involvement in the design of a portion of the completed subsurface project (to be distinguished from responsibility for the design of the entire project under the design-build delivery method, see Section 10.2.3). For example, a contractor with expertise in the design of a permanent component of a subsurface project, such as a foundation system or tunnel lining system, may be required to undertake the responsibility for that design as part of its contract. Or, after the award of the construction contract, a contractor may propose and obtain approval of its value engineering or other submittal, proposing a design substitution for a permanent component of the completed subsurface project. In these examples, the contractor has, in all probability, assumed legal responsibility for the component of the design that it (or its subconsultant) performed and furnished, thereby resulting in a reallocation to the contractor of the risk of design adequacy (traditionally allocated to the owner and/or design professional) for that component of the project.

In addition, there are several circumstances that may result in a design professional's involvement in the construction means and methods on a subsurface project (with a corresponding risk reallocation to that professional). Such circumstances include the following: (1) the design professional may include in the contract documents the detailed design of specific construction means and methods that the contractor is obligated to use in constructing the project; (2) the owner may require that the design professional design and include, as part of the contract documents, specific construction means and methods (e.g., temporary support, bracing, shoring, or dewatering) due to the owner's concerns or indemnity obligations with regard to potential damage to adjacent property resulting from construction operations; (3) data, interpretive or other reports, and/or the contract documents prepared by the geotechnical engineer or other professionals may include "recommendations," "suggestions," or "assumptions" about particular construction means and methods anticipated to be selected, designed, and implemented by the contractor; and/or (4) the design professional may develop and include in the contract documents performance criteria, standards, or requirements to be utilized by the contractor in the selection and design of particular construction means and methods. In these examples, the design professional has, in all probability, assumed responsibility (to a greater or lesser degree) for the design of those particular construction means and methods, thereby resulting in a reallocation to the design professional of risks associated with those construction means and methods (traditionally allocated to the contractor).

Subsurface projects pose a significant challenge to the traditional risk allocation principles for design adequacy and construction means and methods and, for the reasons discussed above, frequently result in risk reallocation for those activities among the project participants. As will be discussed in greater detail below, there is nothing inherently problematic about these types of risk reallocations, provided that the responsibility for these risks are clearly, unambiguously, and fairly allocated and defined in the pertinent contract provisions and communicated to all project participants. Risk assumption and allocation—whether assigned on traditional principles or on a "reallocation" basis—should always be undertaken in a conscious, voluntary manner and in accordance with certain fundamental principles such as the ability to control risk and fairness. See generally R. J. Smith, "Risk Identification and Allocation: Saving Money by Improving Contracts and Contracting Practices," *Int'l. Constr. L. Rev.* 40 (Lloyds of London Press, Ltd., 1995); D. J. Hatem, "Risk Allocation And Dispute Resolution On Construction Projects: Roles And Challenges For Legal Counsel," *The CA/T Professional Liability Reporter*, Vol. 2, No. 4 (Burns & Levinson LLP, Boston, MA, June 1997).

(A) Reallocation of Risk for Design Adequacy In the traditional design-bid-build delivery approach, there are several circumstances which may result in a reallocation of the risk of design adequacy from the traditional standards set forth in Section 10.2.2.1(A), some of which arise by contract document provisions and others of which arise (more or less consciously) through the performance or conduct of the owner, design professional and/or contractor.

In general terms, there are certain basic principles which should determine whether risk should be reallocated under these "nontraditional" circumstances. Simply put, responsibility for risk should follow actual control of, or ability to control the risk. As Justin Sweet has stated:

> As a basic principle, responsibility for a defect rests on the party to the construction contract who essentially controls and represents that it possesses skill in that phase of the overall Construction Process that substantially caused the defect. Usually defects caused by design are the responsibility of the owner in a traditional construction project and the contractor who both designs and builds. Control does not mean the *power* to make design choices. Usually every owner has the power to determine design choices. For example, an owner may require a particular tile to be used, a power within its contract rights. But the control needed to invoke the basic principle means a skilled choice, either one made by an owner who has professional skill in tile selection or an advisor such as an architect with those skills.
>
> This principle recognizes that the owner who supplies the design is responsible for design that does not accomplish the owner's objective and may not have a claim against the design professional. Usually the standard to which the design professional is held is whether she would have performed as would have other design professionals similarly situated.
>
> Under this principle, the owner, though faultless, bears the cost of correcting defects. The owner has the principal economic stake in the project and will benefit from a successful project. There is no reason why it cannot be responsible for project failures even though it is blameless and cannot transfer the loss.
>
> . . .
>
> Risk allocation is based on the probable intention of the parties, the greater skill possessed or supplied by the owner, the contractor's lack of discretion, and the owner's being in the best position to avoid the harm, as well as the owner's ability to spread, absorb, or shift the risk to the design professional.

J. Sweet, *Legal Aspects of Architecture, Engineering and the Construction Process*, 5th ed. (St. Paul, MN: West Publishing Co., 1994), pp. 515–16 (Reprinted by permission; copyright 1994 West Publishing Company, St. Paul, MN. All other rights reserved).

Some of the "nontraditional" circumstances that may result in risk reallocation are described in more detail in the following subsections.

Substantial involvement of contractor in design—The traditional rule that the owner bears the risk of "defective" design is predicated on the rationale that the owner (typically, by and through its design professional) prepares and furnishes the design and construction documents. However, contractors may have an involvement in the design of a permanent component of the completed structure. Generally, the more extensive and substantial the involvement of the *contractor* in the development and finalization of the design and related documents, the greater the degree of discretion given to the contractor in the development and finalization of the design, and the greater the expertise of the contractor as a reason for delegating a design role to the contractor, the greater the probability that the allocation of risk for defective design will be allocated to the contractor or, alternatively, will be shared in some fashion between the owner and contractor. For any such "reallocation" to occur, however, the contractor must have *significant* involvement in the development and finalization of the design.[59] For example, in *Johns-Manville Corp. v. United States*, 13 Cl. Ct. 72 (1987), *vacated for lack of jurisdiction*, 855 F.2d 1571 (Fed. Cir. 1988), the Claims Court stated:

> [For the owner's implied warranty obligation to apply, it must be determined] that the design specifications [were] devised and drafted by the Government. Because of the often extensive interplay between contractors and the Government during the development, modification, and purchase of products, it must be determined whether design specifications are truly the Government's own, or whether the contractor has assisted in their drafting and provided such expertise as to negate the Government's implied warranty, because the Government specifications, in effect, endorse the contractor's product or input.
>
> . . .
>
> In *Austin Co. v. United States*, 161 Ct. Cl. 76, 314 F.2d 518, *cert. denied*, 375 U.S. 830 (1963), the contractor's implied warranty claim failed because its proposed technical modifications were incorporated into the Government's specifications. The contractor's voluntary preparation of the specifications for what was essentially a new product indicated to the court that the contractor had assumed the risk of impossibility of performance, and the court found no implied warranty by the Government. In *Bethlehem Corp. v. United States*, 199 Ct. Cl. 247, 462 F.2d 1400 (1972) (per curiam), the court also denied a warranty of specifications claim. Although the Government had drafted the specifications for an environmental test chamber, it did so in reliance on extensive information provided by the contractor and with the contractor's full knowledge that the Government was relying on its expertise. The court held that the contractor had assumed the risk of non-performance.

Johns-Manville, 13 Cl. Ct. at 119-20.

In circumstances in which a contractor develops its own design for a permanent component of the project (especially where the contractor is given the option to do so by virtue of its expertise, qualification, and/or specialized experience), generally, the contractor alone should be responsible for that design.[60]

"Design" versus "Performance" Distinction—In general, the allocation of risk to the owner for "defective" design exists primarily in circumstances in which the owner furnishes a detailed, fully developed, and completed design. In this specific context, the allocation of risk principles often turns on the distinction between "design" and "performance" specifications, the former typically resulting in allocation of risk to the owner and the latter typically resulting in allocation of risk to the contractor. See *Universal Contracting and Brick Pointing Co., Inc. v. United States*, 19 Cl. Ct. 785, 793-95 (1990).

One court has described the distinction between a "design" and "performance" specification as follows:

> The difference between performance specifications and design specifications is well established. Performance specifications "set forth an objective or standard to be achieved, and the successful bidder is expected to exercise his ingenuity in achieving that objective or standard of performance, selecting the means and assuming a corresponding responsibility for that selection." Design specifications, on the other hand, describe in precise detail the materials to be employed and the manner in which the work is to be performed. The contractor has no discretion to deviate from the specifications, but is "required to follow them as one would a road map." "Detailed design specifications contain an implied warranty that if they are followed, an acceptable result will be produced." [Citations omitted.]

Blake Construction Co., Inc. v. United States, 987 F.2d 743, 745 (Fed. Cir. 1993), *cert. denied*, 510 U.S. 963 (1993).[61]

The significance of the distinction between "design" and "performance" specifications is important in the context of risk allocation for "defective" design. Specifically, the preparer and issuer of a design specification (typically, the owner, by and through the design professional consultant) has the risk associated with defects in that design; see *Universal Contracting*, 19 Cl. Ct. 785 (1990). In contrast, since performance specifications typically give the contractor the responsibility and discretion to develop and finalize the detail of the design, the contractor generally bears the risk of defective design developed and finalized by the contractor pursuant to a performance specification. *Aleutian Constructors v. United States*, 24 Cl. Ct. 372, 378-81 (1991).[62]

In determining whether a particular specification is a "design" or "performance" specification (and, hence, whether the risk of "defective" design should be allocated to the owner or contractor), the courts and administrative boards generally look to the overall contractual scheme and other factors for any reliable evidence as to how the parties intended to allocate the risk for defective design. It was recently observed that such evidence includes: "(1) the historical context in which the parties inserted the relevant specification into the contract; (2) the broad purposes of the contract; (3) the structure of the contract; (4) the contractor's course of performance; and (5) communications between the

Government and its contractor concerning the work affected by the disputed specification." K. Golden and J. Thomas, "The Spearin Doctrine: The False Dichotomy Between Design and Performance Specifications," 25 *Public Contract L.J.* 47, 53 (1995). In determining whether to allocate risk for defective design to a contractor who developed that design in accordance with a performance specification, courts are influenced by the degree of design discretion delegated to or undertaken by the contractor and the expertise or qualification of the contractor with regard to that field of design.[63]

Risk may be imposed under performance specifications in a variety of ways. The case of *Florida Board of Regents v. Mycon Corp.*, 651 So. 2d. 149 (Fla. Dist. Ct. App. 1995), involved what appeared to be a "brand name or equal" specification for architectural concrete. Under the specification, the contractor was to "provide a skin plate with a smooth, noncorded 'true-radius' forming surface, equal to that manufactured by Symons." The contractor used the referenced Symons forming system. The owner concluded, however, that the construction failed to achieve compliance with the specifications. The contractor argued that the use of the specified Symons system was subject to the owner's implied warranty of the adequacy of the specifications, so the contractor could not be liable if the finished product did not meet specifications. The Florida Court of Appeals disagreed.

The court found that the specification was not a design specification and *did not require* the use of the Symons forming system. By allowing the contractor to use an "and/or equal" system, the contractor bore the burden of achieving the specified finish through whatever system the contractor deemed appropriate. It was the court's view that the brand name "or equal" provision was a performance specification, which did not excuse the contractor's failure to comply with the specification. The court noted that if only one brand-name product was specified, so that the contractor had no discretion but to use that one product, then the owner's implied warranty obligation would apply and the contractor would be entitled to relief.

Under certain circumstances, the owner may be responsible for the accuracy of preliminary design criteria or conceptual information that it furnishes to the contractor, even though the contractor has responsibility for the development and finalization of the design. For example, in *M.A. Mortenson Co.*, 93-3 B.C.A. (CCH) ¶26,189 (June 30, 1993), the board found that the government was liable for the estimated quantities of material indicated in conceptual design drawings, even though the contract to be awarded was a design-build contract under which the contractor had the responsibility for completing the design and preparing the construction drawings. The board reasoned that the contractor had no choice but to rely on conceptual drawings in preparing a firm price for the construction aspect of the work.[64]

Other factors and considerations influencing risk reallocation for defective design—The mere characterization of a specification as "design" or "performance" in nature is not necessarily determinative or controlling on the issue of allocation of risk for defective design.

For example, even in circumstances in which the owner issues a fully detailed and developed design, the owner's otherwise applicable implied warranty obligation to the contractor (see Section 10.2.2.1), may be negated by contract provisions and/or actual performance or conduct of the parties.[65] In some cases, courts have held that a contractor's actual pre-award knowledge of design or other defects in the contract documents may prevent the contractor from subsequently recovering any extra costs allegedly resulting from those defects.[66] In addition, implied warranties regarding accuracy, completeness, and/or constructability of a design may be expressly disclaimed or negated by provisions in the contract documents.[67]

The issue of the effectiveness of a design warranty disclaimer was presented in *Massachusetts Bay Transportation Authority v. United States*, 21 Cl. Ct. 252 (1990). As part of the renovation of the South Station Project in Boston, the Massachusetts Bay Transportation Authority (MBTA) and the Federal Railroad Administration (FRA) entered into various agreements under which the FRA agreed to procure (from design professionals retained by it) and furnish the renovation design, and partially fund the project, and the MBTA agreed to renovate the project in accordance with that design. As a result of various issues that arose during the construction of the project, the MBTA sued the FRA, claiming that the design furnished (and prepared by design professionals engaged by) the FRA was deficient and that the FRA breached its implied warranty of design adequacy by furnishing that (allegedly) defective design to the MBTA. The Claims Court dismissed the MBTA's defective design claims on the basis of a provision in one of the agreements between the MBTA and the FRA, under which the FRA disclaimed any warranty of design adequacy. Specifically, the court stated:

> The Government does not impliedly or expressly warrant the accuracy of its specifications in the face of an unambiguous disclaimer in the contract.
>
> . . .
>
> FRA unambiguously disclaimed all express and implied warranties in §222(a) of its construction agreement with [MBTA]. The operative language states:
>
> Title to the Project Design Documents shall pass to MBTA upon acceptance by MBTA. MBTA acknowledges that the Project Design Documents are being prepared by an A-E [Architect-Engineer] as a contractor to FRA, not as FRA's agent. FRA makes no warranties, express or implied, concerning the Project Design Documents. No FRA or MBTA approval given under this Agreement shall be construed as a warranty of any kind.

As such, the above-quoted contract language served to effectively preclude any "defective design" breach of warranty claim by the MBTA against the FRA.[68]

(B) Reallocation of Risk for Construction Means and Methods As stated above, construction contractors typically are exclusively responsible for the design and implementation of construction means and methods.

Typically, construction contract documents, despite their volume and detail, are not intended to address and do not address the specific construction means and methods to be utilized by the construction contractor. As a general matter, because construction means and methods are not part of the design professional's design concept, nor of the construction documents prepared by the design professional, the design professional's construction administration duties (i.e., construction observation, shop drawings/submittal review and approval, and so on) do not involve or create any liability or obligation on the part of the design professional with respect to the means and methods of the construction contractor.

Under some circumstances, however, the design professional may include design or other information (and/or specific directions—either conceptual, performance- or criteria-based, and/or detailed in nature) in the contract documents with respect to construction means and methods-related issues. In those circumstances, the design professional may be held responsible for the design and/or adequacy of the specific construction means and methods-related design or other criteria or information contained in the contract documents, especially if the contractor is required to utilize specific construction means and methods designed and specified by the design professional. S. Stein, *Construction Law*, ¶3.01[2] and nn. 50, 57 (New York: Matthew Bender, 1997).

In some situations, an issue may arise as to whether a *general* reference in the contract documents to particular and/or suggested construction means and methods, absent any detailed design or other specific information, results in any responsibility of the design professional in connection with the implementation of the particular construction means and methods. The recent cases of *Burns v. Black & Veatch Architects, Inc.*, 854 S.W.2d 450 (Mo. Ct. App. 1993), and *Case v. Midwest Mechanical Contractors, Inc.*, 876 S.W.2d 51 (Mo. Ct. App. 1994), are illustrative.

In *Black & Veatch*, the plaintiffs sued, among others, an architectural/engineering firm, Black & Veatch, for various claims arising out of injuries sustained by a construction worker when a trench in which he was working collapsed on him. Black & Veatch's agreement with the project owner obligated it to perform architectural services, including design of the completed project, preparation of construction documents, and observation of construction to determine whether the work was being performed in accordance with the requirements of the contract documents. The agreement also contained the following paragraph:

> The Architect shall not have control over or charge of and shall not be responsible for construction means, methods, techniques, sequences or procedures, or for safety precautions and programs in connection with the Work, since these are solely the Contractor's responsibility under the Contract for Construction unless such means or methods have been specified by Architect for the performance of the Work.

Black & Veatch, 854 S.W.2d at 453.

The project owner's agreement with the construction contractor contained standard language imposing on the contractor the obligation to "supervise and direct" the work and sole responsibility for safety and construction means and methods "unless Contract Documents give other specific instructions concerning these matters." The plaintiffs contended that the project manual (prepared by Black & Veatch and a part of the contract documents) had specified particular construction means and methods with regard to excavation and shoring and, therefore, Black & Veatch assumed responsibility for the adequacy and proper implementation of those specified construction means and methods. The trial court rejected that contention and granted Black & Veatch's summary judgment motion, dismissing all of the plaintiffs' claims against Black & Veatch. On appeal, the Missouri Court of Appeals affirmed, stating:

> The conditions contained in the Project Manual are general conditions, instructing the contractor . . . to provide shoring, bracing, etc. where needed. They do not constitute specific means and methods. The incontroverted testimony of [the construction superintendent] establishes that the specifications provided by Black & Veatch do not specify means and methods of excavating or protecting the trench in which [the plaintiff] was injured. The method used by [the contractor] was one way to construct the foundation. The evidence disclosed that other methods were available to [the contractor].
>
> Missouri courts have not addressed the exact issue of whether an architect has mandated means and methods if the contractor is free to choose between two or more methods of performing work. We agree, however, with the reasoning of the Montana Supreme Court in its interpretation of a similar provision.
>
> The plans and specifications also provided for shoring where necessary. From that, it is clear to this Court that there were provisions in the plans and specifications that safety precautions should be taken to assure the safety of the employees working on the project. Those provisions in the contract mean that it was the duty of the contractor to take all necessary safety precautions. *Wells v. Stanley J. Thill & Associates, Inc.*, 153 Mont. 28, 452 P.2d 1015, 1018 (Mont. 1969).
>
> Appellants' expert witness, . . . while acknowledging that he has no formal education beyond high school and that he does not consider himself an expert in the field of architecture or structural engineering, testified that, in his opinion, Black & Veatch should have prepared drawings of a trench protection system. In other words, he believes Black & Veatch should have provided means and methods even though he agrees that the contract does not require Black & Veatch to perform that function.
>
> [The expert] did agree "that the generally accepted practice in the construction industry is that architects and engineers do not provide trench shoring [plans]" because to do so would infringe on the contractor's right to establish means and methods of construction.

Id. at 454.

Case v. Midwest Mechanical Contractors, Inc., 876 S.W.2d 51 (Mo. Ct. App. 1994), involved an injury by a construction worker during construction of a state correctional facility in Missouri. The worker sued, among others, the architect for the project for failing to provide for a railing or barrier along the roof edge (i.e., construction means and methods or safety precautions incident to the performance of the work) to prevent the worker from falling off the roof during construction operations. The court affirmed summary judgment in favor of the architect, stating:

> Although [the architect] agreed to review the safety programs of the general contractor and their trade contractors and make appropriate recommendations, the contract nevertheless placed the responsibility for the safety of those persons upon those contractors. [The architect] never agreed to supervise the construction of the Correctional Center and never expressly agreed to assume responsibility for safety precautions to be taken during construction. It is a far cry from being responsible to review the safety programs of others to specifically assuming responsibility for safety precautions to be taken.

Case at 53.

Of course, a design professional may contractually undertake responsibility for the design and/or implementation of construction means and methods and, in those circumstances, would be responsible for negligence in the performance of that undertaking.[69] Similarly, if a design professional, by actual conduct, becomes involved in the design or implementation of construction means and methods, he or she may be subject to liability based on alleged deficiencies in those construction means and methods. See, e.g., *Balagna v. Shawnee County*, 668 P.2d 157 (Kan. 1983).[70] Although many construction managers generally undertake greater involvement in construction means and methods than design professionals in the traditional project delivery system,[71] the actual terms of the construction manager's agreement, as well as the actual conduct of the construction manager, are critically important factors in determining whether, in any given circumstance, a construction manager may be held legally responsible for deficient construction means and methods.[72]

Depending on the degree and level of involvement of the design professional in construction means and methods and/or safety matters, the design professional may be exposed to fines and penalties under OSHA and related statutes, on the basis that the design professional is engaged in construction work and/or substantial supervision of construction work.[73]

In two recent decisions, *Secretary of Labor v. Foit-Albert Associates, Architects & Engineers, P.C.*, OSHRC Docket No. 92-654, 1997 WL 197016 (O.S.H.R.C. April 21, 1977), and *Secretary of Labor v. CH2M Hill Central, Inc.*, OSHRC Docket No. 89-1712, 1997 WL 197011 (O.S.H.R.C. April 21, 1997), *appeal docketed*, No. 97-2504 and No. 97-2603 (7th Cir. 1997). The Occupational Safety and Health Review Commission further clarified the circumstances under which design and construction management professionals (who do not directly or indirectly provide construction trade labor) are subject to OSHA's construction industry standards and regulations (OSHA Construction Regulations), 29 CFR §1926. *Foit-Albert* and *CH2M Hill*, taken together, make clear that design and construction management professionals' scope of contractual obligations and their actual conduct (or ability to exercise control) with respect to construction means and methods, and safety issues will be the critical factors in determining whether they will be subject to the OSHA Construction Regulations.

In *Foit-Albert*, the Review Commission held that an engineering firm retained by an architect to review construction documents, review and prepare change orders for approval by the project owner, and inspect the work of trade contractors for conformance with the contract documents was not subject to the OSHA Construction Regulations. In reaching that determination, the Review Commission noted that the engineering firm did not have contractual authority to advise or issue directives concerning construction means and methods, or to stop the work. The Secretary of Labor argued that, despite these contractual scope limitations, the engineering firm was subject to the OSHA Construction Regulations because the engineering firm actually had involved itself in matters relating to construction means and methods and safety issues in a manner that exceeded those contractual scope limitations. Specifically, the Secretary pointed to evidence that established that:

1. During inspection of the construction work, the engineering firm determined and reported to the contractor that in the cast-in-place concrete process, the concrete was not being poured at the proper time. Specifically, the engineering firm determined and reported that the contractor had begun the pour prematurely in a certain area (area 4A) where other trades were still placing rebar and completing the electrical and plumbing installation; this presented a hazard because workers unfamiliar with concrete placement were in close proximity to the concrete bucket as it moved across the floor, and it also precluded the engineering firm from properly inspecting the rebar and electrical and plumbing connections. In addition, during the work, the engineering firm's inspector noted several deviations from the approved shoring design (done by an engineer retained by the contractor) for support of the concrete formwork, and requested that the contractor "review the configuration of approved shoring details on the project." Specifically, the engineering firm's representative did not think that there was sufficient structural support for a horizontal spandrel beam because the scaffolding did not match the drawings of the shoring contractor's engineer. During a subsequent pour of concrete in the next area (area 4B), a collapse occurred.
2. In addition to inspecting and noting some deficiencies in the shoring process, the engineering firm refused to accept certain work because it did not conform to contract documents. Specifically, the engineering firm refused

to allow a concrete truck to make a delivery because its inspector observed water in the concrete.

At the outset, the Review Commission stated that the mere characterization of one's status as "engineer" or "architect" was not controlling in terms of the applicability of the OSHA Construction Regulations. Rather, the proper focus was on the professional's "relationship to the construction work, including its contractual responsibilities and the nature of the activities which it actually performs at the site." The Review Commission further noted that the engineering firm "had no contractual or actual authority to direct activities of the trade contractors, although during the course of its assigned work duties it notified them of safety hazards of which it became aware." The Review Commission noted that the engineering firm's contractual role was to review the contract documents and project specifications, to prepare change orders as necessary, and to inspect the work of the trade contractors to ensure that it conformed to the contracts and specifications. The Review Commission further observed that the engineering firm was not engaged in an extensive range of activities with respect to construction safety issues, nor did it have overall managerial or administrative responsibility for the project. The Review Commission stated that "a non-trade employer whose contractual authority is limited to inspection for conformity to contract specifications may nevertheless be held subject to the construction standards if, notwithstanding the restrictions of its contract, it in fact exercises control over safety programs or safety precautions." Applying that standard, the Review Commission noted that:

> In this case, the pour that collapsed, 4B, was commenced without [the engineering firm's] knowledge at a time when [the engineering firm] was occupied in other inspection activities. The fact that in the normal course of work on the project, [the engineering firm] did not have the opportunity to observe a critical aspect of the work is consistent with the contractual evidence that [the engineering firm's] role in the project included little if any actual responsibility to control or direct the manner of performance of that work.
>
> On the other hand, we recognize, as the Secretary emphasizes in her brief, that notwithstanding the limited scope of its contractual responsibilities, [the engineering firm] did warn the trade contractors of safety hazards in the shoring and of hazards in the manner in which the concrete pour was being conducted in area 4A. However, we conclude that these activities fail to demonstrate a control over safety issues sufficient to bring [the engineering firm] within the scope of the construction standards.
>
> At least to some extent [the engineering firm's] concerns were based on discrepancies between the shoring as installed by the shoring subcontractor and the engineer's drawings. (Footnote omitted.) To the extent that [the engineering firm] raised safety issues beyond mere compliance with design specifications, we cannot conclude on these facts that by so doing [the engineering firm] exercised substantial supervision over construction work ... The inspection of work being

> performed on a job site may implicate matters relating to safety, and our decision here is not to be interpreted as a finding that no such relationship exists on the facts of this case. In a case such as this where an accident occurred and injuries resulted, it is reasonable to assume that a competent inspection would detect hazardous working conditions or potential safety hazards. Our decision merely holds that by making known its concerns about safety [the engineering firm] did not manifest the ability to control or direct matters of safety. . . . [The engineering firm's] warnings about the shoring and the conduct of the 4A pour were an incidence of [the engineering firm's] inspection responsibility and did not arise to the level of supervisory responsibility for the implementation of safety measures and safety precautions at the site.

Based on that analysis, the Review Commission held that the engineering firm was not subject to the OSHA Construction Regulations.

In *CH2M Hill*, the Review Commission held that an engineering firm performing services on a subsurface project was subject to the OSHA Construction Regulations. Essentially, the Review Commission ruled that design and construction management professionals are subject to the OSHA Construction Regulations when they (1) possess broad responsibilities in relation to construction activities, including both contractual and de facto authority relating directly to the work of trade contractors, and (2) are directly and substantially engaged in activities that are integrally connected with safety issues.

The project at issue in *CH2M Hill* involved the Water Pollution Abatement Project owned by the Milwaukee Metropolitan Sewerage District. The project consisted of numerous tunnels, shafts, sewers, and other systems to collect and convey both storm drainage and sewerage to two waste water treatment plants. At issue was a portion of the tunnel part of the project known as the "crosstown collector" (CT), one of several 10-foot-diameter sewers. The District entered into an agreement with CH2M Hill. Because of the size of the project, the District contracted separately with various general contractors for the construction of portions of the project. S.A. Healy Company was the general contractor hired by the District to construct CT-7 tunnel.

The citations at issue in this case arose out of an inspection and investigation conducted by the Secretary of Labor following a methane gas explosion in the CT-7 tunnel, in which three employees of Healy died. Both Healy and CH2M Hill were cited for violations of construction standards based on allegations that the electrical equipment and circuits in the CT-7 tunnel were not approved for hazardous locations, that employees were not trained in the explosive and toxic gas hazards associated with tunnel construction, and that the ventilation in the tunnel was inadequate.

In evaluating this issue, the Review Commission started its inquiry by reviewing CH2M Hill's contractual responsibilities, which (under a master agreement) included program management, planning, engineering, construction management, and related services. The pertinent contract document provisions made it clear that CH2M Hill was not responsible for the construction means, methods, techniques, or procedures, or for safety precautions and programs in connection with the work; rather, the contractor was solely responsible for construction means and methods, and safety precautions and programs. CH2M Hill had no authority to stop work in the event of nonconformance with the contract documents. Although CH2M Hill had explicit authority to give direction to trade contractors with respect to compliance with the contract documents and scheduling of the work, CH2M Hill had no responsibility over construction means and methods and safety.

In addition, the record before the Review Commission contained evidence of CH2M Hill's role in assisting the District in connection with the evaluation and resolution of a differing site conditions claim. Soils in the project area were generally known as gas-producing, and methane was encountered in early 1993 in a boring drilled near the alignment of the CT-7 tunnel. Methane also was encountered in the construction of the CT-8 tunnel, which was the first tunnel constructed. The contractor on the CT-8 tunnel eventually filed a differing site conditions claim due to the existence of the methane. After receiving a recommendation from CH2M Hill that the differing site conditions claim had merit, the project owner acknowledged that the occurrence of methane in the CT-8 tunnel qualified as a differing site condition.

Following CH2M Hill's investigation into the incidence of methane in the CT-8 tunnel, a meeting took place in connection with the CT-7 tunnel project. At that meeting, a representative of CH2M Hill proposed four modifications to the CT-7 tunnel contract: increased ventilation in the tunnel, elimination of sources of ignition on machinery, provision of additional methane monitoring, and making personal safety equipment available to employees along with an evacuation plan. In addition, with respect to the tunnel machinery, minutes of that meeting stated that "[a] directive will be issued requiring the contractor to comply with the requirements of OSHA relating to the explosion proofing of machinery. This is intended to apply primarily to the mining machine and the mock haulage equipment."

In reviewing CH2M Hill's contractual authority and obligations for the project, the Review Commission noted that its precedent had not required that a professional be empowered to direct "means and methods" of construction in order to be subject to the OSHA Construction Regulations. Summarizing that precedent, the Review Commission observed:

> Accordingly, in determining whether the construction standards are applicable to an employer performing non-trade or professional services at a construction worksite, we look to two factors: the extent to which the employer is involved in the multitude of different sorts of activities that are necessary for the completion of the typical construction project and the degree to which it is empowered to direct or control the actions of the trade contractors.

Applying those standards to the facts of the *CH2M Hill* case, the Review Commission noted:

1. CH2M Hill had "broad and comprehensive responsibility in many aspects of a very complex and extensive construction project." In this regard, the Review Commission noted CH2M Hill's administrative responsibilities under the master agreement that encompassed a wide variety of managerial matters, such as scheduling, coordination of construction activities, preparation and interpretation of the contract documents, and modifications, including negotiation directly with trade contractors, claims processing, and dispute resolution.
2. CH2M Hill had authority to direct work activities, and had the contractual power to reject nonconforming work based on its own judgment as to when such action would be necessary to effectuate the construction contracts; and CH2M Hill had authority to implement its contractual responsibilities through the control of scheduling and the allocation of resources. The Review Commission rejected CH2M Hill's defense that it lacked the contractual authority to generally direct "means and methods" or how the contractors perform their work. The Review Commission noted that CH2M Hill possessed authority to interpret scheduling requirements, to establish priorities, to "enforce such measures as may be necessary to maintain" schedules, and to obtain "prompt" compliance by the trade contractors with its scheduling decisions. The Review Commission considered insignificant the fact that CH2M Hill lacked control over construction means and methods and safety practices, and had no capability to direct trade contractors to perform their work in any particular way, stating that "... our precedent ... has never required this authority as a prerequisite to considering an employer a 'construction manager,' and indeed we have never found such authority present in any of our cases."
3. Although the Review Commission acknowledged that the contract documents placed exclusive responsibility for safety upon Healy (the general contractor), it noted that "it is equally true that health and safety responsibility may be shared by more than one employer at a work site." The Review Commission noted that the question before it pertained to the relationship between contractual limitations on CH2M Hill's responsibility for safety and such other contractual terms that granted responsibility and authority. In addressing this issue, the Review Commission stated:

> In order to answer that question we will review in some detail the circumstances of the Differing Site Condition (DSC) contract modification. Broadly stated, from CH2M's perspective, the function of the modification was solely to determine whether or not Healy would be reimbursed for actions related to the potential methane gas hazard at CT-7. In the Secretary's view, the contract modification was not only about money but necessarily implicated safety. We agree with the Secretary.
>
> Acting under the differing site condition and contract modification mechanisms, CH2M implemented a contract specification directed specifically toward, and with the intent of eliminating, a substantial safety hazard at

> the site, the occurrence of methane gas. CH2M initiated a safety meeting with Healy and gave explicit safety instruction to the trade contractors, who in turn understood that CH2M was providing guidance and direction. CH2M's manager of the geotechnical office ... initially formulated the contract modification, and it was communicated ... through the resident engineer as a mandatory requirement.... In fact, the [district] and its approving official ... relied on CH2M's expertise to determine what contractual provisions or specifications were necessary to meet the changed conditions encountered during the course of the work.

Accordingly, the Review Commission held that CH2M Hill was subject to the OSHA Construction Regulations. In summarizing its conclusion, the Review Commission stated:

> The contracts here contain express language providing that the trade contractor, not CH2M, would have sole responsibility for safety precautions and programs. On the other hand, CH2M's many affirmative responsibilities and its authority clearly implicated safety, as did the exercise of its authority. In finding the construction standards applicable, we do not do so based on a presumption that a broad scope of duties necessarily implies responsibility for safety as well. Rather, we reach our conclusion based on an extensive record which illuminates specifically how safety concerns and safety issues were resolved in actual practice on the work site in question. As we have stated, in terms of its de facto actions, CH2M effectively was the nerve center through which means were developed and implemented for allowing the work to be conducted in light of a major safety hazard for a tunneling operation, the presence of methane gas. In terms of its contractual authority, CH2M was required to review and approve necessary actions, including Healy's safety program.

Significantly, a substantial portion of the record before the Review Commission in *CH2M Hill* related to the methane gas differing site condition. As such, *CH2M Hill* demonstrates that the actual performance or conduct of the design and/or construction management professional pertaining to site conditions may form the basis of citations under the OSHA Construction Regulations. In addition, the performance or conduct of the professional of an administrative or management nature—such as the negotiation of a contract modification for a time or cost equitable adjustment due to a differing site condition—may be used as a springboard for civil liability and/or OSHA exposure for safety-related incidents. The relevant factors in determining the exposure of the professional for alleged violations of the OSHA Construction Regulations appear to be the scope of the professional's contractual and actual responsibilities and conduct relating to scheduling, control, or coordination of construction work, and the professional's involvement in construction means and methods. In *CH2M Hill*, the Review Commission ruled that the professional's involvement in construction means and methods, as related to the methane gas differing site condition, led to its direct involvement in safety matters and that, by so doing, the professional "effectively was the nerve center through which means were developed and implemented for allowing the work to be conducted in light of a major safety hazard for a tunneling operation, the presence of methane gas." Given the substantial potential for interaction and interdependence between (1) the design of the completed project and (2) the contractor's selected construction means and methods, the Review Commission's decision in *CH2M Hill* requires serious consideration by the design and construction management professional involved in subsurface projects.[74]

Juno Industries, Inc. v. Heery Int'l, 646 So. 2d 818 (Fla. Dist. Ct. App. 1994), involved personal injury actions by two construction workers who were employees of a general contractor at a construction project at Disney World. Although the design specifications for the project, prepared by an engineering firm, called for a particular polyethylene pipe to be water tested, the pipe was air tested instead. The change in specifications was proposed by the general contractor and reviewed by the site representative of the engineering firm, who did not object to the submittal. During testing, the pipe came apart at a fused weld and whipped around, striking the plaintiffs, killing one and injuring the other. The plaintiffs brought suit against Disney, the general contractor, and the engineering firm.

The appeals court reversed the trial court's finding of summary judgment in favor of the engineering firm, ruling that there was a jury question as to whether the engineering firm assumed a duty regarding the means and methods of construction when its representative reviewed but failed to object to the submittal. The court stated that, by exercising control over the medium to be used for the test, it could be argued that the engineering firm exercised control over the method of construction, and, therefore, the case should not have been decided on summary judgment.

The court reasoned that, based on the contract provisions, there was a material issue of fact as to whether the engineer's contractual duty to review "submittals" or "other contractor-provided information" to determine compliance with the contract documents included within its scope a duty to review the general contractor's proposal to pressure-test the pipe with air rather than water, making summary judgment improper. If the contract required the engineering firm's approval of submittals for changes in the method of testing the pipe, then the engineering firm owed the owner and the workers affected by those changes its professional expertise to determine that the changes met engineering standards. Even if there was no contractual duty to review the change in testing methods, there was a question of fact as to whether the representative, by his actions on behalf of the engineering firm, assumed a duty to review the change in method for testing the pipe when he did not object to the change.

The discussion in Chapter 6 demonstrates the ways and circumstances in which design and construction management professionals become involved in construction means/methods and the factors that frequently and typically result in that type of involvement in connection with subsurface projects. For the reasons more particularly discussed in that chapter, the risk of departure from the traditional role demarcations (under which the design professional typically is not involved, by contract or conduct, in construction means and methods and in which the construction contractor is exclusively responsible for construction means and methods) is substantially increased on subsurface projects. For that reason, design and construction management professionals must pay special and careful attention to the need for clear contractual definition and confirmatory documentation of any involvement that they may have in construction means and methods. In addition, in their actual performance and conduct, those professionals should adhere to their scope of services, and other relevant contractual limitations on involvement in construction means and methods and safety issues. Absent such a clear definition and documentation, the professional faces a substantially heightened risk of professional liability exposure.

10.2.2.3 Application of Risk Allocation Principles for Design Adequacy in the Specific Context of Subsurface Conditions In the specific context of subsurface projects, contractors frequently assert claims for differing site conditions that are based primarily or alternatively on allegations of defective design ("combined differing site conditions and defective design claims").[75] In other words, contractors claim that they are entitled to economic or time adjustments based on site conditions, asserting (simultaneously or alternatively) that: (1) the conditions actually encountered materially differ from those reasonably anticipated; and/or (2) incomplete or inaccurate information, data, or interpretations regarding site conditions, or defective design, resulted in the claimed additional cost or time. As noted above, given the combination of such differing site conditions claims with specific allegations of "defective" design or other deficiencies (i.e., incompleteness or inaccuracies) in information, data, or opinions prepared or furnished by the professional, it is not uncommon, in these circumstances, for claims to be asserted by either or both the owner or contractor against the professional.

The traditional risk allocation principles set forth in Section 10.2.2.1 apply in the specific context of combined differing site conditions and defective design claims. For example, in circumstances in which an owner furnishes to a contractor a detailed design and/or data or interpretive reports regarding site conditions intended to be relied on and strictly adhered to by the contractor in the performance of the work, the owner generally will be found to have impliedly warranted the adequacy, completeness, and constructability of the design and the related information and reports.[76] As noted above, allegations of defective design or specifications, when combined with a differing site conditions claim, typically call into question the adequacy of the design professional's performance. Under such circumstances, in addition to asserting claims against the owner, it is not uncommon for the contractor to assert independent negligence and/or negligent misrepresentation claims against the design professional, or for the owner to assert contractual, negligence, indemnification, or cost recovery claims against the professional derivative of, or in addition to, the contractor's underlying differing site conditions claim against the owner.[77]

Several important cases illustrate the application of these risk allocation principles in the specific context of subsurface condition claims. Many of these cases involve an extended analysis of the issue of whether a particular provision of the contract documents constituted a "design" or "performance" specification (see Section 10.2.2.2(A)).

In *Stuyvesant Dredging Co. v. United States*, 834 F.2d 1576 (Fed. Cir. 1987), a contractor did not succeed in advancing a combined differing site conditions and defective design claim against the government, primarily based on the court's determination that the pertinent provisions of the contract documents were merely performance rather than design specifications, thereby negating the establishment of any implied warranty of the government with respect to the adequacy of the design. In *Stuyvesant*, the contractor had been awarded a contract with the Army Corps of Engineers for maintenance dredging of the Corpus Christi Entrance Channel. The pertinent provisions of the contract documents provided that:

> 4-1.1 ... DESCRIPTION OF WORK, is that composing [sic] of shoaling that has occurred since the channel was last dredged.... Bidders are expected to examine the site of the work and the records of previous dredging....
>
> 4-1.3 *In Situ Densities.* The following table details the results of a nuclear density survey conducted in the project area on 4 April 1979. The in-place density readings presented represent the average value of the density readings taken within the range indicated. The averaged values should not be interpreted as indicating the maximum or minimum density of material which may be encountered.

The table showed six places in the channel at which readings had been taken, and density averages had ranged from 1.380 to 1.675 kilograms per liter.

For the first few weeks of the project, the contractor met or exceeded its plan for construction work. Thereafter, the rate of removal fell drastically, because the contractor encountered large quantities of material that were difficult to dredge, and which were of densities "up to, if not more than 1.9 kilograms" per liter. The contractor then submitted a claim to the contracting officer seeking an equitable adjustment due to these site conditions, claiming entitlement under the differing site conditions clause and on the basis of defective design. In support of its claim, the contractor contended that it encountered differing site conditions within the meaning of the differing site conditions clause because paragraph 4-1.3 of the technical provisions (quoted above) represented that the material to be removed had a density ranging from 1.380 to 1.675 kilograms per liter which, according to the contractor, would have been relatively easy to remove, and that, in actuality, the material that the contractor removed had significantly higher density and, therefore, was more difficult to remove. The contracting officer denied the contractor's combined differing site conditions and defective design claim for an equitable adjustment.

The U.S. Court of Appeals for the Federal Circuit affirmed the denial of the claim. In addressing the differing site conditions portion of that claim, the court noted that, in order to prevail on a claim for differing site conditions, the contractor must prove, by a preponderance of the evidence, that the conditions indicated in the contract differ materially from those encountered during performance. In particular, the court stated:

> Technical provision 4-1.3 did not indicate, much less warrant, that the density of the material to be removed from the channel would range from 1.380 to 1.675 kilograms per liter. To the extent that [the contractor] relied upon that provision as so representing, [the contractor's] reliance was unreasonable. Moreover . . . the density of the material is only one of several factors that determines the difficulty of dredging.
>
> The provision stated that the figures given "represent the average value of the density readings taken within the range indicated" and warned that "[t]he averaged values should not be interpreted as indicating the maximum or minimum density of material which may be encountered." By definition, averages are composed of figures both greater and less than the average, so that [the contractor] must have known that necessarily it would encounter greater densities than the averages. Moreover, the averages were shown for only six specified locations in a channel that was 600–700 feet wide and almost four miles long. They provided a totally inadequate basis for any generalization regarding the character of the material to be removed from the channel.
>
> Moreover, technical provision 4-1.1 stated that "[b]idders are expected to examine the site of the work and the records of previous dredging" that were available at the two designated Corps offices "and after investigation decide for themselves the character of the materials." Such an express instruction to bidders obligated them to perform the necessary investigation to "decide for themselves the character of the materials to be removed," and is inconsistent with the notion that [the contractor] acted reasonably in assuming that the averages indicated the character of the material to be removed.

In support of the defective design claim, the contractor had contended that technical provision 4-1.3 was a design specification, that the government therefore warranted the correctness of the information contained therein, and that the average densities listed in technical provision 4-1.3 inadequately and erroneously described the character of the material. The court rejected this contention, stating that:

> Design specifications explicitly state how the contract is to be performed and permit no deviations. Performance specifications, on the other hand, specify the results to be obtained, and leave it to the contractor to determine how to achieve those results. . . . Detailed design specifications contain an implied warranty that if they are followed, an acceptable result will be produced. . . .
>
> The Claims Court ruled that technical provision 4-1.3 was not "a detailed [design] type specification for which the government might be liable under the defective specification theory." . . . Instead it was a "performance type specification for which the government does not warrant accuracy or adequacy."

> That determination was correct. Technical provision 4-1 did not instruct [the contractor] how it should perform the dredging of the channel. It merely stated the result to be achieved, namely, that the channel was to be dredged to its acceptable prism. [The contractor] had complete discretion to determine how it would perform that work. Its only obligation was to accomplish the designated result.

A similar result, based on similar reasoning, was reached in *John Massman Contracting Co. v. United States*, 23 Cl. Ct. 24 (1991). In *Massman*, a contractor entered into a fixed-price contract with the Army Corps of Engineers (COE) for scour protection work at Lock and Dam No. 10 on the Mississippi River. The tables included in the contract documents stated guidelines established by the COE that governed, based on river flow, the number of dam gates that could be closed simultaneously. Contract provision 1A-8.1 stated that the number of gates that could be closed at a time was limited, but that the COE would close as many gates as possible, in accordance with its guidelines, to facilitate the contractor's scour protection work. The contract documents did not identify the flow volumes at which scour protection work could be performed efficiently, nor did they specify the number of gate closures required for performance of the work. Instead, the contract noted that these tables, along with a third table indicating the approximate average monthly flow for Lock and Dam No. 10, were to be used by the contractor only as guides for scheduling operations.

The contract contained a standard differing site conditions clause. The contractor asserted a combined differing site conditions and defective design claim on the basis that the contract documents indicated what the river flow conditions during the contract period would be and that a reasonable contractor would consider the table included in the contract as representing the approximate average monthly flow as a reliable indication of the expected river flow. The contractor contended that the encountered conditions differed materially from the approximate average monthly flow information. The Claims Court affirmed the rejection of the contractor's claim, stating that:

> [T]he table provided by the Corps merely was historical data given to the contractor as a general guideline for scheduling operations. The tables specifically indicated that the contractor was to use them only as guides for scheduling purposes. As a matter of common knowledge, weather conditions fluctuate from year to year, oftentimes wildly. No reasonable contractor would presume that such information constituted indications of river flow conditions for the specific years the contract was to be performed, as [the COE] reasonably could not be expected to predict weather conditions for the upcoming contract period. Where reasonably there are no representations as to what conditions are, there can be no differing site condition.

In *J. L. Simmons Co. v. United States*, 412 F.2d 1360 (Ct. Cl. 1969), which also involved a combined defective design and differing site conditions claim, the Claims Court allowed a contractor's claim against the government based on alleged deficiencies in the design of the pile foundation system for a project.

After reviewing the pertinent portions of the contract documents, the court concluded that the pile foundation specifications were "design" and not "performance" in nature, based on the "most definite and precise requirements with respect to the piles to be installed." Accordingly, as the specifications were "design" specifications, the court readily concluded that the government impliedly warranted the accuracy and constructability of the pile foundation system.

During construction, the specified "pedestal" piles (which had been approved by the government) failed to satisfy the testing requirements and settlement limitations prescribed by the specifications. Tests performed on the piles indicated that the government-designed 40-ton pedestal piles, even when driven far beyond the order of magnitude contemplated by the original specifications, would not support the design loads. The tests revealed that the pedestal piles were not dependable at the designed load because reasonable driving would not provide support penetration; because there was an absence of beneficial side friction; and because soil considerations would not permit proper forming of pedestals. In response, the government then sought a substitute for the pedestal piles originally specified, and ultimately selected a composite-type pile. The contractor sought additional compensation and a time extension due to the alleged deficiencies in the original design of the pile foundation. The court held that the government was responsible for the adequacy of the design specification which it issued and, therefore, impliedly warranted that the installation of the specified pedestal piles by the specified methods would produce a satisfactory foundation.

Of particular note, the contractor in *J. L. Simmons* contended that the design deficiency resulted from the government's specification of high-capacity displacement piles in such heavy concentrations that damaging mass movements of the incompressible clay soils were inevitable and inherent in the design. The contractor further contended that there was nothing in the specifications to warn bidders that serious movements might occur, nor any provision for methods to prevent movements or to control them if they should occur. The contractor maintained that it was justified in assuming that the design was workable and constructable, and that it performed its work strictly in accordance with the plans and specifications, and was therefore in no way responsible for damages that resulted from its execution of the government's design.

In contrast, the government contended that the design was entirely workable and constructable and that the contractor's increased costs of performance were the direct result of its own faults and failures in execution of the design. The government contended that displacement of soils by the driving of piles was well known in the industry at the time of bidding, and that standard methods of controlling such displacement were practiced by competent pile driving contractors. The government further maintained that the difficulties the contractor encountered were the consequences of its failure to employ such controls and its failure to keep a close watch on the work as it progressed.

The court considered these competing contentions in detail, and also considered what alternative means and methods the contractor might have adopted to avoid the impact of this defective design. The court concluded that, while the soil conditions indicated by the borings were typical of conditions found in the project location and, therefore, no changed conditions were encountered, neither the government nor the contractor had any conception that either the original or the changed design plan would produce disastrous mass movements. The court also concluded that other possible methods of control that may have been available to the contractor (even though not specified or contemplated)—such as the driving of piles prior to general excavation—would not have been feasible because excavation after piles were in place was slow and expensive and involved much hard labor; because there was evidence that the bearing values of piles may be affected by excavating around them after they are driven; and because test piles driven through a considerable depth of overburden that is later to be removed afford an uncertain indication of what actual bearing capacity might later result.

In short, the court concluded that it was evident that the government did not anticipate that displacement of this nature would be a serious problem given the specified design and, therefore, it was unreasonable to expect that the contractor could have anticipated or utilized any construction means/methods to control displacement to the level encountered.

In *Raimonde Drilling Corp.*, 86-3 B.C.A. (CCH) ¶19,282 (Sept. 12, 1986), the Corps of Engineers Board of Contract Appeals held that a contractor was not entitled to an equitable adjustment because the contract documents were "performance" and not "design" in nature. In *Raimonde*, a contractor sought an equitable adjustment in the amount of $143,555.85, allegedly incurred during the performance of a $28,380 contract to obtain 309 samples of underwater materials. In substance, the scope of work was to consist of boring 11 holes to determine the type, nature, and characteristics of the subsurface materials. Drawings showed the "approximate locations" of drill holes, with actual locations to be established in the field by the government's contracting officer. The specifications stated that equipment to be furnished by the contractor

> shall include power driven core drilling machinery of an approved type or types complete with the special devices and accessories enumerated and described hereinafter.... The equipment shall also include a seaworthy floating or raised platform or barge which will be towed to the boring locations by a boat with ample power to move and maneuver the drill platform or barge. Once on location, adequate holding system will be provided to maintain the location of the barge and its contents.... The Contracting Officer will inspect the drilling and floating equipment prior to award.

During the course of the work, Raimonde encountered difficulties in drilling certain holes due to rough seas, which caused excessive lateral movement by the barge, which, in turn, caused three of the spuds to bend. Raimonde determined that the remaining work could not be accomplished with the barge that it had been using, and that a more stable platform would be required. Ultimately, Raimonde changed from the flotation barge method to a jack-up platform barge. Raimonde presented a claim for an equitable adjustment to the government, which the government rejected.

On appeal before the board, Raimonde contended that the government's inspection and approval of Raimonde's equipment made the government liable for extra cost when this equipment failed to perform all of the contract work, and that the specifications should have called for the jack-barge that was actually used to complete a substantial portion of the work. In addition, the contractor contended that the specifications were misleading in that they "represented that the work could be performed with a floating barge at all times of the year."

The board rejected these contentions, stating:

> [The contractor] asserts that it relied [on the equipment section of the specifications] and that [those] specifications represented that the work could be accomplished by any one of three methods: viz., floating platform, raised platform or barge, and, since all the work could not be performed by the method initially chosen by [the contractor]—i.e., by floating barge—the specifications were defective.... However, this is not a situation of a detailed Government design to construct, so that failure of the specifications to produce the expected or desired end result is the Government's responsibility. We conclude that *Spearin* is not applicable to the present situation.
>
> ... The enumeration of equipment in the specification describes the equipment in generic terms and the specific requirement is for adequate holding ability to obtain the desired samples. The specification cannot reasonably be interpreted as listing all equipment that can perform the work, but must be read to list equipment that may work under favorable conditions of wind, weather and sea. Thus a proper reading of the specification leaves to each bidder the choice of such equipment and the risk that the least costly equipment may not be adequate if the weather turns bad. Thus there was no representation that the equipment listed would be adequate under all weather conditions. The requirement for Government inspection and approval of equipment was merely for the purpose of safety of personnel and cannot be read to include any warranty that the floating barge could perform all the borings under all weather conditions. The specification should not be read to be a representation that work in all seasons could be accomplished with a floating barge (or, for that matter, any of the equipment listed.) Surely, there is no representation that boring can be accomplished from a jack-up barge when a hurricane's eye passes over the inlet.

The case *Appeal of Gulf Construction Group, Inc.*, 95-1 B.C.A. (CCH) ¶27,502 (Feb. 8, 1995), *motion to reconsider denied*, 95-2 B.C.A. (CCH) ¶27,649 (April 12, 1995), also involved a combined differing site conditions and defective design claim. Gulf Construction Group, Inc. entered into a contract in the amount of $561,859.87 with the Air Force for the demolition of an existing metal seawall and installation of a replacement seawall consisting of precast concrete sheet piles; the contract contained a differing site conditions clause. The specifications permitted installation of the piles by means of jetting, excavation, driving, or a combination thereof. (The evidence established that the typical method for installation of such sheet piles is jetting.) The specifications also provided: "Subsurface Data: Visual observation near the base line indicates beach sand at all locations." The contract did not include data from site borings, and none had been taken prior to award. The drawings called for sheet piles of precast concrete 26 feet long, 3 feet wide, and 8 inches thick.

During the performance of the work, Gulf encountered dense material (later determined to be wood and debris) that could not be penetrated with its jetting equipment, and Gulf asserted a combined differing site conditions and defective design claim in the amount of $251,310.91. Both the contracting officer and the Armed Services Board of Contract Appeals rejected the claim. In so doing, the board stated:

> [Gulf] argues that a differing site condition was encountered. In addition, [Gulf] alleges defective specifications. As to the defective specifications allegation, [Gulf] claims it was led to believe that jetting was called for by the specifications, but it became necessary to excavate ...
>
> ... With respect to the differing site condition alleged, there are two categories set out in the clause ... Type I Conditions arise from differences between the conditions encountered and contract representations. Type II Conditions arise from unknown conditions differing materially from those ordinarily encountered and are not dependent on contract representations. [Gulf] has not met its burden with respect to either Type I or Type II Conditions.
>
> The contract, under Subsurface Data, includes only a reference to beach sand visible on the surface ... There were no borings taken ... [Gulf] argues that the reference to beach sand was a contract representation as to subsurface conditions. We disagree. The reference to beach sand is manifestly a description of visible surface conditions. As such, it was unreasonable to rely on that contract representation as an indication of the subsurface or latent conditions to be expected in contract performance ... We cannot, therefore, find the existence of a Type I differing site condition.
>
> As to a Type II Condition, other contractors testified that conditions such as [Gulf] encountered were typical ... Moreover, [Gulf's] own crane operator testified that jetting always involves dropping the pilings to penetrate dense material and that problems of the type encountered are common place ... Based on the evidence of record, we have found as a fact there was no differing site condition encountered. Accordingly, the appeal is denied insofar as it is based on the Differing Site Conditions clause.
>
> With respect to the defective specifications argument, appellant appears to argue that the specifications led it to bid expecting to install the piles by jetting but it was forced to excavate ... A contractor is entitled to an equitable adjustment under the [Differing Site Conditions clause] for increased costs due to defective specifications ... However, the contract at issue clearly gives the contractor a choice of methods that includes excavation and jetting or combinations thereof.

> The decision to excavate ... was [Gulf's] alone. Thus, there was neither a prescribed contract method of performance nor a Government direction to perform by a particular method ... The record is not clear as to why [Gulf] decided to excavate ... There is no probative evidence [Gulf] attempted unsuccessfully to install the pilings by jetting at that location ..., and Government specifications do not warrant the absence of difficulty even where the method of performance is dictated. [Gulf] has not proved that jetting was not practicable ... or that it was directed to excavate.

The *Gulf* case is important as it suggests that a contractor may be able to assert a defective design claim based on impracticable and/or unconstructable construction means and methods in circumstances in which the particular construction means and methods are specified and required by the contract documents.

An important New York decision, *Fruin-Colnon Corp. v. Niagara Frontier Transportation Authority*, 585 N.Y.S.2d 248 (App. Div. 1992), also deals with a contractor's combined differing site conditions and defective design claim. In *Fruin-Colnon*, the contractor entered into a contract with the Niagara Frontier Transportation Authority to excavate and construct twin subway tunnels, each approximately two miles long, as part of the Buffalo Light Rail Rapid Transit System. During the course of performance, several disputes arose between the contractor and the authority. In particular, the contractor alleged that, during excavation, it encountered a differing site condition, entitling it to recover $3,255,150 in excess of the contract price of $38,949,800; a second dispute concerned whether the contractor or the authority bore responsibility for addressing the infiltration of water into the tunnels—the so-called watertightness dispute, for which the contractor was claiming $568,297.80 as reimbursement for remedial grouting; finally, various counterclaims for defective work and liquidated damages were asserted by the authority against the contractor.

Following a trial, the court denied the contractor's differing site conditions claim, and awarded the contractor $568,297.80 on its "watertightness" claim. Both the contractor and the authority appealed.

In addressing the differing site conditions claim, the appellate court upheld the denial of that claim on the basis that the contractor had failed to provide adequate proof that the conditions actually encountered differed materially from those indicated in the contract documents, and that the conditions actually encountered were not reasonably foreseeable. Specifically, the appellate court held that examination of the bid and contract documents in their entirety demonstrated that they accurately represented the subsurface conditions actually encountered and, more importantly, they adequately put the contractor on notice of the possible need to employ steel ribbing as temporary support in portions of the excavation. The appellate court relied on the Engineering Design Rationale (EDR) and the Tunnel Interpretive Report (TIR) in support of its conclusion. The EDR indicated that rock quality generally would be "average to good" for tunneling, but warned of the existence of localized areas of poor rock quality; opened, weathered, and, in some cases, "solutioned" fractures; and intersecting vertical and horizontal joints.

The TIR indicated the existence of intersecting joints forming blocks, and cautioned that such blocks might not be self-supporting as excavation progressed and, depending on local conditions, might require varying levels of primary support. Both the EDR and TIR specifically warned of areas of relatively permeable and solutionized rock and open, intersecting, and water-bearing beams.

Based on a review of those contract document indications, the court concluded that the actual conditions did not materially differ from those which reasonably should have been anticipated by the contractor. In response to the contractor's contention that the documents did not accurately predict the need to place steel ribs ahead of the gripper pads of the tunnel boring machine (TBM), the court, in significant part, disposed of that contention, stating that the contractor's need to place ribs ahead of the gripper pads was the consequence of its ill-advised purchase of a TBM that would not allow for placement of rock bolts within ten feet of the face of the excavation, as originally required by the contract documents.

Regarding the authority's appeal from the award in favor of the contractor on the watertightness claim, the appellate court also affirmed that award. On appeal, the authority contended that the court erred in awarding judgment in favor of the contractor on the watertightness claim, since the contract contained a performance specification that imposed the responsibility for achieving watertightness on the contractor and, consequently, the contractor was not entitled to extra payment for grouting. The contractor contended that the trial court erred in concluding that no differing site condition existed. In affirming the judgment in favor of the contractor on the watertightness claim, the appellate court engaged in an extensive analysis of the authority's contention that the contract documents constituted a "performance" as distinct from "design" specification. In addressing this distinction in the specific context of the watertightness claim, the appellate court stated:

> A performance specification requires a contractor to produce a specific result without specifying the particular method or means of achieving that result. . . . Under a performance specification, only an objective or standard of performance is set forth, and the contractor is free to choose the materials, methods and design necessary to meet the objective or standard of performance. . . . Concomitant with control over the choice of design, materials and methods is the corresponding responsibility to ensure that the end product performs as desired. . . . In other words, the contractual risk of non-performance is upon the contractor. . . . That is in contrast to a design specification, where the owner specifies the design, materials and methods and impliedly warrants their feasibility and sufficiency. . . . A contractor must follow a design specification without deviation and bears no responsibility if the design proves inadequate to achieve the intended result. . . . Whether a provision is a performance specification or a design specification depends upon the language of the contract as a whole. . . . Other factors to consider include the nature and degree of the contractor's involvement in the specification process, and the degree to which contractor is allowed to exercise discretion in carrying out its performance under the contract. . . .

> In arguing that the watertightness requirement was a performance specification that [the contractor] assumed the responsibility of meeting, [the authority] relies primarily on the language of [the watertightness clause]. Read in isolation, [the watertightness clause] appears to be a performance specification; it specifies the end objective (watertightness) and the standards for measuring that objective, but does not specify the methods of achieving watertightness. Nevertheless, the language and structure of the contract as a whole, as well as the parties' usage and course of performance under the contract, support the conclusion that a design specification was created.
>
> Although the watertightness clause itself does not set forth a particular method for achieving watertightness, the contract as a whole establishes complex and exacting standards for design and construction of the tunnel. [The contractor] was to construct an unreinforced, cast-in-place concrete liner of precise dimension. The type and mix of the concrete was precisely specified, as were detailed requirements for placing, curing, protecting, and finishing the concrete. [The contractor] was given no discretion to deviate from those specifications, whether for the purpose of water proofing or otherwise. For example, [the contractor] had no discretion to install an impermeable outer liner to resist the hydrostatic pressure that both parties knew would exist following completion of construction. Other provisions of the contract contemplate that waterproofing would be accomplished by means of fissure grouting, which also was to be carried out pursuant to detailed specifications. Additionally, the payment and warranty provisions of the contract support the conclusion that, as a whole, it created a design specification. The contract explicitly provides that "all measures necessary for achieving the degree of watertightness specified in [the watertightness provision] including remedial treatments to stem leaks," would be paid for at the contract unit prices. It is unlikely that [the authority] would have agreed to pay [the contractor] on a per unit basis if, as [the authority] contends, [the contractor] had assumed the responsibility of achieving watertightness. Further, unlike the general warranty set forth in the contract, the extended watertightness warranty did not provide that [the contractor] would remedy water leaks at its own expense, as it would have if [the contractor] had assumed the responsibility of achieving watertightness.

In addition to its examination of the contract documents as a whole in determining whether a specification was "design" or "performance" in nature, the appellate court also considered significant the course of dealing, communications, and conduct of the authority and the contractor both prior to construction and during and following construction. Specifically, the appellate court stated:

> The parties' course of dealing prior to construction also supports the inference that a performance specification was not intended. Bidders had no input into the design of the tunnel, nor did [the contractor] exercise any independent design judgment after it was awarded the contract. [The Authority] relies heavily on [the contractor's] March 5, 1980 letter submitted in support of its Value Engineering Change Proposal (VECP), a type of bilateral change order. In that letter, [the contractor] asserted, as a reason why it should be permitted to change the design of the tunnel liners to a "full circle pour," that [the contractor] bore the risk of meet-

> ing the watertightness requirement. In relying on the language of that letter, [the authority] overlooks the context in which it was sent. If the contract had created a performance specification with respect to watertightness, it would have been unnecessary for [the contractor] to obtain [the authority's] approval for a design change, and it would have been impermissible for [the authority] to withhold such approval. Similarly, the parties' course of dealing during and following construction illustrates that the contract did not establish a performance specification. As evidence that [the contractor] undertook the responsibility of waterproofing, [the authority] cites that [the contractor], on its own initiative, developed and carried out a course of chemical grouting, a method not mentioned in the contract and for which [the contractor] did not bill [the authority]. It is far more significant, however, that [the authority] routinely denied [the contractor's] numerous requests to implement various other waterproofing methods. Even before the concrete liners were put in place, and while the water diversion system was operating, [the contractor] repeatedly requested permission to grout the numerous water-bearing fissures that were present in the exposed rock. Those requests consistently were denied based on [the authority's] erroneous view that fissure grouting was not intended to achieve watertightness. After the tunnel was constructed and the water diversion system turned off, [the contractor] unsuccessfully requested permission to fissure grout, plug the deep wells with piezometer testing holes and construct a permanent dewatering system. Those requests were denied by [the authority] which maintained an uncooperative and obstructive attitude. [The contractor] nonetheless proceeded to fissure grout under process, based on its proper interpretation of the contract. It would have been unnecessary for [the contractor] to seek [the authority's] consent for such measures, and contractually impermissible for [the authority] to withhold such approval, if [the contractor] were in fact responsible for achieving watertightness and had discretion to chose the means to achieve that objective.

Another important case in this area is *Big Chief Drilling Co. v. United States*, 26 Cl. Ct. 1276 (1992). In *Big Chief*, a contractor asserted a claim against the government for damages allegedly sustained due to "defects in the contract specifications and omissions or mistakes by the Government's on-site representatives during contract performance." The project involved the drilling of ten cavern wells at the Department of Energy Strategic Petroleum Reserve in Texas. The wells were to be drilled in the Big Hill salt dome, which is a geologic formation; the subsurface stratum immediately above the Big Hill salt dome was known as Caprock. For reasons not entirely clear from the record in the case, the government deleted the standard differing site conditions clause prior to the award of the construction contract; thus, the construction contract itself did not contain any differing site conditions clause. In pertinent part, the technical provisions of the contract documents provided:

> 8.1.1 The depth to the top of caprock will be determined by: a decline in the rate of drill bit penetration, and cuttings which contain anhydrite, limestone or gypsum. Concurrence of these factors must be obtained from the Contracting Officer or his designated representative.

. . .

> 9.1 It is anticipated that the top of caprock will be encountered near a depth of 300', and the top of salt will be encountered near a depth of 1750'. Casing shall be installed to minimum depths and the hole drilled to sizes as defined below and as shown in Figures II.3, II.4 and tables G.1 and G.2 (Section G) for Wells A and B respectively.

. . .

> 9.3 Surface Casing: The Contractor shall run and set 30" O.D. Grade B casing in a 36" hole from surface to 15' +/− 5' into the caprock. On the bottom 40' casing joint, weld a stab-in float collar on the top and a guide shoe on the bottom.

The contractor sought an equitable adjustment based on two grounds. First, the contractor contended that under the contract documents, the government had the duty to determine and designate competent Caprock, and that the government's on-site representatives had a duty to determine the location of competent Caprock but failed in this duty, thereby precluding the secure placement and cementing of a "casing shoe" which, the contractor contended, was an essential step in the completion of a cavern well. Second, the contractor argued that the specifications relating to placement of the surface-casing were defective (Section 9.3, particularly) because they defined an imperfect range within which the contractor was required to set and cement the 30" casing shoe/seat in order to establish a secure casing bond with competent Caprock. The government disputed the contractor's claim, on the basis that the directions included in Section 9.3 of the contract did not constitute a defective specification and that those directions were not responsible for the loss of the integrity of the casing shoe. In addition, the government asserted that it did not have an obligation to determine the top of Caprock under the contract documents and that such responsibility rested with the contractor. Finally, the government contended that under a fixed-price contract, especially one in which the differing site conditions clause had been specifically deleted and the site investigation clause remained intact, the risk of unanticipated difficulties, including the risk of subsurface or latent physical conditions different from those anticipated, should rest with the contractor.

The Claims Court, finding in favor of the contractor, determined that Section 9.3 was a design specification and, accordingly, the government warranted the accuracy, adequacy, and constructability of the information contained in that specification provision. In reaching that conclusion, the Claims Court was strongly influenced by what it perceived to be the detailed nature of the information contained in Section 9.3 and in the lack of contractor discretion in the execution of the work defined in that section.

With regard to Section 8.1.1, the contractor contended that the government had the duty to determine the depth of Caprock and that its on-site geologist failed to accurately determine the depth of the Caprock. The contractor contended that as a result of that failure, it encountered the loss of the integrity of the casing shoe. The Claims Court again agreed with the contractor, finding that Section 8.1.1 was clear and unambiguous in requiring that the location of the top of Caprock was to be determined by the government and that the government's failure to make that determination constituted a breach of the government's obligations to the contractor. Alternatively, the Claims Court found that inaccurate or incorrect determinations as to the location of the top of Caprock by the government could also be viewed "as another Government issued defective specification." Specifically, the Claims Court stated:

> Although the precise depth of the top of caprock was not part of the original specifications when the contract was signed, the determination of the top of caprock was included in the contract as one of the duties of the government.
>
> The words "concurrence must be obtained from the contracting officer or his designated representative" clearly require that the ultimate determination as to the location of the top of caprock must be made by that appropriate government employee. Therefore, under the contract, even if the contractor had the initial responsibility to identify the top of caprock under the contract, the [government's] representative had the assigned and required task under the contract to make the determination as to the location of the top of caprock, or, at the very least, ratify, adopt and promulgate that determination. Therefore, this court finds that not only was the failure to properly determine the top of caprock a breach of duty under the terms of the contract on the part of the [government], but the incorrect determination of the top of caprock by the government can be viewed as the government issuing another defective specification, which constitutes a breach of its duty to properly determine the top of caprock under contract specification 8.1.1.

The *Big Chief* case clearly demonstrates another significant point—that acts or omissions of the government (or its design professional, construction manager, or other agent or representative) in the administration of the construction contract in the field may also provide the basis for contractor claims based on defective specification, design inadequacy, or breach of the owner's implied warranty obligation.

The recent decision of *S.A. Healy Co. v. Milwaukee Metropolitan Sewerage District*, 50 F.3d 476 (7th Cir. 1995), also demonstrates the impact that a design professional and/or construction manager's conduct and statements in the administration of the construction contract may have on the resolution of a differing site conditions claim. Healy, a contractor, entered into a construction contract with the Milwaukee Metropolitan Sewerage District to construct a vertical shaft 300 feet deep and 13 feet in diameter to carry liquid wastes to storage and treatment facilities. During the course of the work, Healy encountered "unexpected difficulties" in the construction of the dropshaft as a result of a heavy flow of groundwater into the underground work site. Healy submitted a differing site conditions claim to the district, which, in turn, the district rejected. After a trial, a jury awarded Healy $1.5 million. The central issue in the case was whether risk of groundwater inflow in excess of 25 gallons per minute (gpm) rested with the district or with Healy.

Inflow of groundwater was recognized from the beginning of the project as a potentially serious impediment to the construction of the shaft. The contract required Healy to perform consolidation grouting (to control groundwater inflow) and to construct several pumping test wells to determine the adequacy of grouting. The contract stated that "after completion of grouting from the surface, the total quantity of groundwater pumped from each drop shaft pumping test well shall not exceed 25 gpm.... [I]f the pumping tests do not meet the above criteria, additional grouting may be directed by the Engineer [retained by the district to monitor the contractor's performance] followed by additional pumping tests." Healy was not authorized to commence excavation of the dropshaft until the engineer determined that the grouting was adequate.

Healy conducted tests, which demonstrated that the amount of groundwater inflow exceeded 25 gpm, so the engineer directed Healy to do more grouting, which Healy did at the district's expense. Once tests demonstrated that the inflow of groundwater had fallen to 19 gpm, the engineer authorized Healy to commence excavation. After the commencement of excavation, Healy encountered inflows of groundwater that exceeded 25 gpm, and the excess groundwater slowed the project and increased Healy's performance costs. Healy then submitted a differing site conditions claim to the district for those increased performance costs.

In support of its claim, Healy contended that the contract "indicated" 25 gpm to be the maximum inflow of groundwater that Healy should expect to encounter, so that anything significantly above that level would "differ materially" from the contractual indication and entitle Healy to an equitable adjustment. Healy based this interpretation not only on the language of the contract, but also on testimony as to what the parties had said in the negotiations that led up to the execution of the contract.

In response, the district maintained that the contract was unambiguous and, therefore, any extrinsic evidence (testimony of the parties and/or experts regarding what the contract language meant) should not have been admitted to illuminate the meaning of that contract provision. More particularly, the district contended that the contract explicitly placed the risk of unexpected impediments to construction on the contractor, absent the limited circumstances that may fall within the differing site conditions provision. However, the district contended that the differing site conditions clause was inapplicable because nowhere did the contract "indicate" that Healy should not expect to encounter more than 25 gpm of groundwater inflow. The district maintained that the 25 gpm number referred solely to the pumping test wells, and that once the wells passed the 25 gpm groundwater test, Healy was obliged to go forward with the excavation and became responsible for coping at its own expense with any excess groundwater it encountered.

The court held that the contract language was ambiguous and subject to varied interpretations. As such, extrinsic evidence (i.e., testimony) would be admissible to "disambiguate" the ambiguous contract language. In addition to admitting into evidence testimony of Healy personnel regarding statements made during the negotiation of the Healy contract with respect to the risk of water inflow during construction, the court admitted into evidence a statement by the resident engineer that Healy's claim for a price adjustment under the differing site condition clause "probably has merit." The latter statement was made after Healy had submitted its differing site conditions claim to the engineer in accordance with the contract documents but prior to a formal decision with respect to that claim. The court held that evidence of the engineer's statement was admissible because no dispute had arisen as of the date that the statement was made; therefore, evidentiary rules precluding admissibility of evidence as to attempts to compromise a dispute were irrelevant since no dispute, within the meaning of the contract documents, existed at the time the engineer made his statement. Considering all of this extrinsic evidence, along with the contract language, the court rejected the district's position regarding contract interpretation and articulated its own conception of a more possible interpretation, as follows:

> We do not understand the sewage authority to disagree with the proposition that inflows of groundwater can greatly increase the cost of excavating an underground shaft. Since it was a fixed-price contract, the contractor would naturally be anxious about the possibility of encountering unexpectedly large groundwater inflows during the course of construction; and since the construction was underground, the contractor could not determine the volume of groundwater by simple inspection before beginning to excavate. The provision on test wells suggests—no stronger word is possible—that the figure of 25 gpm represented the parties' estimate of the maximum flow of groundwater consistent with performance of the contract at the fixed contract price. If the tests revealed a higher flow, the sewage authority would be responsible for the cost of such additional grouting as might be required to reduce the flow beneath that level. Since the test wells were much shallower than the projected dropshaft itself, the parties could not be confident that grouting sufficient to reduce the groundwater inflow to the test wells below 25 gpm would be sufficient to reduce the inflow below that level throughout the length of the dropshaft. But here the differing site condition clause clicked in; if the inflow was unexpectedly heavy, the contractor would be entitled to an adjustment of the contract price.

The court held that the jury was warranted, based on the evidence, in awarding Healy $1.5 million on its differing site conditions claim.

The Court of Federal Claims' decision in *Kit-San-Azusa, J.V. v. United States*, 32 Fed. Cl. 647 (1995), *aff'd in part, modified in part and remanded*, 86 F.3d 1175 (Fed. Cir. 1996), provides useful guidance for analyzing combined differing site conditions and defective design claims. There, the joint venture contract of Kit-San-Azusa (KSA) alleged a variety of differing site conditions and defective design claims.

In its first claim, KSA contended that the amount of cobble and boulder it encountered could not have been anticipated from the borings and specifications. The specifications indicated that much of the building site was composed of "a heterogenous mixture of silt, sand, gravel, cobbles and boulders." The government took a substantial number of soil borings at the site. Those borings were referred to in the specifications which informed the contractor that "glacial outwash ... was encountered at the site in all drill holes and consists of a heterogenous mixture of silt, sand, gravel, cobbles and boulders." The drilling notes confirm that description. In those notes, the drillers indicate boulder in three of the five sets of notes available. However, when those drilling notes were translated into the drilling logs furnished to the contractor, no reference was made to boulders. Five of the drilling logs simply referenced "cobble," or "scattered cobble," and other soils.

The court noted that, while the contractor could not ignore the site description, it was entitled to place greater reliance on boring logs if they were numerous and well-spaced, as was the case on this project. The specifications' reference to boulders, even at the site-specific level, was due to the general classification of the soil. The court stated that the government placed too much emphasis on the general classification of the soil and not enough on what was specifically found, at least insofar as what was disclosed to the contractor in the boring logs. In granting an equitable adjustment to the contractor, the court stated:

> The Court concludes that to the extent there was boulder, and that the boulders interfered with construction, KSA encountered a differing site condition. To hold to the contrary means that a contractor has to assume from the possibility of boulder in a general type of geologic stratum, that boulder could be encountered at any point, despite the lack of indications in the boring logs. Such an assumption would be nearly impossible to factor into a bid. Richard Galster, a private consulting engineering geologist, testified for the Government that one cannot extrapolate more than a foot from a boring log. Under that approach, the site could have no boulders, or, according to the Government, it could be virtually solid boulder. Would there be no differing site condition so long as there was no boulder within a few feet of any of the boring holes? Clearly at some point, the degree of difficulty due to boulder becomes a differing site condition. That point was reached here.

In addressing the contractor's differing site conditions claim with respect to cobble, the court stated:

> The question of cobble is admittedly different, because cobble did show up in some of the boring holes. There was testimony that nested cobble could result in problems similar to those caused by boulders. However, there was also testimony, which the Court accepts, that, unless the cobble is closely packed, a vibrating hammer will be able to cause the sheet pile to penetrate. Moreover, the contract drawings and specifications called for use of sheet pile to create a cofferdam. This is a warranty of sorts that sheet pile may be driven without use of extraordinary effort, despite the presence of some cobble.

> [KSA] did in fact encounter substantial difficulties at various points in driving sheet pile . . . It tried extraordinary means to install sheets and to install wells . . . The Court concludes that the presence of either boulder or cobble sufficient to impede sheet pile driving constituted a differing site condition.

The court, however, noted that there were other causes of KSA's difficulties—such as late delivery of sheet pile, delivery of the wrong sheets, omission of corner sheets, delivery of the sheet pile with concrete and dirt in the grooves, and delivery of sheets in singles instead of doubles. The court observed that all of those errors affected production and, therefore, would need to be taken into account in the amount of compensation given to the contractor on account of the boulder and cobble differing site condition.

KSA also made a claim of a differing site condition with respect to installation of dewatering wells. KSA bid on the assumption that it could use a water jet system to install deep watering wells. In a prior phase of the project, KSA had used that method with success. There was testimony at trial that a modest amount of cobble should not interfere with installation of deep dewatering wells even when using only a jetting system. On the second phase, however, KSA had significant problems sinking its dewatering wells. One of the holes had to be moved because even dynamite was inadequate to penetrate the cobble. The court found that these conditions constituted a differing site condition entitling KSA to an equitable adjustment.

The contract directed that excavation had to be "in the dry." Before excavation could commence, the ground had to be dewatered to one meter below the deepest excavation. Prior to excavation for the intake, KSA had dewatered to the required level. However, small amounts of what appeared to be water from surface runoff collected in the excavation point for the intake structure. KSA then requested permission to "unwater." The government's resident engineer denied permission to unwater and directed installation of a well-point dewatering system. The well-point system was expensive and cumbersome and, in the court's opinion, neither practically necessary, nor compelled by the contract.

The specifications made a distinction between "unwatering" and "dewatering." Unwatering applied to removing surface water; dewatering applied to eliminating groundwater. Unwatering can be accomplished using a sump pump; dewatering must be accomplished with deep wells or a well-point system. The court found that there was sufficient and adequate evidence and documentary proof that the water in question was surface water and, therefore, that the government's resident engineer exceeded his contractual power in requiring dewatering and that the resident engineer acted beyond the scope of his authority in instructing KSA to install a well-point dewatering system.

KSA asserted a claim for "twisted H-piles." The court denied recovery on this claim. In particular, the specifications warned KSA that it may have to predrill holes for the H-pile beams supporting the intake structure. KSA did not do so. As a result, KSA ended up with crooked H-piles and additional expense for a larger concrete pad. The court noted the absence of any evidence that H-piles, unlike sheet piles, could be vibrated through cobble. Drilling at the location in question was described in the contract documents as "hard to very hard" in places. Not having tried pre-drilling, KSA was not entitled to any entitlement under this differing site conditions claim.

KSA asserted a claim in connection with the failure of the tie-back anchoring system. KSA had hired an engineer to design a shoring system for the sheet piling. KSA's engineer selected a helical anchoring system designed and manufactured by A.B. Chance Company. One of the north wall tie-back anchors was exposed to a tension load equal to 150 percent of the design working load. The anchor failed. More helical plates were added to another anchor, and it failed, as well. KSA and its engineer came up with a totally different design system, using concrete deadmen. KSA argued that the soils were different than shown in the borings, and that, but for those differences, the original design would have worked. KSA advanced little evidence to support its differing site conditions claim that boulders and cobbles were the cause of the failure of the anchor system. The court noted the absence of any evidence and the lack of any direct causation between boulders and cobbles and the failure of the tie-back system. In addition, the government's expert in soil mechanics testified that the tie-back design was deficient in at least four specific respects (detailed in the court's opinion) and that the notes accompanying the design stated that if conditions at the site are generally similar to those reflected in boring DH-100, they would "pose installation problems." Boring DH-100 reflected, among other things, cobble, poorly grated sand, and silt. The court noted that there was no reason to assume that the addition of some boulders would have pushed the anchor tie-back system into failure when the designer recognized up front that the type of soils advertised in the drawings would cause problems. Thus, the court found that the anchor system failed for reasons other than a differing site condition.

KSA asserted another claim based on unanticipated export/import of film material. KSA had stockpiled the material it excavated while preparing the pumping plan site. KSA apparently planned to use that material for backfilling and compaction as construction progressed. In addition, KSA planned to use that material without it being reworked. KSA was required to import suitable material from off-site, and then sought additional compensation for screening of on-site soil (to remove cobbles) and importation of suitable off-site fill. The court rejected this claim, stating:

> KSA never justified the reasonableness of its underlying assumption that it would be able to replace the excavated material. Five of the boring logs show cobble. Because cobble in any quantity was unacceptable as backfill [the specifications specifically precluded use of cobble backfill], the contractor should have reckoned up front with the need of screening or importation of fresh material. It matters not that some of the boring logs did not indicate the presence of cobble. Because some of the logs indicated cobble, an unquantified amount of cobble was foreseeable. The difficulty in predicting the precise amount of cobble in the soil should have

> warned KSA of the possibility that the soil would not be suitable for backfill. In any event, the soil, for purposes of this claim, was not sufficiently different from that indicated by the boring logs to constitute a differing site condition. KSA may not recover the costs of screening or importing fill.

KSA asserted a claim for cold weather protection and loss of efficiency. In support of this claim, KSA contended that it had anticipated that it would be finished with major concrete work well before the winter of 1985–86. Instead, KSA finished three months later. In addition, due to that delay, KSA contended that it suffered a 35 percent loss of efficiency.

While the court accepted KSA's estimation of 35 percent loss of efficiency, it denied entitlement on this claim because KSA had failed to prove its underlying premise that, but for the differing site conditions and other government-caused delays, the work would have been completed on schedule. The court noted that the largest single period of delay—75 days—was attributable to the failure and redesign of the anchor system. That period was more than coextensive with the delay period attributable to subsurface conditions during installation of dewatering wells. Since the anchor tie-back system failure could not be attributed to the government, it followed that any government-caused delays during that 75-day period could not have been the sole cause for delaying the work. In addition, the court noted the following:

> Finally, KSA must cope with the fact that its managerial personnel have on numerous occasions and under oath, blamed various suppliers and its first project manager for most of its problems, including problems that occurred during the period in question.

Shortly after removal of sheet piling, KSA noticed that cracks had developed in some of the wing walls of the pump plant. KSA commissioned a number of surveys of the slab and surrounding area, which showed that the slab settled dramatically over the winter due to soil problems. The overall settlement was a relatively uniform two inches. KSA notified the government that it would cease all work in the project until the government proposed a solution. The government acknowledged some settlement, but contended that settlement was due to poor construction practices. KSA sought reimbursement of the costs of monitoring the slab for signs of settlement, the costs and delay attendant to remediating the problem, and other related costs.

In denying recovery, the court noted that it was not persuaded that the government either caused or assumed the risk of any settlement problems. Specifically, the court noted that there was no evidence that the government's design was defective, nor was there any proof that the soils were of such a nature that they could not be properly compacted. In particular, the court stated:

> The Court is unwilling to find, from the mere fact of settlement, that settlement problems were due to factors for which the Government was responsible. . . . KSA

> argues that the law establishes a warranty that the contractor's work ... will be suitable absent defective workmanship, which was not shown here. KSA presented evidence of correct construction, and the Government presented no evidence of defective construction. Thus, proceeds the argument, KSA is entitled to recover its costs in making the plant conform to the specifications and to producing a suitable result. [KSA's] argument is self-defeating. Although it is true that the law imposes an implied warranty that construction according to contract specifications will produce a satisfactory result, [KSA] does not argue that it was required to deviate from the specifications to produce a satisfactory result. Rather, [KSA] argues that it needed to perform extra work to bring the pumping plant into conformity with the specifications. A contractor is not entitled to additional compensation simply because it encounters unforeseen difficulties. It has no right to complain if the structure it ultimately constructed was essentially the same as the one it contracted to construct ... If the specifications were defective, KSA might have had a claim for costs incurred in attempting to achieve conformity ... KSA has not shown, however, that the specifications were defective and is therefore not entitled to recovery on that theory ...

In *Brant Const. Co. v. Metropolitan Water Reclamation District*, 967 F.2d 244 (7th Cir. 1992), the court affirmed the dismissal of a contractor's claims for additional compensation, which was based on allegations that plans provided by the Water Reclamation District of Greater Chicago were defective as they resulted in a gross underestimation of excavation quantities. The project in *Brant* required the contractor to construct a 6,000-foot-long earth dam embankment with compaction, and to conduct extensive foundation removal beneath it. The district (the project owner) provided bidders with plans and specifications, as well as soil borings and soil analyses on which the plans were based. Bidders were also provided with the opportunity to conduct further tests on their own.

In bidding the project, Brant Construction submitted a unit price (per cubic yard) for "common excavation." The bid for common excavation was derived by estimating the amount and cost of two types of excavation for the project—scraper excavation and dragline excavation, the latter of which was at least twice as costly as the former.

The district originally anticipated common excavation of approximately 518,718 cubic yards of material, but later revised the estimate to 596,192 cubic yards prior to any bidding. Of the latter amount, the district estimated that only 18,006 cubic yards of dragline excavation would be required. However, after the contractor conducted excavation to the specified level set forth in the plans (the "neat lines"), the district's chief engineer found the soil at that level unsuitable and ordered further excavation beyond the neat lines—i.e., "overexcavation." Ultimately, the contractor was required to overexcavate 67,804 cubic yards of soil, roughly 7 percent more than the approximate quantities set forth in the contract. Most of the overexcavation involved soils that had to be removed by the more costly drag line method of excavation.

The parties did not dispute the district's responsibility to compensate the contractor for the overexcavated material; rather, they disputed the appropriate measure of that compensation. The district maintained that it satisfied its obligation because it paid the basic contract unit price for the quantity of material actually excavated, including overexcavation. However, the contractor sought an additional $187,270.02 for the actual cost of the overexcavation, which it estimated to be $4.00 per cubic yard of soil removed—more than twice the unit price provided by the contract.

The contractor contended that the district had breached an implied warranty that the plans it furnished would be adequate to determine the quantity of material to be removed. The court, however, rejected this contention on the grounds that the district specifically stated that the plans were approximations and disclaimed that the quantity of material to be removed would be the ultimate quantity; accordingly, actual reliance by the contractor on the estimated quantities in submitting its bid would be immaterial "because such reliance would appear to have been unreasonable."

In *W. H. Lyman Construction Co. v. Village of Gurnee*, 403 N.E.2d 1325 (Ill. App. Ct. 1980), a contractor employed on a public sewer project encountered high subsurface hydrostatic pressures—a condition that was unanticipated by the project's contract documents, and which resulted in the contractor's inability to seal manhole bases, as designed, by the means permitted in the plans and specifications. Following completion of the project, the contractor filed suit against the village of Gurnee (the owner of the project) and the engineering firm that prepared the plans and specifications for the project, seeking compensation for extra costs incurred as a result of this alleged differing site condition. The contractor asserted claims against the village and the engineer for, among other things, breach of implied warranty of accuracy and sufficiency of the plans and specifications. Specifically, the contractor alleged that the plans and specifications included a manhole base which, when constructed and sealed according to the plans, was inadequate to withstand the subsurface hydrostatic pressures encountered.

The lower court dismissed the contractor's breach of warranty claim for failure to state a cause of action. The appellate court, however, reversed the lower court's decision, concluding that the complaint adequately alleged a cause of action against the village and the engineering firm for breach of implied warranty of accuracy and sufficiency of the plans and specifications with regard to plans and specifications for the manhole base. In support of its decision, the court stated:

> [W]hen a contractor agrees to construct something for a village—or the government, or an owner—and he is required to follow plans and specifications supplied by the village, and those plans are faulty thereby requiring the contractor to incur additional costs, the village must reimburse the contractor for damages incurred because of the breach of the implied warranty that the plans will be sufficient. This implied warranty has been described as akin to the condition implied in every construction contract that neither party will do anything to hinder the performance of the other party. If faulty specifications prevent or delay completion of the con-

> tract, the contractor is entitled to recover damages for the defendant's breach of the design professional's implied warranty. [Citation and internal quotation marks omitted.]

In its defense, the village argued that the contractor's breach of warranty claim failed, as a matter of law, because the project's contract documents contained the following exculpatory clause:

> 8. Infiltration Testing and Limits—In specifying the various types of pipe and joints which will be allowed and the methods of making same, the Engineer or Owner assumes no responsibility regarding the Contractor's ability to meet the infiltration limits as set forth, as it shall be the Contractor's sole responsibility to meet these infiltration limits and, if in the opinion of the Contractor, he cannot comply with the infiltration limits as set forth and still meet the specifications regarding the type of sewer pipe, the type of joints, or the method of making the joints, then the Contractor must indicate this in writing when submitting his proposal.

The appellate court, however, rejected the village's argument, stating that the exculpatory provision was an "impermissible attempt on the part of the village to shift the responsibility for the sufficiency and adequacy of the plans to the contractor, without providing the contractor the corresponding benefit of having something to say about the plan which he is strictly bound to follow."

Hardwick Brothers Co. v. United States, 36 Fed. Cl. 347 (1996), also involved a combined differing site conditions and defective design claim. The project at issue involved the construction of a system of levees, which was part of a three-phased comprehensive flood control project north of the Missouri River. Hardwick Brothers Company, a general contractor, entered into a construction contract with the Army Corps of Engineers (COE) for the construction of the project. In the litigation, Hardwick asserted approximately $4 million in claims, a significant portion of which related to alleged differing site conditions and defective design pertaining to "extremely wet conditions." Hardwick alleged that it was not sufficiently warned, or was materially misled, by positive representations in the contract documents regarding the potential job-site wetness and that, as a result, it was entitled to recover, on the alternative grounds of the differing site conditions clause and the defective design contained in those documents.

In support of its differing site conditions claims, Hardwick contended that the contract documents were inaccurate or insufficient in the following respects: soils at the site had a materially worse bearing capacity than indicated by the soil classes shown on the government's borings and, therefore, had unusual instability and workability problems when wet; the groundwater levels encountered during construction were substantially higher than indicated, and were even artesian; the numbers and sizes of swampy areas shown were deficient; the contract documents failed to warn that one lake level and the Missouri River levels could fluctuate significantly from the indicated levels; the contours shown were materially in error, thereby resulting in the ground in certain critical areas being lower than depicted, which, in turn, misled Hardwick as to potential inundation and drainage problems at the site; and the contract documents failed to contain sufficient indications that the project would require abnormal means of construction, such as floatation dosers and other equipment, thereby implying that operators with only the usual skills and standard earth-moving equipment could construct the levee.

In rejecting the differing site conditions claim, the court observed that the contract documents did not contain any affirmative representations that the groundwater, river, and lake levels were static or would not change from the levels shown on the contract documents. The court emphasized that in order to prevail on a Type I differing site conditions claim, a contractor must prove by a preponderance of the evidence that the conditions indicated in the contract documents differed materially from those it encountered during performance. The court found that the contract documents did not represent an express or implied representation or guarantee that water levels indicated would remain the same throughout the period of contract performance.

As stated above, Hardwick claimed that the contract documents were defective or incomplete with respect to job wetness conditions and, as a result, the COE breached implied warranties as to the adequacy of those documents. Specifically, Hardwick alleged that the contract documents were erroneous with respect to the topographic information and soil bearing data (showing areas within five feet of the surface and water tables), swamp and ponding areas, lake levels, and the types of equipment needed to perform the work, and that the contract documents contained various omissions, including omissions with respect to known and unknown backwater conditions, the slope of the river, river flows, hydrologic charts, and the exact location of the old Missouri River Channel.

The COE maintained that Hardwick was not entitled to any recovery on a theory of defective design, since the contract documents in question were merely performance (and not design) specifications. In addition, the COE maintained that it owed no implied warranty obligations to Hardwick for the same reasons that Hardwick's differing site conditions claim failed—i.e., the contract documents contain no positive or affirmative representations regarding any of the various factors relied on by the plaintiff in support of its claim, such as water levels, backwater, or rates of drainage. In addition, the COE maintained that the defective design claim failed with respect to the omitted information identified by Hardwick, since all of that information was either available to Hardwick, should have been known to it as an experienced levee contractor, or could have been derived by its own independent investigation.

In rejecting the defective design claim, the court discussed, in some detail, the distinction between design and performance specifications, and concluded that the specifications in question were, "in the main," performance specifications. In making that finding, the court stated:

> The Specifications did not direct the various types of equipment to be used or describe the way in which the contractor was to build the Project. The Specifications also left the solution or problems involving methods of dewatering, drainage and construction techniques up to the contractor. Further, it was given the option to use the uncompacted fill alternate. . . . In short, within its engineering judgment and expertise, Hardwick was given the discretion to chose the means and methods to construct the levee.

The court also found that the defective design claim could not succeed because the contract documents contained no positive or affirmative representations regarding river flows, lake levels, precipitation, drainage, ponding, groundwater, or backwater effect. In addition, the contract documents specifically informed bidders that they were to make their own assumptions regarding the effects of upriver releases and weather conditions on the work. The court further concluded that the drainage design in the contract documents was not a representation that the drainage would be at any particular rate, or that water would not remain for a period of time in ditches adjacent to the levee after heavy rains.

The court agreed with the COE that the alleged omissions from the contract documents largely reflected information relating to water level fluctuations that the contractor either knew or should have known as an experienced levee contractor. As such, the alleged omissions did not support a defective design claim. Specifically, the court found that Hardwick, as an experienced contractor in levee construction, should have known that the project would, or very likely could, incur all of the wetness conditions that it actually encountered—flooding (as defined in the contract), backwater, fluctuations, groundwater, lake levels, the existence and size of piezometric mound under the lake, river levels, Palmer Creek levels, inundation of large segments of the job site at less than bank-full conditions, wet soils, poor drainage, the existence of old river channels in the levee's alignment and high levels of precipitation. The court noted that if Hardwick did not know these things, as an experienced contractor, it could have discovered them simply by securing the services of a competent hydrologist. The court further concluded that Hardwick's difficulties were, in large part, attributable to its failure to take prudent investigatory actions, in conjunction with its unreasonable assumptions with regard to the static nature of various described water levels at the job site. Simply put, if Hardwick was misled, "it was not due to the alleged defects, but resulted from its own unreasonable assumptions and actions." Accordingly, the court found that Hardwick failed to prove that it was "misled by any defects in or omissions from the specification."

Finally, *Neal & Company, Inc. v. United States*, 36 Fed. Cl. 600 (1996), also involved a combined differing site conditions and defective design claim. In *Neal*, while the contractor did not prove entitlement on the basis of a differing site condition, it was able to recover under the theory of defective design.

The *Neal* case arose out of a construction contract between Neal & Company, Inc. (Neal) and the U.S. Coast Guard. The project at issue involved the construction of 30 units of base housing for the Coast Guard in Kodiak Island, Alaska. During construction, Neal encountered performance problems and delays due to the presence of "excessive water" on the site. The source of that water became the subject of a significant dispute between Neal and the Coast Guard, which ultimately led to the assertion of the differing site conditions and defective design claim.

According to Neal, the water was primarily subsurface moisture—water seeping out of the sides of excavations—and the infiltration of water was contrary to indications contained in the contract documents. Alternatively, Neal contended that to the extent that the water was surface runoff (for which it would otherwise be responsible under the contract documents), the water accumulated, according to Neal, as a result of a defective drainage design (furnished by the Coast Guard) that the contractor was required to implement. It was Neal's position that, by following the Coast Guard's grading plan, excessive water accumulated on the surface during construction. Finally, Neal maintained that under either theory—differing site conditions or defective design—these factors, in conjunction with excessive overexcavation, affected the construction schedule and resulted in delay.

More particularly, Neal contended that the excess water "seeped" in from the soils lining the sides of the excavations, and that this was "groundwater" or "subsurface water" (not surface water), and thus not the responsibility of Neal under the contract. According to Neal, the water source was unforeseeable and, thus, constituted a differing site condition. Neal also alleged that the surface water could not be properly controlled because of a defective design furnished by the Coast Guard.

In response, the Coast Guard maintained that most of the excessive water encountered was surface runoff, the control of which was Neal's contractual responsibility; the logs and geotechnical information furnished to Neal were sufficient for Neal to have foreseen the water seepage; the site design was adequate to handle the water that was encountered; and the problems encountered were the result of Neal's failure to properly grade or protect the site.

The court first addressed the differing site conditions claim. In so doing, the court noted that, in order to prevail on a Type I differing site conditions claim, the contractor must prove that the Coast Guard explicitly, or implicitly, indicated within the contract documents subsurface conditions materially different from those actually encountered. In that regard, Neal alleged that it interpreted the boring logs and the contract specifications to indicate a lack of subsurface water on the site, and that this interpretation was reasonable. The Coast Guard, however, maintained that Neal's interpretation was unreasonable, given the information available to bidders. Significantly, the Coast Guard noted that Neal did not review geotechnical reports and other related information available to Neal prior to execution of the construction contract. The court agreed with the Coast Guard, finding that Neal's failure in this regard was not reasonable.

In addition, Neal contended that the limited reference to water within the boring samples and the geotechnical report, in conjunction with the absence of information regarding subsurface water, was the equivalent of an affirmative representation by the Coast Guard that there would be no water problem on the site. The court, however, rejected this argument, stating that "[w]hile the bid documents may not have been as thorough as they could have been with respect to water conditions, the court concludes that the specifications failed to affirmatively indicate by implication, or otherwise show, the extent of subsurface water flow." The court emphasized that Neal's failure to review the geotechnical report was the result of its own "negligence." The court indicated that while that report may not have provided the most thorough investigation and analysis, it did provide prospective bidders with additional information regarding the amount of permeable soil and fill on the site, and the fact that the report indicated "wet" soil or "water seep" in three test pits should (or could) have provided Neal with enough information to do a more thorough prebid site investigation. Finally, the court noted that the geotechnical report, in conjunction with the boring logs, provided the bidders with enough information to, at the very least, warn them of the possibility of water seepage from the soft permeable ash and peat layers. Based on the foregoing, the court found that Neal failed to establish an affirmative representation by the Coast Guard that there would not be subsurface water flow into the site; nor had Neal established that the actual conditions differed materially from information represented in the contract documents. Accordingly, the court rejected Neal's claim, concluding that Neal failed to prove the existence of any entitlement on the theory of differing site condition.

Regarding Neal's defective design claim, Neal alleged that the Coast Guard design that it was required to implement contained significant site drainage flaws, and that the plan was defective in several respects. Specifically, Neal alleged that the design did not provide for adequate surface grading to channel water away from the houses and construction, and that the Coast Guard caused undue delay to the project by not responding in a timely manner to Neal's design clarification requests regarding proposed drainage solutions. In response, the Coast Guard maintained that the project could have been constructed as designed without difficulty, and that the excessive water and drainage problems encountered were the result of the poor construction practices of Neal and its earthwork subcontractor.

Upon reviewing the evidence with respect to the defective design claim, as well as basic principles of government contract law, the court observed that the Coast Guard, in furnishing the design, warranted the performability and constructability of the design. The court concluded that the contract documents contained significant grade and design errors and that the poor drainage design made it virtually impossible for the contractor to control surface water.

Thus, while Neal was not able to prove entitlement on the basis of a differing site condition due to the excess water, the court found that the same excess water conditions justified entitlement to recovery on a theory of defective design.

The various cases discussed in this Section demonstrate that contractor claims for differing site conditions are often combined with or alternatively predicated on allegations of defective design. In some instances, the allegations of defective design are triggered by inadequate information disclosure in the bid and/or contract documents; inaccurate, ambiguous, unclear, or otherwise inadequate expression of anticipated site conditions in the contract documents and/or in the data or interpretive reports furnished by the owner; inaccurate or incomplete design in the contract documents; and/or confusion or ambiguity as to whether specifications are design or performance in nature (and, therefore, uncertainty and conflict about whether the owner or contractor is responsible for design adequacy or construction means and methods). In addition, some of the cases demonstrate that the owner may be held responsible for contractor claims due to statements or actions of the design and/or construction management professional during the administration of the construction contract.

As discussed in more detail in Sections 10.1.1.2 and 10.2.2.1, and 10.2.5, in circumstances in which an owner is found legally liable (or otherwise obligated) to pay the contractor with respect to a claim based on allegations of defective design and/or statements or conduct of the owner's professional in the administration of the construction contract, there is a substantial probability that the owner will pursue a contractual, negligence, indemnification, or cost recovery claim against that professional.[78] In addition, as discussed in Sections 10.1.2 and 10.2.5, in circumstances involving those allegations, the professional is also exposed to a significant risk of third-party claims by any involved contractor or subcontractor.

As previously noted, in circumstances in which it is adjudicated that the contractor is entitled to an equitable adjustment (and/or in which the owner agrees to pay the contractor) based on a claim involving (in whole or in part) allegations of defective design, it is probable that the owner will pursue some form of cost recovery claim (based on the legal theories of negligence, breach of contract, or indemnification) against the professional that prepared the design. Similarly, in situations in which the contractor's claim was based and compensated, in whole or in part, on allegations involving some act, omission, statement, or conduct of the professional during the construction phase, the owner may also pursue such cost recovery claims against the professional. In the context of defending those cost recovery claims, the important distinctions between the professional's more limited negligence-based professional standard of care obligation (see Section 10.1.1.2) and the owner's implied warranty obligation must be emphasized. Again, defective design is not necessarily negligent design (i.e., defective design does not necessarily represent a departure from the professional standard of care), and the owner's implied warranty obligation is broader than the design professional's professional standard of care obligation (see Section 10.2.2.1(A)). Similarly, the appropriateness of acts, conducts, and statements of the design and construction management professional in the administration of the construction contract is also governed by the professional standard of care and, typically, expert testimony is required to establish a departure from that standard. (See Section 10.1.1.2 and Notes 14 and 15, *supra*.)

10.2.3 Design-Build Subsurface Projects: Risk Allocation Principles and Professional Liability Risk

In general, a design-build project is one in which a single entity is contractually responsible (typically, to the project owner) for both the design and construction of the project. There are several variations in the contractual arrangements under which design-build projects may be undertaken. In some of the more common of these alternate arrangements, the design professional may be: the prime entity responsible for both design and construction; part of a joint venture with the construction contractor and, in that capacity, jointly and severally responsible with the contractor as a prime for both design and construction; or a subconsultant to a construction contractor and responsible for the design and certain (contractually defined) services during the construction phase.

Virtually all of the same professional liability and risk allocation issues existing in the context of the traditional design-bid-build delivery method are present in design-build projects. For example, a clear definition of the scope of the design professional's services is essential and, especially with respect to the "build" component of the design-build undertaking, every reasonable effort should be made to fairly allocate risk for site conditions, such as through the inclusion of a differing site conditions clause in the construction documents. In addition, the design professional should endeavor to include contractual provisions defining performance and indemnity obligations in accordance with the negligence-based professional standard of care. The designer-builder who is part of the design-build team on a subsurface project bears all of the professional liability risk discussed in the preceding sections of this chapter, as well as additional professional liability risk (discussed below) due to the nature of the design-build undertaking.[79]

What is so different about the risk of professional liability for the design professional involved in a design-build subsurface project? The primary areas of increased professional liability for the professional involved in a design-build project are: (1) the potential that a warranty standard may be substituted for the negligence-based professional standard of care as the basis for judging the adequacy of the professional's performance; (2) the general diminution or elimination of the owner's implied warranty obligation and the transference of the risk associated with that obligation to the professional; and (3) the potential for an unduly restrictive role of the professional during the construction phase. In addition, the issue of increased professional liability should be approached from the standpoint of two distinct and independent types of involvement of the design professional: (1) the design professional who is prime, subconsultant, or, in some other capacity, a part of the design-build team; and (2) the design professional retained by the owner on a design-build project to perform predesign subsurface investigation and prepare associated preliminary data and/or interpretive reports; prepare preliminary or conceptual design for use in the design-build procurement process; and/or to represent (independent of the designer-builder) the interests of the owner in the review of the designer-builder's design submissions, shop drawings, or other submittals and/or in observing the performance of the construction work undertaken by the designer-builder.

10.2.3.1 Professional Standard of Care—Warranty Liability The professional liability risk of the design professional involved in the design-build delivery method will vary, depending on the type of contractual arrangement, the scope of services, and other risk allocation contractual provisions included in the design professional's agreement. However, regardless of all these potentially variable considerations and factors, there are certain common and constant issues regarding the potential for increased professional liability risk in the design-build method that must be considered.

There are several ways to approach this subject, but the most appropriate starting point is from the "top down"—i.e., starting with the expectations of the owner in the design-build method. In significant part, this is probably the most appropriate approach, since the owner's expectations in the design-build method (and those that motivate and underlie its selection of that method) will drive the owner's subsequent contractual discussions with the designer-builder regarding risk tolerance and allocation with respect to *all* aspects of the design-build undertaking, including those relating to the design professional's standard of performance. In addition, it is those same owner expectations that will form the basis of the owner's assessment of the professional's actual performance on the design-build project. As most professionals know or may intuitively sense, disappointed owner expectations (whether realistic or not) frequently lead to professional liability claims.

The design-build method requires that the delivered and completed project comply with the owner's program and expectations of performance. From the standpoint of the designer-builder, the legal obligation to meet those expectations and requirements amounts to a warranty obligation. Under the warranty obligation, the risk of inability to achieve the owner's required performance standards, as well as the risk of unanticipated or unforeseen events, shifts to the designer-builder. In the event that the completed project fails to meet those performance standards and requirements, the owner will look to the designer-builder for single-point responsibility.

In general, from the expectation standpoint of the owner, it does not matter whether the failure to meet its program or performance requirements is attributable to design or construction deficiencies. Under the design-build method, the owner (in legal, contractual, and practical effect) expects to be shielded from the economic, time, and performance failure risk, or other potentially "adverse" consequences associated with inadequate or deficient design, regardless of whether any of those consequences are the result of a departure from the professional standard of care.[80] Specifically, the design professional's adherence to the professional standard of care may not be sufficient to satisfy the design professional's warranty obligation to the owner. In addition, in sharp contrast to the owner's implied warranty obligation with respect to the adequacy of design furnished through it under the design-bid-build method (see Section

10.2.2.1(A)), the owner on a design-build project typically has no such implied warranty obligation to either the designer-builder or subcontractors with respect to design, because the owner neither controls the preparation of nor furnishes that design.

Many owners believe and expect that the design-build method will eliminate most, if not all, project risk for them, including any risk associated with design adequacy. While it may not be realistic for an owner to expect that utilization of the design-build method will eliminate virtually all project risk for the owner, it is accurate to state that the design-build method generally will result in a significant reduction in owner project risk and a transference of that risk to, and corresponding increase in the liability exposure of, the designer-builder for design and construction defects, or any combination thereof.

In order to meet these owner expectations, the designer-builder will be legally required to provide—either expressly or impliedly—certain warranties to the owner. The warranty liability of the designer-builder to the owner with respect to the construction or workmanship aspects of the design-build undertaking (as distinct from design) is similar to the warranty liability of the construction contractor under the traditional design-bid-build delivery method. More specifically, the standard industry construction contracts contain express warranties by the contractor, such as that materials used in the construction will be new, that workmanship will be of good quality and without defects, and that the work will conform to the contract documents.[81] In addition to express warranties, contractors typically are found to have impliedly warranted their work.[82]

While contractors generally are accustomed to providing express and implied warranties with respect to materials and workmanship, traditionally, those warranties have not extended to adequacy of design (since, typically, under the design-bid-build method, the design is furnished by the owner and prepared by a design professional under direct agreement with the owner).[83] In fact, in the design-bid-build method, the *owner* impliedly warrants the adequacy of design to the contractor (see Section 10.2.2.1(A)).

In addition, as discussed above, in the design-bid-build method, design professionals typically do not warrant that their design will meet specific or implied performance objectives of the owner. Similarly, design professionals generally are not subject to implied warranty obligations that the design they furnish will be "fit for its intended purpose." As a general matter, the design professional's design adequacy obligation in the design-bid-build method typically is governed by the negligence-based professional standard of care and not the higher obligations imposed by express or implied warranties (unless, of course, the design professional has agreed to contractually assume warranty obligations) (see Section 10.1.1.2.)

Based on the preceding discussion, it is clear that the designer-builder is exposed to significantly greater liability exposure for project-related defects (both design and construction) than if the project were designed and constructed under the design-bid-build method by a design professional and contractor independently and respectively retained by the owner. In contrast to the professional standard of care liability exposure of the design professional applicable under the design-bid-build method (see Section 10.1.1.2), the designer-builder may be liable under warranty law for design defects, including inadequate performance of the design, irrespective of whether the design was prepared in accordance with the professional standard of care and/or whether the construction work was performed in accordance with the contract documents. In addition, the owner owes no implied warranty obligation to the designer-builder with respect to design adequacy. Further, the designer-builder impliedly warrants the adequacy of the design to its subcontractors (i.e., under the design-bid-build method, the designer-builder assumes the position of the project owner, who furnishes design to others).

Generally, irrespective of involvement in the design-bid-build or design-build methods, design professionals should restrict the standard of their performance obligation to the negligence-based professional standard of care (see Sections 10.1.1.2 and 10.2.2.1(A)). While this recommendation and the underlying legal principles should not vary in the design-build project, the expectations and factors identified above may result in pressure on the design professional, either as prime designer-builder or subconsultant to a prime designer-builder, to agree to a modified and higher performance standard, which may constitute a warranty.

As a threshold matter, some contractors who are prime in the design-build relationship may attempt to transfer liability risk for design adequacy or failure of design to meet owner performance expectations by contractually enabling (through assignment of rights or otherwise) and/or requiring that the owner pursue claims for design adequacy or performance failure directly against the design professional. Although this option may be legally achievable, it may be perceived by many owners to undermine and circumvent the objective of single-point responsibility, which, for many owners, is one of the main ingredients and the "value-added" of the design-build method. A more viable approach for the contractor as the prime in the design-build relationship is to include contractual provisions in the prime agreement with the owner, which (1) limit liability for design deficiencies/performance failures to a departure from the negligence-based professional standard of care; and/or (2) limit owner remedies in the event of a failure to meet that standard of care to either reperformance or redesign to remedy the deficiency or, alternatively, to monetary recovery in a limited amount.[84]

The design professional, whether in the contractual position of a subconsultant to the prime designer-builder or as a prime or part of a prime design-build entity, should always endeavor to limit its liability exposure for design adequacy—performance failure—to a departure from the negligence-based professional standard of care. This limitation is entirely consistent with the scope of coverage afforded under most professional liability insurance policies issued to design professionals (see Note 3).

In addition to standard of care and risk allocation (e.g., indemnification or limitation of liability) issues between the design professional and others in the design-build arrangement, it is important that professional liability risk be identified, defined, and clearly allocated between the design professional and the designer-builder and the project owner.

Certainly, risk should be clearly allocated between the design professional and other members of the design-build team in both the teaming agreement (typically entered into prior to a team proposal) and in the agreement between those team participants (if selected). Although the designer-builder may be unwilling to limit the design professional's risk for design adequacy and/or performance failure to the professional standard of care if the designer-builder has risk exposure to the owner in excess of that standard, it is strongly recommended that the design professional insist on such a limited standard and/or include a limitation of liability or other provisions limiting the remedy for defective design to redesign rather than reimbursement of corrective, reconstruction or consequential damages due to defective design. In part, the rationale supporting that contractual approach is based on the relative reward (fee) of the design professional and the designer-builder, which must be taken into account in the risk allocation for any increased liability risk exposure for the design professional posed by the design-build method.

In addition to defining and limiting the risk of design adequacy/performance failure to the negligence-based professional standard of care, the design professional and contractor—i.e., the design-build team—should discuss other related issues, such as whether the design professional's professional liability insurance limits are adequate and, if not, who will pay the cost of the additional premium for increased limits of project insurance; how the deductible or self-insured retention obligation under the professional liability policy will be allocated between design professionals and the contractor; and how professional liability in excess of the available professional liability insurance will be allocated between them. The contractor must understand, however, that even if he or she contributes to the payment of the premium and/or to the deductible/self-insured-retention obligation, the professional liability insurer in all probability will maintain the right to control the conduct of the defense of any claims made under the policy, and settlement of any such claims will require the consent of the design professional.

Despite the increased risk of professional liability exposure, there are mitigating factors in connection with utilization of the design-build method that may serve to correspondingly mitigate the professional liability exposure, such as:

- Improved relations among design-build team members over those that typically exist in the traditional design-bid-method.
- Economic incentives of the design-build team to avoid conflict and dispute, especially among themselves and also with the owner.
- Substantially increased contractor involvement in and review of the design and contract documents (through development), particularly with a view

toward constructability issues and identifying and resolving ambiguities, omissions, and inconsistencies prior to the start of construction, thereby reducing the likelihood of the disruptive cost and time impact from such issues during the construction process.

- A less "defensive" approach to design; the design professional need not prepare overly conservative design and contract documents in anticipation of the "low bidder"—who may bid or otherwise approach the project from an overly optimistic and/or adversarial perspective.

10.2.3.2 Owner Involvement in Design of Design-Build Projects

Despite a general transference of design adequacy risk from owner to designer-builder, there may be circumstances that result in owner retention of that risk or the reallocation of that risk to the owner. These circumstances include situations where an owner furnishes specifications or criteria (whether prescriptive or performance) and/or the conceptual design to the designer-builder, either as part of the procurement process or otherwise. The *more detailed* the specifications and design furnished by the owner, and the *less discretion* given to the designer-builder to deviate from those specifications or design, the more likely that the owner will retain some level of risk and responsibility for the adequacy of that design.[85] Similarly, the more involved the owner becomes in reviewing, commenting on and/or directing changes to the design developed by the designer-builder, the greater the likelihood that the owner will assume some degree of risk for design adequacy. The degree of owner involvement during construction in the submittal review and/or proposed substitution review/approval process may also affect the design risk exposure of the owner. Finally, the extent to which the owner furnishes geotechnical data and/or interprative reports to the designer-builder with the expectation that the designer-builder (and/or the design professional subconsultant) will rely on that information in the design and/or construction of the project may also result in a level of design adequacy risk for the owner (and/or its independently retained geotechnical engineer), assuming, of course, reasonable reliance on those data or reports.

Simply put, the greater degree of involvement of the owner in the design development process and/or in the review, evaluation, and direction of the design-builder's design, the greater the risk of owner responsibility for design adequacy.[86]

10.2.3.3 Professional Liability Risk During Construction Phase

Much has been written about the professional liability risk and claims associated with the design professional's role during the construction phase of the design-bid-build project.[87] In addition to the professional liability risks inherent in the design professional's construction phase role on the design-bid-build project, design professionals involved in design-build projects face the potential of increased professional liability exposure, depending on their scope of services and contractual arrangement.[88]

In general, the more involved the design professional is in the design, direction, and/or supervision of construction means and methods, the greater the potential liability exposure of the design professional for constructability and/or safety-related problems that arise during the construction phase (see Section 10.2.2.2(B)). In addition, involvement in construction means and methods and/or supervision of construction may result in liability under OSHA regulations and environmental laws.[89]

In addition to these objectively definable increased risks of "construction phase" professional liability exposures for the design professional involved in design-build projects, other liability exposures may result from disappointed owner expectations as to whose interests the design professional (who is part of the design-build team) represents during the construction phase of a design-build project. Specifically, despite the (contractual) fact that the contractor or a separate design-build entity has retained the design professional, or despite the existence of a design-build entity in which the financial and other interests of the designer and builder are merged, an owner may expect that the design professional will act as its agent or representative in administering the construction, as is the case in the traditional design-bid-build method, particularly with respect to the review and approval of contractor payment requisitions, contractor and subcontractor shop drawings, and other submittals; the review and approval of contractor-proposed submissions or value engineering proposals; and the observation of construction for conformance with the contract documents prepared by the owner, as well as contract documents prepared by the designer-builder. These expectations may not match the practical or economic reality, as well as the accountability and allegiance established by the existence of a direct contractual relationship between the design professional and someone other than the project owner, and, for that reason, may subject the design professional to the risk of claims based on conflicting loyalties.[90]

During the construction phase, the role of the design professional who is part of the design-build team may be significantly curtailed from the scope of services provided by the design professional (independently retained by the owner) in the traditional design-bid-build method. For example, the designer-builder may not retain its design professional to provide traditional construction administration services, such as shop drawing or submittal review, construction observation, approval of substitutions, and/or review and approval of payment requisitions. These limitations on the scope of the design professional's services, in most situations, result in a corresponding increase in professional liability exposure. In circumstances in which an owner expects that the design-build design professional will serve as its agent and representative, the professional liability exposure is yet further increased when the design professional's role during the construction phase is either precluded or significantly restricted (when compared with the role of the design professional on the traditional design-bid-build method).

The Eighth Circuit decision, *C.L. Maddox, Inc. v. Benham Group, et al.*, 88 F.3d 592 (8th Cir. 1996), demonstrates the importance of a well-drafted scope of services provision in the contractual arrangement of a design professional involved in a design-build project. *Maddox* involved a dispute between a designer-builder and its design professional arising out of a project that involved the extensive remodeling of an electrical power plant's coal processing system. In March 1990, the owner of the electrical power plant began soliciting bids from general contractors for this design-build project. C.L. Maddox, Inc. (Maddox), a contractor, responded to the solicitation.

To assist it in preparing its formal proposal, Maddox entered into an agreement with The Benham Group, Inc. (Benham), an engineering firm, pursuant to which Benham was to perform studies for Maddox of equipment layouts and equipment sizing, supply Maddox with information and assistance necessary to prepare a final construction cost for the project, and develop a final lump-sum engineering cost. Relying heavily on the information provided by Benham, Maddox prepared and submitted its proposal to the owner. Maddox was awarded the project.

In September 1990, Maddox engaged Benham to perform much of the design services required for the project. Pursuant to the terms of the agreement, in addition to agreeing to prepare the construction drawings for the project, Benham agreed to furnish Maddox with all cost estimating required for the project, and agreed that it would keep Maddox "informed of the progress and quality of the Work," and that it would "endeavor to guard Maddox against defects and deficiencies in the Work of [Maddox]."

The project did not progress well, as it was fraught with delays and cost overruns. Consequently, Maddox filed suit against Benham for breach of contract, seeking to recover from Benham over $5 million in damages allegedly due to bidding errors, design errors, and time delays, as well as Benham's alleged breach of its contractual duty to "guard" Maddox against Maddox's own deficiencies on the project.

At trial, Maddox acknowledged that approximately $1.2 million of the damages that it sought was attributable to errors by Maddox, or by the owner of the project. Maddox, however, argued that Benham was liable for this amount because Benham's subcontract with Maddox required Benham to "guard" Maddox against Maddox's own deficiencies. The jury agreed with Maddox's position, and returned a verdict in favor of Maddox and against Benham on all of Maddox's claims, awarding Maddox the full amount of damages sought.

On appeal, the Eighth Circuit reversed the portion of the jury's verdict which held Benham liable for breaching its contractual duty to "guard" Maddox against its own deficiencies. The Eighth Circuit noted that, although Benham agreed to "endeavor to guard [Maddox] against defects and designs in the Work of [Maddox]," the subcontract also specifically provided that Benham was not responsible for the "acts or omissions of Maddox":

> [Benham] shall not have control or charge of and shall not be responsible for construction means, methods, techniques, sequences or procedures ... for the acts or omissions of [Maddox], [Maddox's] subcontractors or any other persons performing any of the Work, or for the failure of any of them to carry out the Work in accordance with the Construction Documents.

Thus, as the scope of Benham's responsibilities under its agreement with Maddox expressly negated any responsibility on Benham's part for Maddox's acts or omissions, the Eighth Circuit refused to shift to Benham the risk that Maddox would not properly perform its obligations under its contract with the owner, and reversed the jury's award on that issue.[91]

The preceding discussion highlights the importance of clearly defining expectations regarding the design professional's representative role, especially during the construction phase of a design-build project. The risk of professional liability attributable to disappointed or unrealistic expectations and allegations of "conflict of interest" is real and substantial in the design-build method. This issue must be addressed and resolved early in the establishment of the design-build relationship.

10.2.3.4 "Bridging": Professional Liability Implications As noted in Section 10.2.3, the role of the professional in the design-build method—i.e., whether the professional is part of the design-build team or an independent consultant to the owner—will also affect the potential professional liability exposure. On some design-build projects, owners retain their own design and/or construction management professional (the "owner's consultant"), who is independent of the designer-builder, to serve as their advisor and representative with respect to a number of activities. Some commentators have used the term "bridging" to describe the role of the owner's consultant.

In substantial part, the perceived need for such an owner's consultant in the design-build method arises because the design professional who is part of the design-build team typically does not function as a representative or agent of the owner (as is the case in the traditional design-bid-build method). As one noted commentator on the design and construction process has observed:

> [Design-build] has weaknesses. The absence of an *independent* design professional selected by the owner can deprive the owner of the widest opportunities for good design. An unsophisticated owner often lacks the skill to determine whether the contractor is doing the job well or as promised. This can reflect itself not only in substandard work but also in excessive payments being made early in the project or slow payment or nonpayments of subcontractors.

J. Sweet, *Legal Aspects of Architecture, Engineering and the Construction Process*, 5th ed. (St. Paul, MN: West Publishing Co. 1994), p. 368.

The concept of the owner's consultant is intended to provide the owner with independent professional representation in the design and construction of a design-build project. While there are a variety of services that the owner's consultant may provide, this section will focus on those that pertain to the points of interface of that consultant with the designer-builder, especially with the design professional who is part of the design-build team.

In general, the owner's consultant serves as the owner's agent and representative in defining and communicating the owner's program; developing conceptual design, preliminary design development and performance criteria, and design standards; preparing a request for proposal (RFP) or other solicitation for the design-build procurement, and generally assisting in the design-build procurement and selection process; and interfacing with the designer-builder in the design development, contract document preparation, and construction phases of the project. In the latter respect, the owner's consultant may be involved in the review of the designer-builder's design development submissions and contract documents, review of shop drawings or other submittals during construction, and observation of construction for conformance with the owner's programatic intent and the compliance with the designer-builder's contract documents, and/or review approval of payment requisitions of the designer-builder.

This "bridging" approach—developed by George Heery in approximately 1976—represents a hybrid between the traditional design-bid-build method and the design-build method. Under the bridging approach, the owner achieves the advantages of single-point responsibility from the designer-builder for design and construction, which is the hallmark of the design-build method, while, at the same time, obtaining independent professional representation from the owner's consultant (who, by definition, is not a part of the design-build team).

There are several professional liability issues for design professionals in connection with the bridging approach. These issues will vary, depending primarily on (1) the nature and scope of the design professional's role and contractual scope of services and (2) whether the design professional is serving as the owner's consultant or part of the design-build team.

(A) Design Professional as Owner's Consultant From the standpoint of the design professional who is an owner's consultant on a design-build project, the following issues should be considered:

1. If the owner's consultant is retained to develop conceptual, preliminary, or design development information, including design criteria, standards, or performance requirements, which will become part of the design-build RFP and/or establish a standard or baseline on which the designer-builder's final design must conform, then it is important to define the nature and extent of the owner's consultant's liability for inaccuracies, errors, or omissions in that preliminary work product, especially if the owner's consultant will have no role in the review and evaluation of the designer-builder's design and construction documents and/or no role during the construction phase of the project. In addition, the extent to which the designer-builder is entitled to rely on the accuracy of any preliminary design information or concepts developed and furnished by the owner's consultant should also be addressed, along with any appropriate qualifications or limitations on that reliance. Similarly, it may be prudent to require that the designer-builder, in responding to the RFP, advise of any perceived ambiguities, inaccuracies, or other deficiencies or flaws in the

preliminary design information or concepts developed and furnished by the owner's consultant.

2. If the owner's consultant is to have any role in the review and/or approval of the designer-builder's design and construction documents, the nature and extent of that review/approval role should be clearly defined and qualified (e.g., solely for conformance with preliminary or conceptual design requirements). In order to maintain and preserve the concept of single-point responsibility, it is important that the designer-builder's agreement make clear that the owner's consultant's review and/or approval of the designer-builder's design and construction document submissions will not relieve the designer-builder (and/or the design professional who will ultimately stamp the documents) of final design responsibility. The owner's consultant should not stamp or seal contract documents prepared by the designer-builder or its design professional (or any design professional who is part of the design-build team).
3. While the owner's consultant—in representing the interests of the owner—may attempt to influence the designer-builder's design decisions, the owner's consultant should not impose his or her design preferences or judgments on the designer-builder or otherwise seek to limit the reasonable degree (i.e., as defined contractually or by professional practice standards) of design discretion of the designer-builder. The greater the intrusive and/or dominant role of the owner's consultant in the design development, review, and approval process, the greater the risk of professional liability for the owner's consultant, and the greater the risk that design adequacy risk (and the associated implied warranty obligations) will be reallocated (at least to some degree) to the project owner.
4. During the construction phase, the owner's consultant may be retained to provide a variety of services as agent and representative of the owner, including shop drawing or submittal review, construction observation and/or review and approval of the designer-builder's payment requisitions. If the owner's consultant is retained to review and/or approve the designer-builder's shop drawings or other submittals, it is recommended that any such review be undertaken only after the designer-builder (including the design professional) has reviewed and approved the submissions. The owner's consultant's agreement and the contract documents should make clear that the owner's consultant's review and approval of such submittals is undertaken and performed solely for the benefit of the owner; and the owner's consultant should not engage in communication with any subcontractors, trade contractors, and/or vendors regarding the subject of such submittals. Similarly, with respect to any construction observation services, the owner's consultant should make clear that those services are performed solely for the benefit of the owner and do not relieve the designer-builder of its own independent and

contractual responsibilities for the construction of the project in accordance with the approved contract documents.

(B) Design Professional as Part of the Design-Build Team Professional liability issues for the design professional who is part of the design-build team on a project utilizing the bridging approach arise out of the interface role between the owner's consultant and the design-build design professional. The more important of these issues may be summarized as follows:

1. Although the owner—through the design-build RFP or otherwise—may furnish preliminary or conceptual design criteria, requirements, or standards to the designer-builder, ultimate responsibility for the adequacy of the design rests with the design professional who is part of the design-build team and who will stamp the contract documents. Accordingly, if the design-build design professional perceives the existence of fundamental problems with the preliminary or conceptual criteria, information, requirements, or standards, he or she should identify and take exception to the problems in the response to the RFP or, depending on the nature and scope of the problems, decline to respond. The basic point is that the design-build design professional will be entirely responsible for the final design, and reliance on preliminary or conceptual design or standards prepared by the owner's consultant generally will not relieve the design-build design professional of that responsibility.
2. While the owner's consultant may review and/or approve design development submissions and/or the final design and construction documents prepared by the design-build design professional, that review/approval role generally will not relieve the design-build design professional of responsibility for the adequacy (errors/omissions, completeness, code compliance, constructability) of the final design.
3. During construction, the design-build design professional should have the primary administrative role in addressing design issues, interpreting the contract documents, reviewing shop drawings or other contractor submittals, and communicating with the designer-builder and/or trade contractors. The owner's consultant should not interfere with the design-build design professional's ability to perform that primary role.

10.2.3.5 Professional Liability Risk—Subsurface Conditions In the traditional design-bid-build method, the contractor under the so-called common law rule (see Section 10.2.1) bears the performance, temporal, and economic risk associated with encountering physical conditions at the site. However, in virtually all public—and many private—projects, that risk, to a limited degree, typically is shifted to the owner through the use of a differing site conditions clause under which (assuming satisfaction of certain requirements, such as notice) the owner assumes the risk of encountering conditions at the site that differ materially from those indicated in the contract documents or those of an unusual or unknown nature not ordinarily encountered in the type of work being constructed. (See Section 10.2.1.1). Certainly, every reasonable effort should be undertaken, in both design-bid-build and design-build projects, to fairly allocate the risk of subsurface conditions between owner and other project participants involved in the design and construction of the project (see Section 10.2.2.2).

One commentator has raised certain issues about the use of design-build in certain subsurface projects, such as underground tunnels, that affect the ability of the designer-builder to establish price, noting the significant number of "unknowns," the need for extensive geotechnical investigation to determine price, and difficulty of specifying and controlling construction and underground structures that may respond to subsurface conditions in the absence of 100 percent design documents. The same commentator has questioned whether construction quality standards will be compromised on a design-build project in the absence of a clear definition of those standards in the design-build agreement.[92]

What is so different about subsurface conditions risks for the design professional involved in a design-build project? Several important factors may affect design professional liability risk for subsurface conditions on design-build projects, such as:

1. To what extent, if any, is the designer-builder entitled to rely on pre-design site investigation data and/or interpretative reports furnished to it by the owner and prepared by a geotechnical engineer independently retained by the owner? In some instances, the owner may retain a geotechnical engineer to perform pre-design site investigation and prepare appropriate data and/or interpretive reports. In some cases, the data and/or reports are made available to the designer-builder as part of the design-build RFP, or otherwise. In addition, when those data and/or reports are made available, the owner may disclaim the accuracy and/or completeness of the information and preclude the designer-builder's reliance on it. From the standpoint of the designer-builder, the data and/or interpretative reports may contain information and/or assumptions that will be important to both the design and construction means and methods anticipated to be employed in the construction process, as well as other information that will form the basis of the designer-builder's estimated (bid) cost for the project. For those reasons, it is especially prudent for the designer-builder to ascertain the extent to which it may rely on such data and/or reports and, in addition, to confirm, as part of the design-build contractual undertaking, that certain baseline assumptions have been established as part of the design-build agreement; the baseline assumptions should be predicated on the data and/or reports, and the design-build agreement should provide that a material (or otherwise defined) variation from those baseline assumptions may affect the cost and/or time and result in an equitable adjustment. However, owner use of disclaimers of data and/or interpretive reports that may be furnished to the designer-builder is generally

consistent with the owner expectation that the designer-builder will have single-point responsibility and substantially all risk for all aspects of the design and construction process. Accordingly, it is expected that many owners will utilize disclaimers in furnishing such data and reports, and the designer-builder should be aware of the substantial contingent risk associated with the use of such disclaimers (see Section 10.2.1.1, and Notes 39-41).

2. Regardless of whether the owner furnishes to the designer-builder any subsurface data and/or geotechnical interpretive reports based on pre-design site investigation, in all probability it will be necessary for the design-build team to include a geotechnical engineer. It is generally recognized that geotechnical engineers serve an important "observational" role during the construction process, and the performance of that role may result in the need or advisability for changes in design assumptions, or in anticipated construction means and methods indicated and/or recommended in preliminary data or interpretive reports that may have been furnished by the owner. The extent to which the geotechnical engineer and/or other design professionals who are part of the design-build team may be required to communicate or otherwise interface with the owner's independent geotechnical engineer (if one exists) should be clearly ascertained and defined in the designer-builder's agreement with the owner. In any event, the designer-builder will need to retain a geotechnical engineer as part of the design-build team, and that will result in some increased professional liability exposure for the designer-builder for subsurface conditions risk (in contrast to the corresponding more limited risk associated with the traditional design-bid-build method in which the owner typically retains the geotechnical engineer and in which the design professional is entitled to rely on the completeness and accuracy of reports furnished to it). In addition, the designer-builder will be responsible for coordination of data and reports produced by its geotechnical engineer with the design prepared by other design professional members of the design-build team.

The case of *Nicholson & Loup, Inc. v. Carl E. Woodward, Inc.*, 596 So. 2d 374 (La. Ct. App. 1992) illustrates some of the professional liability issues related to site conditions for design professionals involved in the design-build method. In *Nicholson*, the owner of a supermarket designed and constructed under the design-build method asserted claims against the design-build entity that had contracted with the owner to design and construct the supermarket; the in-house architect employee of the designer-builder; and the geotechnical engineering firm that had been retained by the project owner to perform pre-design site investigation and to prepare a preliminary interpretive report for use in the design of the project. The primary liability issue in *Nicholson* was whether any of the design professionals was responsible for design deficiencies that resulted in unacceptable differential settlement in the floor slab of the supermarket.

Prior to entering into the contract with the designer-builder, the owner had retained a geotechnical engineer to perform a "subsoil investigation" of the site of the proposed supermarket. The owner's geotechnical engineer did two 70-foot soil borings on diagonally opposite corners of the site and prepared a seven-page report. The report listed pile capacities for numerous different-length piles and indicated that a combination of 40-foot piles supporting the interior space and 70-foot piles supporting the exterior and other weight-bearing walls would be appropriate. In addition, the report recommended that all loads be pile-supported and indicated that settlement would not be a problem in pile-supported locations. Although the report discussed the fill material required for grade or drainage purposes, it did not indicate that the introduction of additional fill could cause any settlement problems.

The report prepared by the owner's geotechnical engineer was forwarded to and used by the designer-builder's in-house design professionals in the design of the supermarket. The owner's geotechnical engineer had no involvement during the construction phase.

A topographical survey of the site indicated that it was approximately five feet lower than an adjacent highway. The owner told the designer-builder that it wanted the supermarket to be "even with the highway." In compliance with that request, the designer-builder introduced and placed four to six feet of fill on the site prior to construction of the supermarket. The owner's geotechnical engineer was not consulted with respect to the introduction or placement of that fill.

As constructed, the supermarket was supported by a combination of 40-foot and 70-foot piles, as described in the report prepared by the owner's geotechnical engineer.

Shortly after the supermarket opened, the owner noticed problems with the supermarket building, which were eventually determined to be caused by differential settlement in the building slab. In some places, the differential settlement in the building slab was as much as six inches in a 15- to 20-foot span, with the average differential settlement between four and five inches.

There was substantial conflicting expert testimony at trial; however, *all* of the expert testimony indicated that the differential settlement was caused by the following: (1) the soft compressible soils at the site; (2) the different length piles; and, (3) the addition of four to six feet of fill at the site. The evidence further indicated that the presence of the four to six feet of fill caused the underlying soils to consolidate and that the 40-foot piles moved down with the soil due to a phenomenon known as "down-drag." The 70-foot piles did not move down with the soil to the same degree, because the tips of those piles sat on top of the sand layer. As a result of the differential settlement of the 40- and 70-foot piles, the floor slab settled differentially. Finally, the testimony indicated that the differential settlement would not have occurred if the building had been constructed with uniform length piles, because all piles would have moved down uniformly; and that differential settlement would not have occurred if the fill material had not been placed because it was the weight of the fill that caused the upper layers of the soil to consolidate.

The trial court found all of the defendants—i.e., the designer-builder, the owner's geotechnical engineer, and the designer-builder's in-house architect—liable, and the Louisiana Appellate Court affirmed that judgment.

The owner's geotechnical engineer had argued that it should not be liable because it had no obligation to mention whether fill could or should be introduced and placed at the site—particularly since no one ever informed the owner's geotechnical engineer that the designer intended to add a large amount of fill to the site—and because either the designer-builder or its design professional team should have been able to determine, based on the owner's geotechnical engineer's report, that the introduction of a large amount of fill would cause problems. The court rejected that argument on the basis that the expert testimony was virtually unanimous that the geotechnical engineer's report should have addressed the effect of fill on the existing soils at the site and that the recommendation of different length piles was inappropriate under the circumstances.

The designer-builder had argued that it should not be liable since there was no evidence of any construction (as distinct from design) defects. Although the court agreed that there was no evidence of construction defects, the court ruled that the distinction between design and construction defects was irrelevant and misplaced, since the designer-builder was contractually obligated to both design and build.

In support of their argument that they should not be liable for defective design, the designer-builder and its in-house architect argued that the soil report prepared by the owner's geotechnical engineer was so defective that it was impossible "to properly design the building." The court, however, rejected that argument on the basis of expert testimony that the designer-builder's structural engineer should have contacted the owner's geotechnical engineer prior to the start of construction regarding the proposed introduction of fill, and that a reasonably prudent structural engineer should have been on notice that down-drag could be a problem, given the proposed introduction of fill and the differing lengths of piles contemplated in both the soils report and the proposed design. In addition, further expert testimony supported the view that the designer should have contacted the owner's geotechnical engineer if the designer intended to introduce a "significant amount of fill," and that the designer should have communicated with the owner's geotechnical engineer, especially concerning "any variation . . . not spelled out" in the report or if the designer found anything in the soil report that was vague or ambiguous. Moreover, the expert of the owner's geotechnical engineer testified that no one ever contacted the geotechnical engineer about the introduction of additional fill to the site and, had it been so contacted, it would have advised the designer-builder of possible problems with down-drag and would have provided the designer-builder's design team with alternatives.

The *Nicholson* decision illustrates that the designer-builder is held fully responsible for both design and construction of the project. In addition, *Nichol-*

son demonstrates the importance of interaction and communication between the design professionals who are part of the design-build team, and the owner's geotechnical engineer (as well as other design professionals retained directly by the owner). *Nicholson* further demonstrates the need for involvement of the owner's geotechnical engineer during the design development and construction phases, especially if the expectation is that the owner's pre-design data and/or reports will be the only substantive information and/or interpretive reports to be utilized in the design and construction process.

The case of *Appeals of Pitt-Desmoines, Inc.*, 96-1 BCA (CCH) ¶27,941 (Sept. 26, 1995) (*PDM*), illustrates other professional liability issues related to site conditions in the design-build context, and the responsibilities of the designer-builder entity during the conceptual, preproposal phase of a project as opposed to the design phase of a project. In *PDM*, the government awarded PDM a design-build contract to develop, fabricate, and install a large acoustical tank in an existing building known as "Building 5" at a naval research laboratory. After the contract award, PDM discovered that Building 5 was much heavier than indicated in the contract documents, which necessitated a costly redesign of the building's support system. PDM also discovered that an adjacent building, known as "Building 43" (which was also owned by the government), was in such a deteriorated condition that PDM's dewatering plan, and its performance of its work in accordance with its auger cast pile design for Building 5, risked collapsing Building 43. Consequently, PDM asserted differing site conditions claims against the government based on these site conditions, which it discovered after the contract award, but during the design phase of its services.

The Board of Contract Appeals found that there was no evidence that PDM miscalculated or misread any of the bid documents, and that PDM acted reasonably during the preproposal phase of the project. The board, however, found that PDM did not conduct a reasonable post-award geotechnical investigation. The evidence showed that shortly after it was awarded the contract, PDM hired an independent consultant to perform a geotechnical investigation; the investigation, however, did not adequately address the impact of PDM's auger cast pile design and dewatering plan on the adjacent buildings. It was not until three months after the geotechnical investigation that PDM discovered the deteriorated condition of Building 43, and the potential risk of damaging or collapsing Building 43 as a result of its performance of work on Building 5.

The board held that although PDM, as a designer-builder, had no duty during the preproposal phase of project to conduct technical investigations to determine the accuracy of drawings provided by the government with its original request for proposal solicitations, PDM had a duty during the design phase to perform a reasonable, timely post-award site investigation, which included a geotechnical investigation. According to the board, a reasonable geotechnical investigation would have encompassed an investigation of the foundations, conditions, and ages of the existing building on the project site and any adjacent buildings that could come within the influence of the work. The board concluded that PDM did not conduct a reasonable post-award site investigation, and, therefore, it was not entitled to damages incurred between the time it should have discovered the subsurface conditions of Building 43 and the time that it actually discovered the conditions. Thus, although PDM was entitled to an equitable adjustment due to differing site conditions, PDM's recovery was limited in a manner that took into account the higher responsibilities of a designer-builder in connection with geotechnical investigations and other related inquires of subsurface conditions. That is, PDM was not allowed to recover for cost incurred *during the design phase* of contract performance *after* the point in time when PDM should have discovered the differing subsurface conditions had PDM conducted a reasonable post-award investigation.

One approach in dealing with the subject of risk allocation for subsurface site conditions on a design-build project would be for the owner, and/or the designer-builder together with the owner, to establish a set of baseline geotechnical assumptions for the design and construction of the project. The baseline geotechnical assumptions would be established utilizing either existing site data or reports and/or other reasonably reliable site-specific information. In this regard, professional liability and other considerations involved in the use of the baseline concept are not different from those involved in the traditional design-bid-build method.[93]

10.2.4 Relationship of Risk Allocation for Subsurface Conditions Between Owner and Contractor to Risk of Professional Liability

Subsurface conditions involve a high degree of risk potential for all project participants—especially and primarily the owner and contractor but also, to a significant degree, for the design and construction management professional. Although the owner and contractor are the primary "stakeholders" in the risk allocation process for subsurface conditions, the failure of those parties to identify and fairly allocate risk between them will serve to significantly increase the potential of liability risk exposure for the design and construction management professional.[94]

Much has been written about the salutary goal of fairly allocating risk between the owner and contractor in general, and in the context of subsurface conditions in particular.[95] Subsurface conditions afford an opportunity for owners to unfairly impose on contractors the risk of all unforeseen and unforeseeable conditions (for example, by failing to include a differing site conditions clause in the contract documents), the burden of defining and performing all required preconstruction site investigations, the obligation to correctly interpret any and all subsurface reports, data, or other information furnished or made available to the contractor (through the indiscriminate and wholesale utilization of disclaimer-type provisions), and/or the assumption of all cost and performance problems due to time-related impacts occasioned by the discovery of differing site conditions (through the inclusion of no damage for delay clauses). Alternatively, contractors may underbid subsurface projects in the unrealistic or overly optimistic expectation that conditions more favorable than those indicated in the contract documents will be encountered; may bid a project without conducting any site inspection or in disregard of disclosed or available information regarding site conditions; or may enter into a contract that does not contain a differing site conditions provision and carry no contingency for the risks associated with differing site conditions.

Risk in connection with the design and construction of subsurface projects does not automatically lend itself to a neat and precise allocation between owner and contractor. Decisions, actions and inactions, and contract terms between the owner and contractor may have a significant impact—legally or pragmatically—not only on themselves, but also on the professional liability risk of the professional involved in the subsurface project. For example, an owner's decision about the scope of preconstruction subsurface investigation may materially affect the risk of defective design, and thereby enhance potential professional liability exposure, as well as the owner's liability exposure for differing site conditions claims by the contractor, and the contractor's potential liability for increased cost and time for performance. Similarly, the owner's decision not to include a differing site conditions clause in the contract documents, or to include a differing site conditions clause but disclaim the accuracy or completeness of subsurface information furnished or made available to the contractor, may leave the contractor without a contractual remedy for a valid differing site conditions claim and thereby invite (if not encourage, out of a sense of frustration) a third-party claim by the contractor against the professional based on the theories of negligence and/or negligent misrepresentation.[96]

Unquestionably, most of the important decisions that may materially affect the potential of professional liability risks are *owner* decisions. In light of the critical importance and centrality of owner decisions regarding issues that affect risk allocation for subsurface conditions—among owner, contractor, and professional—professionals should become proactive in educating owners concerning the relevant risk allocation and associated liability factors and considerations.

Although owners are the primary decision makers in the process of identifying and allocating risk for subsurface conditions, the preceding discussion demonstrates that both owners and contractors have the ability, by their conduct and agreements, to allocate and accept such risk as they consider appropriate. When judged with the benefit of hindsight, any particular agreement between the owner and contractor with respect to risk allocation for subsurface conditions (such as where a contractor enters into an agreement either with no differing site conditions clause and/or with disclaimers of site information (see Section 10.2.1)) may appear unfair in light of actual conditions encountered or the actual performance, time, or cost burdens encountered by the contractor due to those conditions. However, the agreement is the agreement, and that should end the matter between the contracting parties (i.e., the owner and contractor). Moreover, contracting parties should not be allowed to circumvent limitations on recovery imposed by their agreements through the pursuit of third-party claims against design and construction management professionals. The absence of any adequate and effective contractual remedy by the contractor against the owner generally should not give rise to negligence, negligent misrepresentation, or other third-party claims against the professional; to allow such claims to be maintained against the professional would allow the contractor to circumvent and avoid the contractual terms and limitations under which it bid and agreed to accept risk on the project.[97]

10.2.5 Professional Liability Claims in the Specific Context of Subsurface Projects

As evidenced by the preceding discussion, the discovery of alleged and/or actual differing site conditions frequently generates disappointed financial and performance expectations and, even more unfortunately, results in disputes, claims, and litigation between owners and contractors, all of which are resolved with varying degrees of "success" and "fairness," depending on the particular factual circumstances, including the nature and quality of information disclosed; the existence or not of a differing site conditions clause; the use and/or clarity of baselines for the definition of contractual assumptions and the application of a differing site conditions clause; the inclusion of disclaimer-type provisions in the contract documents; the reasonableness of the contractor's assumptions in bidding the project; and the dispute resolution mechanisms available to the parties.

It is equally accurate to observe that claims are frequently asserted by owners and contractors, subcontractors, and other third parties against design and construction management professionals arising out of the discovery of actual or alleged differing site conditions. As one commentator has observed:

> Other parties may be liable as well. The architect or engineer who prepared the plans and specifications is often named as a defendant in differing site condition cases. Because there is usually no contract between the contractor and the architect, the contractor's action is typically based upon negligence or some other tort, and cases are split regarding the liability of the architect in these circumstances. If the Owner is held liable to the contractor as a result of an unforeseen differing site condition, the owner may seek indemnity from the architect, who prepared the plans and specifications or who communicated in correct data to the contractor.
>
> The ultimate liability for damages resulting from an unforeseen site condition may also rest upon consultants hired by the owner, architect or contractor, who should have discovered the conditions at the site, but did not. These parties may be soil consultants, testing agencies, or geotechnical engineers. Depending upon the jurisdiction, they may be held directly liable to parties with whom they have no privity of contract. In other jurisdictions, the parties may be subject to indemnity claims brought by those parties found liable to the owner or contractor.[98]

There are a number of reported appellate court decisions involving claims against professionals arising out of the time and economic impacts associated with, or caused by, the encountering of differing site conditions. In general terms, claims involving differing site conditions have been asserted against design and construction management professionals by owners and third parties, including contractors, subcontractors, and construction lenders, based on the following allegations: failure to undertake an adequate pre-construction investigation/survey of existing site conditions;[99] failure to accurately and/or completely describe existing site conditions in bid and contract documents;[100] failure to disclose information concerning existing site conditions that was generally (or otherwise) not available to bidders, or that was not available within the time frame permitted for preparation of the bid;[101] negligence in the drafting of bid or contract documents with respect to the circumstances under which a contractor would be entitled to time or economic relief due to encountering differing site conditions;[102] failure to prepare accurate quantity or other estimates;[103] and failure to properly and/or timely address and resolve differing site conditions claims in accordance with the professional's obligations under the contract documents, or in accordance with the professional standard of care.[104]

Design and construction management professionals may also be exposed to professional liability claims that arise from purely administrative services performed during the construction phase of the contract (i.e., tortious interference with a contractual relationship).[105] Under most standard general conditions of construction contracts, the design and/or construction management professional has the obligation to fairly, promptly, and impartially administer the construction contract. This obligation includes the interpretation of contract documents and resolution of disputes regarding differing site conditions claims between the owner and contractor.[106] In the performance of this role, the design professional must act fairly and impartially, and without unreasonable delay, in addressing and resolving claims involving differing site conditions.[107]

10.3 RECOMMENDATIONS FOR MANAGING THE RISK OF PROFESSIONAL LIABILITY ON SUBSURFACE PROJECTS

Each of the preceding chapters of this book has addressed and made recommendations concerning specific topics related to risk management in connection with subsurface projects. This section will address specific recommendations for managing the risk of professional liability on subsurface projects, based on some of the particular liability concerns previously discussed in this chapter.

10.3.1 Clear and Adequate Scope Definition

The discussions concerning contract-based liability in Section 10.1.1.1 and negligence-based liability in Section 10.1.1.2 highlight the importance of a clear and precise definition of the scope of the professional's service obligation and the need for a client-authorized scope of services adequate (in the reasonable judgment of the professional) to achieve the goals and objectives of the specific project. A clear and precise scope definition may serve as an important front-line defense to both client and third-party claims; an ambiguous, unspecific or inadequate scope definition may provide the basis for liability to the client and third parties. (See Section 10.1.1.1.) The scope definition should clearly state what services will be furnished, which services are offered as "additional services," any particular service exclusions, and any appropriate and pertinent limitations and qualifications with respect to the service engagement.

Typically, the project owner makes the decision regarding the scope and nature of the subsurface investigation to be performed in connection with the planning and design of a subsurface project. However, the professional consultant can and should play an important role in recommending the appropriate scope and advising the owner to authorize an adequate investigation. Curtailment of the scope of a recommended subsurface investigation to "reduce the project costs" is often short-sighted and will lead to ultimate increases in the project costs due to the need to redesign during construction, the increased risk of discovering differing site conditions (which may have been ascertained through an adequate preconstruction investigation), and the associated delays, disruptions, and claims. In addition, inappropriate limitations on the scope of subsurface investigation may require that the design professional qualify, limit, or disclaim the accuracy, completeness, or reliability of information or interpretive opinions which, in turn, could also result in higher construction bids due to the inclusion of a contingency to address the risk associated with the qualified or disclaimed information or opinions. Inadequate subsurface information based on limited investigation may also result in an overly conservative (i.e., more expensive to construct) design.[108]

If the client disagrees with the professional's recommendations regarding the scope of subsurface investigation, the professional should discuss (and later document the discussion) with the client the potential cost, time, and other impacts that the limitation may have on the design and construction of the subsurface project, including any concerns with respect to the quality of the final and completed project, as well as constructability issues and/or the increased risk of differing site conditions claims during construction. In addition, consideration should be given to the future time or cost impact associated with a limited subsurface investigation, including time and cost impact of redesign during the construction of the project required to address site conditions that could have been anticipated and/or known as a result of a more complete and adequate preconstruction subsurface investigation program.

The professional should render an evaluation of the subsurface conditions, and advise the owner of any reservations, qualifications, or limitations that are predicated on a scope of investigation less than what the professional had recommended. Any such qualifications should be included in the agreement between the professional and the client, as well as specifically in any opinions, data, reports, or other work product furnished by the professional that may be relied on by third parties, including bidders and the construction contractor. In addition, if the professional recommends, as part of the professional engagement, that subsurface information be re-evaluated and/or updated at periodic junctures, that recommendation should be included in the agreement between the professional and the client.

In circumstances in which a client does not accept a recommended scope, the professional should document (in a record of negotiation, or otherwise) the recommendation and the client's rejection of that recommendation.[109] However, if the client arbitrarily refuses to authorize a scope that the professional considers minimally required to fulfill legal, professional registration, and/or ethical standards, it may be professionally required or advisable to decline the engagement.

10.3.2 Standard of Care and Indemnification

Design and construction management professionals should not agree to perform their services in accordance with a standard that exceeds the negligence-based professional standard of care (see Section 10.1.1.2). Contractual standards that require that the professional perform its services in accordance with the "highest," "best," or other superlative level will, in legal effect, amount to a warranty of perfection, which will result in a significant enhancement of liability risk for the professional, as well as liability that is not covered by professional liability insurance (see Section 10.1.2.1). Similarly, the design and construction management professional should not agree to defend and/or indemnify the project owner (or others) for liability, claims, damages, or other risks that are not predicated on and caused by actual or (at a minimum) alleged negligence of the professional (see Section 10.1.1.1).

The professional should educate the owner as to the distinction between (1) the owner's implied warranty obligation to the contractor regarding the accuracy, constructability, and completeness of information and design (typically prepared by the professional and) furnished by the owner to the contractor and (2) the design professional's more limited negligence-based professional standard of care obligation (see Section 10.2.2.1(A)). It is important that the owner understand that its implied warranty obligation is not coterminous with, and is more encompassing than, the professional's professional standard of care obligation, and that there may well be circumstances in which a contractor claim based on inaccurate, incomplete, or defective data, opinions, or design prepared by the professional may fall within the owner's implied warranty obligation, but not constitute a departure from the professional standard of care (see Section 10.2.2.1(A)).[110] This is an especially important point, given that differing site conditions claims frequently are combined with allegations of defective designs or specifications (see Section 10.2.2.3), which may lead to client and/or third-party claims against the professional.

10.3.3 Risk Identification and Fair Allocation

There is a direct correlation on subsurface projects between (1) the owner's adequate identification and fair allocation of risk between it and the contractor; and

(2) the degree of professional liability risk for the design and construction management professional. Design and construction management professionals have the opportunity and the obligation to serve a critically important role in educating owners regarding the need for adequate identification and fair allocation of risk on subsurface projects. The major components of an effective process for adequate and fair risk identification and allocation include the commissioning and performance of an appropriate preconstruction subsurface investigation (see Section 10.3.1); the disclosure to bidders of all pertinent information and data regarding known or anticipated site conditions; the limited and selective use of disclaimers; the inclusion of a differing site conditions clause (see Section 10.2.1); the involvement of the professional during the construction phase; and the inclusion in the contract documents of a fair procedure for the submission, evaluation, and timely, fair, and impartial resolution of differing site conditions claims.

As noted above, it is recommended that *pertinent* (i.e., not necessarily *all*) information regarding site conditions be disclosed; concealment of any relevant information is strongly discouraged (and may lead to contractor claims based on fraud, misrepresentation, impossibility of performance, and/or superior knowledge) (see Section 10.2.1). The disclosure, without appropriate limitations or disclaimers, of information that is either irrelevant to the final design (e.g., conceptual studies or preliminary design submissions) and/or not relevant to the construction of the project may not only be unhelpful and distracting, but may also be potentially confusing and misleading to the bidders and the contractor. In addition, this type of information may provide a seed or opportunity for contractor claims based on allegations of contradictory, conflicting, confusing, and misleading information. Although concealment of this type of information is not recommended, the disclosure of that information should be accompanied by appropriate limitations, qualifications, and/or disclaimers. A conscientious effort should be undertaken to identify technical or other provisions of information (data or interpretive) that may be relevant to the construction and/or constructability issues.[111]

Disclaimers of information furnished to bidders should be utilized selectively and should specifically reference the particular information disclaimed. General and indiscriminate use of disclaimers should be discouraged.

10.3.4 Definition and Allocation of Roles and Responsibilities Regarding Design and Construction Means and Methods

As discussed in Section 10.2.2.3, differing site conditions claims often involve issues relating to both the professional's design and the construction means and methods selected by the contractor. In some instances, all of the design of the permanent and completed project is performed by the professional; in other instances, portions of that design are performed by the contractor (or a consultant engaged by the contractor). In addition, in some instances, the professional has no role whatsoever in connection with the design and/or implementation of construction means and methods, while in other instances, the professional either designs (or otherwise requires) specific construction means and methods or otherwise directs the implementation of particular construction means and methods. The resolution of differing site conditions claims involving both "design" and "construction means and methods" issues becomes more complicated, and less predictable, and the associated professional liability risk is significantly increased when the respective roles and responsibilities of the professional and contractor regarding (1) design and (2) construction means and methods are not clearly defined, allocated, and documented.

If the construction contractor is to have a role in the development and finalization of the design of a portion of the permanent and completed project, such an expectation should be explicitly stated and the extent of that role defined in the contract documents. Consideration should be given to the inclusion of a prefatory statement as to why design responsibility has been delegated to the contractor (for example, relative experience, qualifications, and expertise of contractors in this type of project; the high degree of interrelatedness between the final design and the contractor's selected construction means and methods). In circumstances in which the owner delegates responsibility for design to the contractor, the contract documents should clearly state that the contractor will fully assume risks associated with defects in that design. In addition, the more design discretion given to the contractor and the more information disclosed by the owner regarding the known and unknown information and anticipated risks, the greater the likelihood that the risk of defective design will be effectively reallocated to the contractor. (See Section 10.2.2.2(A).)

In circumstances in which the design professional develops performance criteria and/or a conceptual design that is furnished to the contractor for the latter's use in developing and finalizing the design, the level of detail in the criteria or concept and the degree of discretion given to the contractor in the design development process will influence the allocation of risk for defective design (see Section 10.2.2.2(A)). In addition, the nature, purpose, and extent of any design and/or construction management professional review of any contractor design submissions should be explicitly defined and stated in the contract documents. Stamp language should be specifically developed and utilized by the design and construction management professional that conforms with the stated purposes and limitations of any such design submittal review.

In any situation in which the contractor (or a design professional engaged by the contractor) is expected to undertake a design role, the contract documents should specify the qualifications and experience required of the contractor's designer; the indemnification obligations of the contractor and/or its designer with respect to claims and liabilities arising out of defective design furnished by either of them; the professional liability or other insurance requirements of the contractor and its designer; the expected role and services to be performed by the contractor's designer during the construction phase (i.e., construction observation, submittal review, issuance of certifications, affidavits, and so forth); and the degree of discretion (and/or immunity) given to the design professional of record in its review of the proposed design of the contractor's designer and the standard that will govern the exercise of that review role; and who has the responsibility for coordinating the contractor's design with the overall project design done by others.

In some situations, design and construction management professionals do become involved in the design of particular construction means and methods. In special circumstances in which the need and/or advisability of design professional involvement in a particular construction means and methods is apparent at the time of the negotiation of the owner/professional agreement, and in which a design professional chooses to become involved in some aspect of a particular construction means and methods, the nature and extent of that involvement and the respective obligations and duties of the design professional and construction contractor with respect to the design and/or implementation of the construction means and methods should be clearly defined in the owner/professional agreement. In addition, the owner/professional agreement should make clear that the professional's defined and specific involvement in the particular construction means and methods does not create or imply any broader obligation of the professional, with respect to any other specific construction means and methods and/or construction means and methods in general; nor does the professional's involvement in the particular construction means and methods negate, diminish, or otherwise affect the construction contractor's exclusive responsibility for all other construction means and methods and safety-related issues. The specific level and degree of responsibility of the design professional for the particular construction means and methods at issue should be specifically defined in the owner/professional agreement and in the general and/or supplementary conditions of the construction contract. Assuming that the design professional chooses to become involved in the design of particular construction means and methods, it is recommended that the design professional also provide observational services for the purpose of determining whether the construction contractor is constructing or otherwise implementing the particular construction means and methods in accordance with the design professional's design of those construction means and methods. In addition, the design professional should review any submittals of the contractor regarding the particular construction means and methods.

In special circumstances in which the need for and/or advisability of the design professional's involvement in a particular construction means and methods is not apparent at the time of the negotiation of the owner/professional agreement (for example, where the circumstances involve issues that arise during construction such as differing site conditions) and in which the design and/or construction management professional chooses to become involved in some aspect of the particular construction means and methods, the nature and extent of that involvement and the respective obligations and duties of the professional and construction contractor with respect to the design and/or implementation of the particular construction means and methods should be clearly defined in a written amendment to the agreement or otherwise documented and communicated to all relevant parties involved in the construction process. The amendment should make clear that the professional's defined and specific involvement in the particular construction means and methods does not create or imply any broader obligation of the professional with respect to any other specific construction means and methods, and/or construction means and methods in general; nor does the professional's involvement in the particular construction means and methods negate, diminish, or otherwise affect the construction contractor's exclusive responsibility for all other construction means and methods and safety-related issues. In addition, in these specific and "unanticipated" circumstances, some written communication should be issued (such as a no-cost change order) to the construction contractor, setting forth similar information to that recommended for inclusion in the amendment to the professional's agreement.

10.3.4.1 The Baseline Approach and Construction Means and Methods For a discussion of risk management and professional liability implications associated with the use of the geotechnical or contractual baseline approach (the "baseline approach") to address construction means and methods and related contractor performance matters, the reader is referred to the Chapter 10 Appendix.

10.3.5 Third-Party Liability

Although the risk of liability to third parties can be significant for design and construction management professionals involved in subsurface projects, there are ways to limit, manage, and control that risk. The discussion in Section 10.1.1.1 highlights the importance of precisely and clearly defining the scope of services to be provided by the professional for the client and, preferably, the purposes for which those services have been requested and will be performed. A precise definition of the scope of services, apart from representing a sound contracting practice, may provide the foundation for the defense to a third party claim on the basis that the claim does not arise out of services that the professional was retained to perform. (See Section 10.1.2.1.) However, in circumstances or transactions in which the professional actually knows or reasonably should foresee that certain third parties may rely on the performance of the professional's services, opinions, or other work product, it is sound practice for the definition of service scope to be communicated to those third parties, especially any limitations or qualifications that vary from the scope of services typically performed by or expected of a professional in similar circumstances or transactions. In addition, a statement in the professional's agreement with the client setting forth the purpose or reason why (and/or for whose benefit) the professional is being retained may provide the basis for a defense to a claim by a third party whose interests are inconsistent and/or otherwise in conflict with those of the professional's client.

In a related context, in circumstances in which the professional may be obligated to give opinions or certifications in discharge of a public or statutory duty (such as a design professional's obligation to provide a building inspector with certain design or construction control affidavits), it is important that the professional's client understand (preferably confirmed in contract provisions) that the discharge of such public duties may require the professional to make statements or disclaimers that may be perceived by the client to be inconsistent with the client's best interest. Similarly, in certain contexts the professional is retained by a client to render opinions or other services that it is expected will, under some law or regulation, be relied on by public officials or agencies. If the client does not authorize the professional to perform the services that will be necessary or typically expected by the public official or agency in the performance of those services and/or the rendering of such opinions, the professional should definitely so advise its client (preferably confirmed by contract provision) and the public official or agency involved as well, by way of disclaimer, limitation, or qualification in the professional's opinion or other work product. Although having discussions and making disclaimers concerning these topics are perceived as sensitive subjects and difficult for a number of professionals in terms of client relations (particularly in dealing with those clients represented by legal counsel, such as counsel for environmental clients), those types of discussions and disclaimers are well advised as an effective risk management mechanism designed to avoid misunderstandings and many of the problems that were relied upon by the federal court in *Aliberti* (see Section 10.1.2.2), in imposing liability on the engineer and construction manager.

Similarly, as part of the contract negotiations with the client, the professional should have a candid discussion about whether it is expected that any data, opinions, or other work product produced as a result of the professional's services will be distributed, or otherwise disseminated or furnished, to anyone other than the client. If not, then the agreement should expressly state that the services are being performed solely for the benefit and use of the client, and that the terms of engagement, including the amount of compensation, have been predicated on the professional's expectation and understanding that no one other than the client would rely on the services, reports, or other work product of the professional. (See *Bronstein v. GZA GeoEnvironmental, Inc.*, 665 A.2d 369 (N.H. 1995), discussed in Section 10.1.2.2.) If the owner wants or expects the professional's work product to be disseminated to third parties, then the professional must consider whether he or she wishes to proceed with the undertaking and, if so, under what terms and compensation, given the greater risk of liability to third parties. In any event, in circumstances in which the professional decides to undertake an engagement knowing that persons beyond the client may rely on those services, an effort should be made to specifically define those persons by name or generically.

In connection with the performance of services and/or preparation of work product that the professional knows may be relied on by certain third parties, case law instructs that the professional, while maintaining loyalty and confidentiality obligations to the client, must also be candid in stating and disclosing any appropriate limitations, qualifications, and/or disclaimers. Along those lines, reports or other statements issued by the professional should include, as an appendix or otherwise, the professional's agreement with the client, as well as any appropriate limitations, qualifications, and/or disclaimers, or other constraints or restrictions (time or cost) imposed on the professional in the performance of professional services. Should the professional be of the opinion that the third party should independently verify (through retention of an independent professional or otherwise) the accuracy and/or completeness of information and/or opinions stated by the professional, that admonition should also be explicitly stated. Similarly, if the professional's report contains information and/or opinions derived from a source other than the professional, that fact and the source should be disclosed.

In the event that a professional is aware of third-party reliance on its services, the professional may wish to issue a third-party reliance letter. In this letter, a professional may limit the scope of services to which the third party is entitled to rely. The letter should also limit the reliance on the professional's services to the third party named in the letter and condition that reliance on the same limitations and qualifications contained in the professional's contract with its client. A professional should refrain from issuing a third-party reliance letter unless it is necessary under the circumstances. In most instances, the letter should be prepared by the professional's counsel. In addition, contract and other appropriate limitations/disclaimer language should be developed for use by design and construction management professionals to manage, limit, and control the risk of third-party liability in a variety of specific contexts, including cost estimating, constructability opinions, environmental opinions, and geotechnical reports.

In summary, while the risk of third-party liability for design and construction management professionals is a real and substantial concern, there are certain professional practice contractual mechanisms available to control, limit, and manage that risk.

10.4 CONCLUSION

The risk of professional liability to both clients and third parties is substantial for design and construction management professionals involved in subsurface projects. That risk exposure is materially affected by how risk for subsurface conditions is allocated between the owner and contractor. There is a direct correlation between the degree to which risk is fairly allocated between owners and contractors and the likelihood of professional liability exposure for the design and construction management professional: The fairer and more balanced the allocation, the less likely the risk of professional liability. While professionals are not in a direct position to control the decision making with respect to risk allocation for subsurface conditions between the owner and contractor, they do have the opportunity, if not the obligation, to advise project owners (the primary most influential decision makers in the risk allocation process) concerning risk allocation for those conditions. It is hoped that the discussion in this chapter will be useful to design and construction management professionals in advising their clients about risk allocation for subsurface conditions.

NOTES

1. The phrase "design and construction management professionals" encompasses persons or entities that provide solely professional (as distinct from actual construction) services and primarily includes civil and geotechnical engineers. As used in this chapter, the term "construction management professional" includes any professional who has knowledge of both pre-construction issues (such as scheduling, the development of the decision to build, the design and the delivery method, the construction cost of design alternatives, construction materials, methods, and labor) and construction issues (such as constructability reviews, scheduling, coordination, and supervision) and who does not take any entrepreneurial risk for cost, timeliness, or quality of construction (as distinct from contractors, subcontractors or trade contractors who contract directly with the owner). See W. Foxhall, *Professional Construction Management and Project Administration*, 2d ed., (New York: Architectural Record Books, 1976), pp. 4–5.
2. In addition to the legal theories discussed in Section 10.1, there are other legal theories upon which clients and third parties may assert claims against design and construction management professionals. These include: third-party beneficiary claims, fraud, misrepresentation, defamation, intentional interference with contractual relationships, and violations of unfair business practice statutes. See A. S. Pralgever, "Common Law Theories of Recovery," Chapter 6 in *Differing Site Condition Claims*, R. F. Cushman and D. Tortorello, eds., (New York: John Wiley & Sons, Inc., 1992). Discussion of the latter legal theories has been omitted from Section 10.1 in the interest of focusing and concentrating the analysis on those legal theories (viz., contract, negligence, and negligent misrepresentation) that are predominantly and most frequently asserted against professionals in the specific context of subsurface projects.
3. In general, most professional liability (errors/omissions) insurance policies issued to design and construction management professionals provide coverage (both defense and indemnification) for only those claims arising out of *negligence* in the performance of professional services. These policies typically exclude coverage for claims resulting from breach of warranty or contractual provisions that impose on the professional a performance standard that exceeds the negligence-based professional standard of care. See O. Shean and D. Patkin, *Construction Insurance: Coverage and Disputes*, Chapter 8 (Charlottesville, VA: The Mitchie Company, 1994); D. J. Hatem, "The Owner as 'Additional Insured' Under the Design Professional's Professional Liability Policy: Can It Be Done and, If So, How?", *The CA/T Professional Liability Reporter*, Vol. 2, No. 1, (Boston: Burns & Levinson LLP, September 1996), pp. 1–8.
4. By contract, a design professional may impose obligations upon himself or herself or make express warranties that would subject him or her to liability beyond the professional standard of care. See, e.g., *Chesapeake Paper Products Co. v. Stone*

& Webster Engineering Corp., 51 F.3d 1229 (4th Cir. 1995) (engineer agreed to be bound to higher standard of care by accepting purchase order stating that engineer warranted that "all the materials and articles covered by this order" will be "free from defects in material and/or workmanship, and merchantable"); *Anthony's Pier Four, Inc. v. Crandall Dry Dock Engineers, Inc.*, 489 N.E.2d 172 (Mass. 1986) (if engineer assured owner that its services "would be fit" for the owner's intended purpose, "[s]uch an assurance would impose a higher standard on [the engineer] than its implied warranty of reasonable care," and such assurances would amount to an express warranty); *Klein v. Catalano*, 437 N.E.2d 514 (Mass. 1982) (where architect provides warranty of a certain result, owner may maintain action for breach of express warranty if result is not met; in that case, however, architect did not guarantee specific result, but merely agreed to provide specifications, thus providing only an implied warranty to comply with required standard of professional care). See generally D. G. Gavin and J. K. Holland, "Dealing with Hazardous Waste and Environmental Issues," in *Design Professional's Handbook of Business and Law* R. F. Cushman & J. Dobbs, eds. (New York: John Wiley & Sons, Inc. 1991).

5. Broad-form indemnification provisions typically are declared void and unenforceable under a variety of so-called anti-indemnification statutes that have been enacted in several states. For a summary of the contents and effects of these types of statutes, see S. Stein, *Construction Law*, ¶13.17, pp. 13-128 to 13-143 (New York: Matthew Bender, 1997). However, in many states, such broad-form indemnification provisions are fully enforceable, despite their apparent unfairness.

6. See Note 3, *supra*.

7. See, e.g., *Ramey Construction Co. v. Apache Tribe of Mescalero Reservation*, 673 F.2d 315, 321 (10th Cir. 1982) (where contract between owner and architect did not provide that architect would undertake responsibility for project management, architect had no duty to coordinate and manage progress of work, and therefore could not be held liable for any damages caused by inadequate coordination or management); *Lutz Engineering Co., Inc. v. Industrial Louvers Inc.*, 585 A.2d 631 (R.I. 1991) (holding that engineering firm owed no duty of care to heating and ventilation subcontractor with respect to deficient products furnished by subcontractor under approved submittals where contract documents provided that engineering firm would review submittals only for general conformance to design of project); *Empire West Companies, Inc. v. Albuquerque Testing Laboratories, Inc.*, 800 P.2d 725 (N.M. 1990) (in an action by contractor against geotechnical firm for breach of contract and warranties, negligent misrepresentation, and professional malpractice, scope-of-services provision in geotechnical firm's agreement with contractor was used as successful bar to contractor's action on the basis that geotechnical firm fulfilled its agreement and exercised its limited duties within the appropriate standard of care). See generally Note, "Architectural Malpractice: A Contract-Based Approach," 92 *Harv. L. Rev.* 1075 (1979); P. Jackson, "The Role of Contract and Architectural and Engineering Malpractice," *Ins. Couns. J.*, (Oct. 1984), p. 571.

8. See also *National Foundation Company v. Post, Buckley, Schuh & Jernigan, Inc.*, 465 S.E.2d 726 (Ga. Ct. App. 1995) (ruling that an architectural/engineering firm was not liable for defective design of shoring wall because the firm had no responsibility for design or other services in connection with temporary support or con-

struction means/methods); *Housing Authority of City of Carrollton v. Ayers*, 88 S.E.2d 368 (Ga. 1955) (holding that owner could not maintain negligence action against architect for "failure to determine soil structure" or for "neglect in examination of site," where the owner contractually assumed the duty to select the site and ascertain the subsoil conditions). See generally S. Stein, *Construction Law*, ¶5.07[1] [b] [viii] (New York: Matthew Bender, 1997).

9. For instance, in *Stanley Consultants, Inc. v. H. Kalicak Const. Co.*, 383 F. Supp. 315 (E. D. Mo. 1974), the court held that an architectural/engineering firm (A/E) was obligated to provide a service beyond what was specifically required of the A/E in the A/E's agreement with the owner of a project. Specifically, the agreement between the A/E and the owner did not require the A/E firm to perform surveys. Notwithstanding, based on expert testimony indicating that an architect exercising reasonable care would have surveyed the area under the circumstances, and, further, based on the fact that the plans contained errors "obvious even to [the] Court's nonexpert eye," the court concluded that the A/E had the duty to survey.
   Similarly, in *Kleb v. Wendling*, 385 N.E.2d 346 (Ill. App. Ct. 1978), the owner of the project filed a breach of contract action against the architect alleging, among other things, that the architect's design was defective. The trial court entered judgment in favor of the architect on the grounds that the owner failed to introduce any evidence of defective design. The appellate court, however, reversed and remanded the court's decision for a rehearing on the issue of whether the architect should have performed test borings on the site prior to the completion of its design. Although the owner never requested—either verbally or in writing—that the architect perform test borings, the court stated that additional evidence was needed to determine whether, as a reasonably competent architect, the architect should have obtained subsurface borings prior to completing its design.

10. The recent Massachusetts Superior Court decision in *Chiacchia v. Lycott Environmental Research, Inc.*, Civil Action No. 91-2358C, slip. op. (Worcester Sup. Ct. Sept. 28, 1995) (appeal pending), demonstrates this point. The plaintiffs in *Chiacchia* engaged Lycott, an environmental engineering firm, to perform an environmental assessment of property that the plaintiffs were interested in purchasing. The agreement was imprecise as to the extent of the assessment Lycott was to perform, identifying the subject property simply as "located on Route 16 in Uxbridge, Massachusetts." Believing that the property located on the northern side of Route 16 was not included in its scope of services, Lycott performed an assessment only of the portion of the property located on the southern side of Route 16. As it turned out, however, the client wanted the entire property to be investigated. After the plaintiffs purchased the entire site, hazardous substances were ultimately found on the property located on the northern side of Route 16. The plaintiffs filed suit against Lycott, maintaining that Lycott knew or reasonably should have known that they expected that the property on both the northern and southern sides of Route 16 would be included in Lycott's assessment, since they were contemplating the purchase of *all* of the subject property. The court ruled in favor of the plaintiffs. Although the agreement did not specify that property on the northern side of Route 16 was to be included in the scope of Lycott's assessment, the court stated that, to the extent there was any ambiguity as to the precise boundaries of the site that Lycott was to assess, Lycott should bear the consequence of that ambiguity since it drafted the agreement. In addition, the court concluded that "it was good and

customary practice," and therefore incumbent upon Lycott, as an environmental professional, to verify and resolve the precise boundaries of the site prior to proceeding with its environmental assessment of the property.

11. Several court decisions have enforced limitation-of-liability clauses that are clearly drafted and negotiated between the client and the professional. See, e.g., *R. W. Beck & Associates v. Sitka, affirmed without opinion*, 951 F.2d 362 (9th Cir. 1991) (limitation of liability clause limited engineer's liability to extent of its compensation from owner); *Fla. Power & Light Co. v. Mid-Valley, Inc.*, 763 F.2d 1316 (11th Cir. 1985) (clause limiting liability to amount of insurance agreed upon by the parties enforced where parties are of equal bargaining power); *Central Maine Power Co. v. Foster Wheeler Corp.*, 684 F. Supp. 724 (D. Me. 1988) (upholding unambiguous limitation of liability clause fixing engineers' liability to 50 percent of $500,000 fee); *Long Island Lighting Co. v. IMO Delaval, Inc.*, 668 F. Supp. 237 (S.D.N.Y. 1987) (upholding agreement limiting liability to proceeds of agreed upon insurance arrangement). See also *Markborough Calif., Inc. v. Superior Court*, 277 Cal. Rptr. 919 (Cal. Ct. App. 1991) (limitation of liability provision in contract between engineer and owner valid where parties had opportunity to request change in contract despite arguable inconspicuousness of provision even if parties—when dealing at arm's length—do not read the contract), *rev. denied*, 1991 Cal. LEXIS 1940 (Cal. 1991); D. J. Hatem, "Indemnification and Limitation of Liability," 18 *Guidelines for Improved Practice*, No. 6 (Chicago: Victor O. Schinnerer & Co. 1988); W. J. Kennedy, *Limitation of Liability in Design Professional Contracts*, EJCDC Doc. No. 1910-9E (Alexandria, VA: EJCDC, 1986), pp. 4–6.

12. For a further discussion of the professional standard of care as it relates to design professionals, see J. Sweet, *Legal Aspects of the Architecture, Engineering and the Construction Process*, Section 14.06, 5th ed. (St. Paul, MN: West Publishing Co., 1994), pp. 273–76; D. J. Hatem, "Errors/Omissions Cost Recovery Claims," *The CA/T Professional Liability Reporter*, Vol. 1, No. 4, (Boston: Burns & Levinson LLP, June 1996) pp. 1–10. For a discussion of the professional standard of care as it relates to construction managers, see M. Loulakis, et al., *Construction Management* (New York: John Wiley & Sons, Inc., 1995), pp. 328–31.

13. *Klein v. Catalano*, 437 N.E.2d 514, 525 (Mass. 1982).

14. See *Nelson v. Commonwealth of Virginia*, 368 S.E.2d 239, 244 (Va. 1988) (in an action by an owner against an architect for alleged breaches of the architect's duty to administer the construction contract, expert testimony was required to define standard of care applicable to architect's administration of its duties and to prove violation of that standard). See also *Chicago College of Osteopathic Medicine v. George A. Fuller Co.*, 719 F.2d 1335, 1346 and n.13 (7th Cir. 1983) (architectural expert needed to explain to jury nature of architect's judgment calls); *Bartak v. Bell-Galyardt & Wells, Inc.*, 629 F.2d 523, 530 (8th Cir. 1980) (expert testimony required to prove architect's negligent planning of project because required technical analysis was beyond competence of ordinary lay people); *Aetna Ins. Co. v. Hellmuth, Obata & Kassabaum, Inc.*, 392 F.2d 472, 477-78 (8th Cir. 1968) ("supervision" as responsibility of architect, if term of art, requires expert testimony to establish standard of care required). See also Robert M. Schoenhaus, Annotation, *Necessity of Expert Testimony to Show Malpractice of Architect*, 3 A.L.R.4th 1023 (1981 & Supp. 1991). For a discussion of the duties and responsibilities of a pro-

fessional in serving as an expert, see *Expert: A Guide to Forensic Engineering and Service as an Expert Witness* (ASFE, 1995).

15. However, expert testimony may not be necessary in circumstances in which the professional's performance amounts to gross negligence. See *Zontelli & Sons, Inc. v. City of Nashwauk*, 373 N.W.2d 744, 755 (Minn. 1985) (ruling that the trial court was justified in concluding, without expert testimony, that engineer was negligent based on factual record, which indicated that engineer "drastically" underestimated quantity of concrete and unsuitable material to be excavated and removed by contractor, and misrepresented that the highway in question was municipal [as opposed to state] highway, thereby misleading bidders with respect to scope of work required to be performed where engineer could have avoided those inaccuracies by simply making preliminary inquiries); *Nicholson & Loup, Inc. v. Carl E. Woodward, Inc.*, 596 So.2d 374 (La. Ct. App. 1992) (concluding that with respect to a negligence claim against a geotechnical engineer, expert testimony on the standard of care exercised by geotechnical engineers in the community was not necessary where a lay person, based on the factual evidence, could "infer" negligence using a "common sense standard."), *writ denied*, 605 So. 2d 1098 (1992). Unquestionably, however, cases such as *Zontelli* and *Nicholson* represent exceptions to the general, and more pervasive, rule requiring expert testimony.

16. The terms of the agreement between the professional and the client may be relevant in determining not only the *contractual* liability of the professional to the client, but also whether the professional was *negligent*. See generally Note, "Architectural Malpractice: A Contract-Based Approach," 92 *Harv. L. Rev.* 1075 (1979); P. Jackson, "The Role of Contract in Architectural and Engineering Malpractice," *Ins. Couns. J.* (Oct. 1984), p. 517.

17. This survey of potential defenses to negligence claims is by no means exhaustive. For a further and more detailed discussion, see J. Acret, *Architects and Engineers*, Chapter 9 (Colorado Springs, CO: Shepard's/McGraw-Hill, 1993).

18. As noted in Note 2, *supra*, other legal theories may be pursued by non-clients against professionals. This section, however, will focus on the two most predominant and frequently asserted types of third-party claims.

19. See *Parent v. Stone & Webster Engineering Corp.*, 556 N.E.2d 1009 (Mass. 1990) (ruling that trial court erred in granting summary judgment in favor of defendant engineering firm, which failed to demonstrate that its contractual relationship with plaintiff's employer, an electrical generating plant operator, did not give rise to a duty of reasonable care toward plaintiff, who was burned by high voltage when testing an electrical distribution panel, where contract between engineering firm and electric company designated engineering firm as project manager and imposed on engineering firm the duty of "participation and assistance" in safety; summary judgment record indicated engineering firm, pursuant to performance of its contractual duties, either did discover or should have discovered that distribution panel carried high voltage and was not properly labeled; contract language also supported proposition that engineering firm should have corrected this hazard or brought it to electric company's attention). See also *McElvy, Jennewein, Stefany, Howard, Inc. v. Arlington Elec., Inc.*, 582 So.2d 47, 48-49 (Fla. Dist. Ct. App. 1991) (duty imposed on architects by architects' contract with owner did not extend to electrical subcontractor where architects undertook only to render good-faith advice and inter-

pretation, and not to render the ultimate decisions on the interpretation of project contracts); *Lutz Engineering Co., Inc. v. Industrial Louvers, Inc.*, 585 A.2d 631, 635-37 (R.I. 1991) (holding that engineering firm owed no duty of care to heating and ventilation subcontractor with respect to deficient products furnished by subcontractor under approved submittals where contract documents provided that engineering firm would review submittals only for general conformance to design of project); *Craig v. Everett M. Brooks Co.*, 222 N.E.2d 752, 755 (Mass. 1967) (contractor asserted negligent misrepresentation claim against engineer based on, among other things, engineer's allegedly erroneous representations in plans regarding soil conditions; court barred that portion of contractor's claim based on finding that scope of engineer's agreement with owner did not impose on engineer duty to provide any information regarding soil conditions); *Magnolia Construction Co. Inc. v. Mississippi Gulf Engineers, Inc.*, 518 So. 2d 1194, 1202-03 (Miss. 1988) (holding that whether engineer assumed a duty of care to contractor is contingent on whether engineer, pursuant to contract with owner, agreed to be field "representative" or field "inspector" for project).

20. For a comprehensive listing of jurisdictions that have adopted some form of the economic loss doctrine (also known as the economic loss rule), see D'Angelo, "The Economic Loss Doctrine: Saving Contract Warranty Law from Drowning in a Sea of Torts," 26 *U. Tol. L. Rev.* 591 at Appendix (1995). See also S. Stein, *Construction Law*, ¶5.05[7] (New York: Matthew Bender, 1997).

21. One commentator has accurately noted:

> Although many claims paid by the owner in settlement or after adjudication give the appearance that the owner bears the risk of unknown subsurface conditions, the burden is distributed amongst all construction participants in many situations. Even though the owner may settle a subsurface claim, the contractor is likely to not to receive full compensation in settlement for all of the increased costs of the changed subsurface conditions. Adjudication by the court process or by arbitration may award less than the full increased costs ... Even though it would appear that subsurface claims are reimbursable by the owner to the contractor, the true result is generally a sharing of the unexpected subsurface conditions.

G. Jones, "The U.S. Perspective on Procedures for Subsurface Ground Conditions Claims," *The Int'l Construction L. Rev.* 155, 171 (Lloyds of London Press, Ltd., 1990).

22. See generally J. Feinman, *Economic Negligence: Liability of Professionals and Businesses to Third Party for Economic Loss*, Chapter 14 (Boston: Little, Brown & Company, 1995).

23. Examples of these types of claims include contractor allegations of economic losses due to a professional's failing to adequately identify the presence of certain subsurface conditions, see, e.g., *Gulf Contracting v. Bibb County*, 795 F.2d 980; 982 (11th Cir. 1986) (general contractor allowed to maintain negligent misrepresentation claim against project architects and engineers for failing to disclose subsurface debris); *Texas Tunneling Co. v. City of Chattanooga*, 204 F. Supp. 821 (E.D. Tenn. 1962) (tunneling subcontractor allowed to recover against engineering firm on neg-

ligent misrepresentation claim, based on findings that tunneling subcontractor justifiably relied on inaccurate geological data negligently prepared by engineering firm, and that engineering firm should have reasonably foreseen that subcontractor would have relied upon said inaccurate data), *aff'd in part, rev'd in part*, 329 F.2d 402 (6th Cir. 1964).

24. *Blake Construction Co., Inc. v. Alley*, 353 S.E.2d 724, 727 (Va. Sup. Ct. 1987) (dismissing general contractor's action against architect, which sought only economic losses against architect, on the basis that architect owed no "duty" to general contractor to perform with due care the services for which the architect was responsible under the owner-architect contract).

25. J. Feinman, *Economic Negligence: Liability of Professionals and Businesses to Third Party for Economic Loss*, Chapter 14 (Boston: Little, Brown & Company 1995) (Reprinted by permission; Copyright 1995 by Little, Brown & Company, assigned to Aspen Law & Business, a division of Aspen Publishers, Inc.). See also *Foster Wheeler Enviresponse, Inc. v. Franklin County Convention Facilities Authority, et al.*, 678 N.E.2d 519 (Ohio 1997) (relying upon the economic loss doctrine to preclude contractor's negligent misrepresentation claim against engineering consultant on the basis that the contractor should not be permitted to circumvent limitations on recovery imposed by the construction contract).

26. Several cases provide examples of third-party negligent misrepresentation claims against professionals involved in subsurface projects. See, e.g., *Raymond International, Inc. v. Baltimore County, Maryland*, 412 A.2d 1296 (Md. Ct. Spec. App. 1980) (allowing contractor's claim grounded on allegations of fraud, negligent misrepresentation and breach of warranty and asserted against both the owner and the engineer, and finding that the engineer's estimates were materially wrong and substantially inaccurate, and were based on information that the engineer reasonably should have known was inaccurate in view of its prior involvement at the site), *cert. denied*, 449 U.S. 1013 (1980); *Lane v. Geiger-Berger Associates*, 608 F.2d 1148 (8th Cir. 1979) (holding that a subcontractor may assert claim against professional based on subcontractor's reliance on representations of professional as to character of fill that would be permitted on project). But see *APAC-Carolina, Inc. v. Greensboro-High Point Airport Auth.*, 431 S.E.2d 508 (N.C. Ct. App. 1993) (rejecting contractor's negligent misrepresentation claim against engineering firm based on finding that contractor's reliance on engineer's plans and estimates was not justifiable because contractor failed to fully and adequately inspect available information, and contract clearly stated any quantities mentioned in contract were merely estimates); *Bilotta Construction Corp. v. Village of Mamaroneck*, 199 A.D.2d 230 (N.Y. App. Div. 1993) (rejecting contractor's negligent misrepresentation against owner and engineer on the grounds that the contractor could not establish justifiable reliance on the alleged misrepresentation of information regarding elevations because the contract documents made clear that data concerning elevations was approximate only, with no guaranty of accuracy; and the contractor was obligated to conduct its own investigation prior to entering into the contract); *L. Loyer Construction Co. v. City of Novi*, 446 N.W.2d 364 (Mich. Ct. App. 1989) (rejecting contractor's misrepresentation claim against owner and engineer arising out of allegations that certain work required to complete the project varied from estimates in the contract documents, based on findings that the owner did not guaranty the representation as to soil conditions; the risk that soil and water conditions might

be different than expected was expressly assumed by the contractor; the quantities listed in the proposal were not guaranteed to be actual quantities; and the contractor had a duty to investigate the specific subsurface conditions, as well as the site generally), *appeal denied*, 435 Mich. 871 (1990); *Costanza Construction Corp. v. City of Rochester*, 147 A.D.2d 929 (N.Y. App. Div. 1989) (contractor asserted claim against owner and engineer based on alleged gross inaccuracies in estimates of cubic yards to be excavated; court rejected contractor's negligent misrepresentation, breach of warranty and equitable adjustment claims ruling that, as a matter of law, the contractor could not establish justifiable reliance on estimates because the contract documents disclaimed the accuracy or completeness of information on the drawings concerning existing conditions, and obligated the contractor to satisfy itself as to site conditions), *appeal dismissed*, 541 N.E.2d 429 (N.Y. 1989); *P & Z Pacific, Inc. v. Panorama Apartments, Inc.*, 372 F.2d 759 (9th Cir. 1967) (holding that subcontractor could not recover from owner or general contractor on negligent misrepresentation or breach of warranty claim based on conditions encountered in project but not revealed in soil report prepared by engineer, because no representation was made that the soil conditions described in reports would be uniform throughout the project). See also *Foster Wheeler Enviresponse, Inc. v. Franklin County Convention Facilities Authority, et al.*, 678 N.E.2d. 519 (Ohio 1997) (Note 41, *infra*).

27. The theory of negligent misrepresentation has its doctrinal foundation in §552 of the *Restatement (Second) of Torts* (1977), which provides:

    §552. Information Negligently Supplied for the Guidance of Others

    (1) One who, in the course of his business, profession or employment, or in any other transaction in which he has a pecuniary interest, supplies false information for the guidance of others in their business transactions, is subject to liability for pecuniary loss caused to them by their justifiable reliance upon the information, if he fails to exercise reasonable care or competence in obtaining or communicating the information.

    (2) Except as stated in Subsection (3), the liability stated in Subsection (1) is limited to loss suffered

        (a) by the person or one of a limited group of persons for whose benefit and guidance he intends to supply the information or knows that the recipient intends to supply it; and

        (b) through reliance upon it in a transaction that he intends the information to influence or knows that the recipient so intends or in a substantially similar transaction.

    (3) The liability of one who is under a public duty to give the information extends to loss suffered by any of the class of persons for whose benefit the duty is created, in any of the transactions in which it is intended to protect them.

    See also S. Stein, *Construction Law*, ¶5.04[4] [c] (New York: Matthew Bender, 1997).

28. In many instances, an owner's unwillingness to fund an otherwise valid differing

site conditions claim may result in the inability of the project to proceed altogether, given physical or other limiting site constraints that render the project infeasible in the absence of the performance of additional redesign and/or work associated with, and required by, the condition in question. Moreover, in certain circumstances, the cost and time associated with the condition may render the project economically infeasible and/or otherwise unattractive to the owner, investors, and/or construction lenders. Similarly, owners and/or their lenders may not carry adequate contingencies for unanticipated occurrences encountered during construction, thus increasing the risk of heightened tension not only between the owner and contractor but also between the design professional, owner, and contractor, if and when such conditions are encountered.

29. See C. Allen Foster, et al., eds., *Construction and Design Law*, Chapter 12, Section 12.2 (Charlottesville, VA: The Mitchie Company, 1989).

30. See Alan S. Pralgever, "Common Law Theories for Recovery," in *Differing Site Conditions Claims*, R. F. Cushman and D. Tortorello, eds. (New York: John Wiley & Sons, Inc., 1992).

31. A differing site conditions clause, when included in the construction contract documents, is intended to provide *limited* relief from the common law rule under which the contractor is generally responsible for bearing the time/cost impacts due to the discovery of differing site conditions. C. Allen Foster et al., eds., *Construction and Design Law*, Chapter 12, Section 12.2b, (Charlottesville, VA: The Mitchie Company, 1989); J. Sweet, *Legal Aspects of the Architecture, Engineering and the Construction Process*, 5th ed., Section 25.02, (St. Paul, MN: West Publishing Co., 1994), pp. 539–40; S. Stein, *Construction Law*, ¶3.03[2] [a] & 5.07[1] [b] [viii] (New York: Matthew Bender, 1997). In order to be entitled to the limited relief provided by such a clause, the contractor must comply with the requirements set forth in the clause with respect to notice and the presentation and proof of such a claim. See generally J. Jansey and D. Kessler, "Differing Site Conditions," *Legal Handbook for Architects, Engineers and Contractors*, A. Dib, ed., Vol. 4 (New York: Clark Boardman Gallagher, 1988).

32. A detailed discussion of differing site condition clauses is contained in Chapter 4. In addition, there are many excellent resource materials regarding the various types of differing site condition clauses, see, e.g., J. Sweet, *On Construction Industry Contracts*, 3rd ed., Vol. 1, Chapter 13 (New York: John Wiley & Sons, Inc., 1996); R. F. Cushman and D. Tortorello, eds. *Differing Site Condition Claims*, Chapter 2 (New York: John Wiley & Sons, Inc., 1992); Cibinic & Nash, *Administration of Government Contracts*, 3rd ed. Chapter 5, (The George Washington University Press, 1995); R. S. Brams and C. Lerner, eds., *Construction Claims Deskbook Management, Documentation, and Presentation of Claims*, (New York: John Wiley & Sons, Inc., 1996); C. Allen Foster et al., eds., *Construction and Design Law*, Chapter 12 (Charlottesville, VA: The Mitchie Company, 1989); Hann & Maloney, "What's Different about Differing Site Condition Claims?" *Defense Couns. J.*, July 1990, p. 304; R. Lane, "Differing Site Conditions," *Construction Briefings*, 2nd Series, No. 92-11 (Washington, DC: Federal Publications, Inc., October 1992).

Although a substantial amount of written material is available on the subject of differing site conditions, there has not been much commentary on the potential liability of design and construction management professionals involved in subsurface projects. See D. J. Hatem, "Differing Site Conditions: Liability Precautions for Design and Construction Management Professionals," *Defense Couns. J.* (October 1994).

33. Although virtually all public and a significant number of private owners include differing site conditions clauses in contract documents, some owners opt not to include them for a variety of reasons, such as concern about price certainty prior to the start of construction. J. Sweet, (Note 32, *supra*), §13.1 p. 399 & n. 4. As one commentator has noted:

> [T]here are valid reasons for transferring the risk of unknown site conditions to the contractor. First, the cost efficiencies of a differing site conditions clause are only seen in the aggregate. On any particular project, the efficiency of the clause will depend upon the conditions actually encountered. An owner that rarely engages in construction may be more concerned with the potential for a catastrophic cost overrun than the incrementally higher construction costs that the differing site conditions clause may cause. Second, some owners, particularly public owners, have limited funds for the construction of a project. Substantially increasing the project budget to accommodate a changed condition may be impractical. Third, placing the risk on the contractor provides the contractor with an incentive to minimize the financial effect of the discovered condition. If the contract has a differing site conditions clause, the contractor may see the changed condition as an opportunity to recoup other losses on the project at the owner's expense. Finally, in a competitive market, empirical evidence indicates that contractors do not quantify the risk of differing site conditions and may undervalue the risk. Under these conditions, elimination of the differing site conditions clause benefits the owner at little or no cost.

H. Ashcraft, "Avoiding and Managing Risk of Differing Site Conditions," Chapter 1, in *Differing Site Condition Claims*, R. Cushman and D. Tortorello, eds., (New York: John Wiley & Sons, Inc., 1992) (Reprinted by permission of John Wiley & Sons, Inc.).

34. C. Allen Foster et al., eds., *Construction and Design Law*, Chapter 12, Sections 12.4f & 12.3a (Charlottesville, VA: The Mitchie Company, 1989) (Reprinted with permission; Copyright 1989, The Mitchie Company, Charlottesville, VA. All rights reserved.)

In *Foster Constr. C.A. & Williams Bros. Co. v. United States*, 435 F.2d 873, 887 (Ct. Cl. 1970), the Court of Claims defined the purpose of a differing site conditions clause, as follows:

> The purpose of the changed conditions clause is ... to take at least some of the gamble on subsurface conditions out of bidding. Bidders need not weigh the cost and ease of making their own borings against the risk of encountering an adverse subsurface, and they need not consider how large a contingency should be added to the bid to cover the risk. They will have no windfalls and no disasters. The Government benefits from more accurate bidding, without inflation for risks which may not eventuate. It pays for difficult subsurface work only when it is encountered and was not indicated in the logs.

See also Anderson, "Changes, Changed Conditions and Extras in Government Contracting," 42 *Ill. L. Rev.* 29, 47 (1947).

In *Iacobelli Construction Inc. v. County of Monroe*, 32 F.3d 19, 23-24 (2nd Cir. 1994), the court summarized the purpose of a differing site conditions clause as follows:

> Once construction begins on a project under a contract that is silent about the risk of unforeseen conditions, a contractor bears the risk of running into conditions that were unforeseen at the time he submitted his bid even though they significantly increase the cost of performance. To prevent contractors from bidding on a worst-case-scenario basis, government construction contracts contain a differing site conditions clause. . . . This clause imposes upon the government the risks for conditions that the contract documents fail to disclose, but leaves upon the contractor the costs of encountering those conditions described in the contract. . . . Because the differing site conditions clause alleviates the need for contractors to insert speculative contingency costs in their bids, it reduces inflated bidding, and the government presumably saves money by getting lower bids. . . . The government may further reduce the contractor's need to pad its bid by furnishing as much information about the site to the bidders as it can.

35. Although many courts and administrative boards have declined to strictly enforce notice obligations so as to preclude recovery for otherwise valid differing site conditions claims, see generally J. Brewer, "Effect of Notice Requirements on Differing Site Condition Clauses," Chapter 3, in *Differing Site Condition Claims*, R. F. Cushman and D. Tortorello, eds. (New York: John Wiley & Sons, Inc., 1992); C. Allen Foster et al., eds., *Construction and Design Law*, Chapter 12, Section 12.3c (Charlottesville, VA: The Mitchie Company, 1989), there is a greater tendency to do so in circumstances in which the owner can demonstrate some prejudice due to the failure to receive timely or adequate notice (for example, where an owner was impaired in its ability to evaluate the claim because the site conditions were disturbed or removed prior to notice; or where the owner could have mitigated or avoided costs through redesign or otherwise, had the contractor given timely notice). See *Blackenship Construction Co. v. North Carolina State Highway Commission*, 222 S.E.2d 452, 462-63 (N.C. Ct. App. 1976) (contractor's oral notification to owner of changed conditions did not satisfy contract's provision for written notification and contract's requirement for, in event of dispute, accurate reviewable record of extra costs associated with changed conditions, where contractor did not prepare records and file claims until after completing project); *Inland Bridge Company, Inc. and Roadbuilders, Inc. v. North Carolina State Highway Commission*, 227 S.E.2d 648, 656 (N.C. Ct. App. 1976) (contractor not entitled to additional compensation for extra work where it failed to comply with provision of contract calling for contractor to notify engineer promptly in writing on differing site conditions); *McDevitt & Street Co. v. Marriott Corp.*, 713 F. Supp. 906 (E.D. Va. 1989) (contractor not entitled to extra compensation for work necessitated by changed conditions where contractor failed to give contractually-required written advance notice, even though architect was aware of changes, since there was no evidence that architect or owner was aware of attendant additional costs), *aff'd in part, rev'd*

in part, 911 F.2d 723 (4th Cir. 1990); *Glynn v. City of Gloucester*, 487 N.E.2d 230 (Mass. App. Ct. 1986) (barring contractor's claim for extra costs incurred as a result of alleged differing subsurface conditions encountered, because contractor failed to provide owner with timely, written notice of claim, as required by contract documents); *Skopek Bros., Inc. v. Webster Housing Authority*, 416 N.E.2d 1006 (Mass. App. Ct. 1981) (holding that contractor did not meet statutory notice requirements and was not entitled to statutory "equitable adjustment" for extra work performed as a result of encountering alleged differing site conditions, where contractor did not notify owner of claim for extra costs until sixteen months after encountering the alleged conditions); *D. Federico Co., Inc. v. Commonwealth*, 415 N.E.2d 855 (Mass. App. Ct. 1981) (barring contractor's claim for extra costs for excavation work performed which greatly exceeded design professional's estimates, because contractor did not give notice of claim "on or before the first working day following commencement of ... such work," as required by contract documents); *Cecio Brothers, Inc. v. Town of Greenwich*, 244 A.2d 404 (Conn. 1968) (holding that subcontractor was not entitled to recover damages from owner on basis of unjust enrichment for extra work performed as a result of encountering alleged differing site conditions, where contractor did not give timely notice of the claim in accordance with the terms of its subcontract; noting that had the subcontractor provided timely notice, contractor and owner would have had opportunity to inquire into claim and try to save added expense by changing site plans or reducing costs in other portions of project).

The fact that the notice requirement is included in a differing site conditions clause that is statutorily or otherwise legally mandated for inclusion on a public project or other project involving public funds may also be an influential factor militating in favor of strict enforcement of that requirement. See generally S. Stein, *Construction Law* ¶5.07 (New York: Matthew Bender, 1997).

36. Typically, the bid and contract documents admonish, require, or otherwise direct the bidders/contractor to visit the site and become familiar with the site conditions. J. Schulz, "Effect of Site Investigation Clauses on Claims," Chapter 5 in *Differing Site Condition Claims*, R. F. Cushman and D. Tortorello, eds. (New York: John Wiley & Sons, Inc., 1992). Under such circumstances, if the bidder/contractor fails to undertake such a site investigation (or a reasonable site investigation, as determined by the circumstances, including the time permitted for the preparation of its bid, see, e.g., *Frederick Snare Corp. v. Maine-New Hampshire I.B.A.*, 41 F. Supp. 638 (D.N.H. 1941)), a court may deny recovery of an otherwise valid differing site conditions claim.

37. In general, a contractor will not prevail on a differing site conditions claim where the information concerning site conditions on which the contractor claims he or she relied is determined to be "incomplete, irrelevant to the site, or otherwise so deficient that a reasonable and prudent Contractor would not rely on [it] ... except in cases where no other conflicting information is available to the Contractor." C. Allen Foster et al., eds., *Construction and Design Law*, Chapter 12, Section 12.3b (Charlottesville, VA: The Mitchie Company, 1989). See *Clark v. United States*, 5 Cl. Ct. 447, 452 (1984) (denying contractor's claim for equitable adjustment, which was based on allegations of differing site conditions, on the ground that contractor was unable to prove that he was "misled" by the contract documents, as there were no indications in the contract documents that could or did induce rea-

sonable reliance by the contractor that subsurface conditions would be more favorable than those encountered), *aff'd*, 770 F.2d 180 (Fed. Cir. 1985); *Round Place, Inc. v. United States*, 31 Fed. Cl. 749 (1994) (precluding contractor's differing site condition claim based upon finding of unreasonable reliance where available information and contractor's actual experience were ignored). See also Cibinic & Nash, *Administration of Government Contracts*, 3rd ed., (The George Washington University Press, 1995), p. 508.

38. Both estimated quantity and variation in quantity provisions may have an impact on the application of a differing site conditions clause. The Federal Court of Claims has rejected the notion that a significant variation from an estimated quantity, standing alone, constitutes a differing site condition. See *Perini Corp. v. United States*, 381 F.2d. 403, 410 (Ct. Cl. 1967). However, estimated quantities may serve as a basis for a differing site condition where the change in quantities could not have been foreseen, i.e., where the contractor bid without an opportunity to verify the quantities. *Schutt Construction Co., Inc. v. United States*, 353 F.2d 1018 (Ct. Cl. 1965). However, if the accuracy of the quantities was disclaimed (see Notes 40 and 41, *infra*), then recovery may be denied. Moreover, estimated quantities may provide contractors with implied indications of subsurface conditions and, therefore, on that independent basis, may provide the foundation of a differing site conditions claim. *Foster Const. C.A. & Williams Bros. Co. v. United States*, 435 F.2d 873 (Ct. Cl. 1970); *Perini Corporation v. United States*, 381 F.2d 403 (Ct. Cl. 1967).

39. As to issues relating to the enforceability of disclaimers regarding site conditions, see R. Lane, "Differing Site Conditions" (*supra*, Note 32), pp. 10–12; F. Cohen and J. Oliviero, "Effect of Exculpatory Clauses," Chapter 4, in *Differing Site Condition Claims*, R. F. Cushman and D. Tortorello, eds. (New York: John Wiley & Sons, Inc., 1992); C. Allen Foster et al., eds., *Construction and Design Law*, Chapter 12, Section 12.4e (Charlottesville, VA: The Mitchie Company, 1989). Disclaimer provisions may take several forms and be included in contract documents that may or may not contain differing site conditions clauses. These disclaimer provisions may contain any and/or all of the following statements: (a) that the accuracy or completeness of site information or estimated quantities is disclaimed and should not be relied on by the bidder/contractor; (b) that information concerning site conditions is not part of the contract documents and is furnished solely for the information and convenience of the bidder/contractor and is not guaranteed or warranted; (c) that the contractor should (must) conduct his or her own independent examination of site conditions and may not rely on the accuracy or completeness of information furnished or included in the contract documents; (d) that estimated quantities are furnished for bid comparison purposes only, and are not guaranteed; and (e) that information concerning site conditions contained in the bid and contract documents are not part of the contract documents, and were prepared solely to assist the owner in connection with the design of the project.

40. Although the effectiveness of disclaimer-type provisions concerning site conditions as a complete barrier to recovery on an otherwise valid differing site conditions claim is the subject of some uncertainty and unpredictability, at a very minimum, many courts will consider such provisions in judging whether the contractor's reliance on the disclaimed site information was *reasonable* under the circumstances. If a court were to find that the contractor did not act reasonably in relying on the

allegedly inaccurate or incomplete site information (which was disclaimed), then the contractor may not be entitled to any recovery, even assuming the existence of a differing site condition clause.

In several cases, courts (sometimes employing different reasoning) have rejected differing site conditions claims because of the inclusion in the agreement of disclaimer-type provisions. See, e.g., *D. Federico Co., Inc. v. Commonwealth*, 415 N.E.2d 855, 857 (Mass. App. Ct. 1981) (holding that where the contract terms specifically exclude warranty of, or reliance on, furnished estimates, contractor may not recover for additional compensation for extra excavating and filling); *In re D. Federico Co., Inc.*, 8 B.R. 888, 895 (Bankr. D. Mass. 1981) (contractor barred from recovery for extra costs due to "numerous and conspicuous" disclaimers by owner for liability for location of underground obstructions); *McDevitt & Street Co. v. Marriott Corp.*, 713 F. Supp. 906, 910-11 (E.D. Va. 1989) (where owner expressly disclaimed any responsibility for data in soils report as being "representative of the conditions, and materials which may be encountered," and stated that test results were made available solely as a matter of convenience and general information, contractor's differing site conditions claim rejected) *aff'd in part, rev'd in part*, 911 F.2d 723 (4th Cir. 1990); *Eastern Tunneling Corp. v. Southgate Sanitation Dist.*, 487 F.Supp. 109, 113–14 (D. Colo. 1979) (rejecting contractor's differing site conditions claim based on finding that the contract documents, which disclaimed any liability for inaccuracies contained in geological data, placed the risk of uncertainty in regard to differing soil conditions on contractor); *Costanza Constr. Corp. v. City of Rochester*, 147 A.D.2d 929 (N.Y. App. Div. 1989) (by signing contract that clearly included owner's disclaimer of any responsibility for accuracy of drawings concerning conditions, including rock, contractor obligated itself to make in-depth site inspection and cannot recover for additional costs, even where actual rock encountered was six times more than specifications estimated), *appeal dismissed*, 541 N.E.2d 429 (N.Y. 1989); *W.H. Lyman Constr. Co. v. Village of Gurnee*, 403 N.E.2d 1325, 1328 (Ill. App. Ct. 1980) (contractor denied recovery because nothing in plans indicated that soil boring logs shown in plans were meant to be "specifically relied upon so as to relieve" contractor of its contractual responsibility to inspect site, including subsurface conditions); *Daniel O'Connell's Sons Inc. v. Commonwealth*, 212 N.E.2d 219, 220–222 (Mass. 1965) (where excavation contractor relied on geological data provided by owner despite owner's express disclaimer of responsibility for data's accuracy and provision that contractor "must satisfy himself by his own investigation and research regarding all such conditions," contractor was not entitled to recover costs it incurred for extra work due to differing site condition); *Green Construction Co. v. Kansas Power & Light Co.*, 1 F.3d 1005 (10th Cir. 1993) (holding that owner did not create an implied warranty by providing some soil information where owner instructed the contractor that the information may not be complete and required the contractor to conduct an independent site and soil investigation).

Similarly, if a contractor reads *only a portion* of a soil report or other site information contained in the contract documents, or otherwise does not act reasonably and in accordance with the norm in the construction industry, a court may reject the contractor's differing site conditions claim. See, e.g., *Gardner-Zemke Co. v. State of New Mexico*, 790 P.2d 1010, 1016 (N.M. 1990) (although court did not reach issue of "exculpatory clauses" in contract, court denied contractor's differing site condition claim where contractor failed to read soil report in its entirety or to inquire about inconsistencies in the report); *Siteman v. Woodward-Clyde & Associates, Inc.*, 503 S.W.2d 141, 147 (Mo. Ct. App. 1973) (affirming judgment in favor of engineering firm on professional malpractice claim asserted against it by owner; owner alleged that based on the contents of an ambiguous/inaccurate report prepared by engineering firm on soil conditions and foundation requirements, owner assumed risk of excavation in exchange for lower contract price with contractor; court properly rejected owner's claim against engineer based on finding that the owner did not rely on said report in negotiating with contractor); *Texas Tunneling Co. v. City of Chattanooga*, 329 F.2d 402 (6th Cir. 1964) (engineering firm that failed to include in its boring log a drawing showing results of test drillings information about percentage of core recoveries from another firm's report was not liable to subcontractor, which could have availed itself of all information at city engineer's office).

Some courts have precluded recovery for differing site conditions claims where there are disclaimer-type provisions in the contract documents on the grounds that the inclusion of the disclaimer negated as a matter of law any express or implied warranty as to the accuracy or completeness of the specifically disclaimed information or estimates, thereby eradicating any warranty obligation that otherwise may have existed in favor of the contractor (see Subsection 10.2.2.1(A)). See, e.g., *Eastern Tunneling Corp. v. Southgate Sanitation Dist.*, *supra*, 487 F. Supp. at 114; *H. W. Lyman Coast Co. v. Village of Gurnee*, *supra*, 403 N.E.2d at 1328; *D. Federico Co. v. Commonwealth*, *supra*, 415 N.E.2d at 857; *In re D. Federico Co.*, *supra*, 8 B.R. at 895; *Brown Brothers, Inc. v. Metropolitan Government of Nashville*, 877 S.W.2d 745 (Tenn. App. 1993) (where plan provided that figures estimating amount of rock to be excavated were provided only for convenience of contractor and warned bidders to "rely exclusively upon their own estimates, investigation and other data," court affirmed that plan did not constitute a warranty as to subsurface conditions); *Brant Const. Co., Inc. v. Metropolitan Water Reclamation Dist.*, 967 F.2d 244 (7th Cir. 1992) (court rejected excavation contractor's contention that owner breached implied warranty that plans owner furnished would be adequate to determine quantity of material to be removed on grounds that owner specifically stated that plans were approximations and disclaimed that the quantity of material to be removed would be ultimate quantity; accordingly, actual reliance by contractor on estimated quantities in submitting bid was immaterial "because such reliance would appear to have been unreasonable"); *Fondedile, S.A. v. C.E. Maguire Inc.*, 610 A.2d 87 (R.I. 1992) (rejecting contractor's breach of express warranty claim, holding that there can be no express warranty where contract documents contained disclaimers that disclaimed responsibility for accuracy of data provided in bid packet); *J.F. White Contracting Co. v. Massachusetts Bay Transportation Authority*, 666 N.E.2d 518 (Mass. App. Ct. 1996) (where contract clearly provided that actual costs incurred by contractor could exceed allowance estimates furnished by owner, no merit to contractor's agreement that it "relied" upon estimates), *rev. denied*, 670 N.E.2d 966 (Mass. 1996). See also *Bilotta Construction Corp v. Village of Mamaroneck*, 199 A.D.2d 230 (N.Y. App. Div. 1993) (rejecting contractor's negligent misrepresentation against owner and engineer on the grounds that the contractor could not establish justifiable reliance on the alleged misrepresentation of information regarding elevations because the contract documents made clear that data concerning elevations was approximate only, with no guaranty of accuracy; and the contractor was obligated to conduct its own investigation prior to entering into the contract); *APAC-Carolina, Inc. v. Greensboro-High Point Airport Auth.*, 431 S.E.2d 508 (N.C. Ct. App. 1993) (rejecting contractor's negligent misrepresentation claim against engineering firm based on finding that contractor's reliance on engineering firm's plans and estimates was not justifiable because contractor failed to fully and adequately inspect available information, and contract clearly stated any quantities mentioned in contract were merely estimates), *rev. denied*, 438 S.E.2d 197 (N.C. 1993); *L. Loyer Construction Co. v. City of Novi*, 446 N.W.2d 364 (Mich. Ct. App. 1989) (rejecting contractor's misrepresentation claim against owner and engineer arising out of allegations that certain work required to complete project varied from estimates in contract documents, based on findings that owner did not guaranty the representation as to soil conditions; risk that soil and water conditions might be different than expected was expressly assumed by contractor; quantities listed in proposal were not guaranteed to be actual quantities; and contractor had duty to investigate the specific subsurface conditions, as well as the site generally), *appeal denied*, 435 Mich. 871 (1990); *Costanza Construction Corp. v. City of Rochester*, 147 A.D.2d 929 (N.Y. App. Div. 1989) (contractor asserted claim against owner and engineer based on alleged gross inaccuracies in estimates of cubic yards to be excavated; court rejected contractor's negligence, breach of warranty and equitable adjustment claims ruling that, as a matter of law, the contractor could not establish justifiable reliance on estimates because the contract documents disclaimed the accuracy or completeness of information on drawings concerning existing conditions, and obligated the contractor to satisfy itself as to site conditions).

Even in circumstances in which differing site conditions clauses are included in the contract documents, disclaimer-type provisions may preclude recovery. In *Millgard Corp. v. McKee/Mays*, 49 F.3d 1070 (5th Cir. 1995), a subcontractor encountered wet soil while sinking piers for a foundation. A soil report accompanying the bid documents signaled dry soil, but the bid documents stated that the soil report was "not part of the contract documents." A provision in the subcontract allowed for an equitable adjustment if the subcontractor encountered conditions at variance with those indicated in the contract documents. Another clause in the contract documents disclaimed all responsibility for the accuracy of the soils investigation information furnished to the contractor, but not made part of the contract documents. The subcontractor, during the bid process, was provided with a copy of the soils investigation report performed by an independent engineering firm retained by the owner. The instructions to bidders warned that the soil report and log of borings was available solely for information and that the report was not a warranty of subsurface conditions, nor was it part of the contract documents. Bidders were expected to examine the site and the subsurface investigation reports and then decide for themselves the character of the materials to be encountered; the owner, engineer and construction manager disclaimed any responsibility for the accuracy, true location and extent of the soils investigation. During construction, the subcontractor encountered a layer of quicksand-like material between 5 and 15 feet thick between the fill area and the sand and gravel layer. These conditions affected 74 percent of the holes drilled and made the subcontractor's drilling plans impracticable. Millgard spent more money that it had expected and this delayed completion of the project. The subcontractor presented a claim under the differing site conditions clause. In denying recovery, the court pointed to the disclaimers referenced above and the statement that the soil report was not part of the contract documents, stating that "[i]f the soil report is not part of the contract documents, it cannot form the basis of a claim that conditions were at variance with the conditions indicated in the Contract Documents." The court further noted that "conditions indicated by the Contract Documents" do not embrace soils reports that, by definition, are explicitly excluded from the definition of "contract documents." The subcontractor argued that giving effect to the literal wording of the disclaimers would gut the differing site conditions clause. The court, however, rejected this contention stating, "the converse holds true—allowing reliance on the soils reports under the concealed condition clause would eviscerate the disclaimers. . . . [T]he disclaimers and the language of the project manual show that the parties placed the risk of underground water on [the subcontractor]. [The subcontractor] took its chances by not boring its own hole and instead relying on the soil reports. The bargain struck by the parties allocated the risk and there it ends. We enforce the contract."

Not all courts have reached the same conclusion as the court in *Millgard* in circumstances in which an apparent conflict exists between disclaimer-type provisions and the inclusion of a differing site conditions clause. J. Beard, "Contract Allocation of Risk of Differing Site Conditions," Chapter 2, p. 34, in *Differing Site Condition Claims*, R. F. Cushman and D. Tortorello, eds. (New York: John Wiley & Sons, Inc., 1992). See also *Raymond International, Inc. v. Baltimore County, Maryland*, 412 A.2d 1296 (Md. Ct. Spec. App. 1980) (allowing contractor's claim for an equitable adjustment based upon alleged unforeseen conditions encountered at site, despite the existence of exculpatory clause requiring contractor to make a "reasonable inspection" of the site for the purpose of determining whether conditions were as depicted in contract documents; court found that contractor was not reasonably able to discover true conditions of site itself and was therefore entitled to rely upon representations made by engineer; court further found that engineer's estimates were materially wrong and substantially inaccurate, and were based upon information that the engineer reasonably should have known was inaccurate in view of its prior involvement at site), *cert. denied*, 449 U.S. 1013 (1980); *Ray D. Lowder, Inc. v. North Carolina State Highway Commission*, 217 S.E.2d 682 (N.C. Ct. App. 1975) (allowing contractor to obtain adjustment in unit price from owner, despite existence of disclaimer, where contractor experienced an excessive overrun in estimated quantities, and where awarding authority did not disclose to contractor the actual data report upon which the estimates were derived), *cert. denied*, 218 S.E.2d 467 (N.C. 1975).

Courts that have declined to enforce disclaimer-type provisions to entirely preclude recovery for an otherwise valid differing site condition have done so on several bases. In circumstances where a disclaimer-type provision is included in contract documents that also include a differing site conditions clause, some courts have reasoned that to give effect to the disclaimer would undermine and eradicate the purpose of the differing site conditions clause (i.e., to afford relief to the contractor if actual physical conditions prove materially different from those indicated). *United Contractors v. United States*, 368 F.2d 585, 598 (Ct. Cl. 1966) ("broad exculpatory clauses . . . cannot be given their full literal reach, and do not relieve the [owner] of liability for changed conditions as the broad language thereof would seem to indicate"). Other courts have generally refused to enforce disclaimers in the following three circumstances: (1) where the contractor independently performed a reasonable investigation that confirmed or supported the disclaimed information, see, e.g., *Morrison-Knudsen Co., Inc. v. United States*, 397 F.2d 826 (Ct. Cl. 1968); *J. F. Shea Inc. v. United States*, 4 Cl. Ct. 46 (1983); (2)

where a court concluded that the contractor's reliance upon the disclaimed information was reasonable and justified based upon convincing evidence that the owner intended and/or expected the contractor to rely upon the information in preparing its bid, see, e.g., *Eastern Tunneling, supra*; *Western States Mechanical Contractors, Inc. v. Sandia Corp.*, 798 P.2d 1062 (N.M. Ct. App. 1990) (holding that a disclaimer of rock estimates contained in soil investigation report was not sufficient to preclude bidder's reliance on rock estimates where separate, positive representations were made in bid materials regarding rock estimates; rather, disclaimer must be included in the bid and contract documents to be effective), *cert. denied*, 798 P.2d 1039 (N.M. 1990); and/or (3) where it is determined that reliance was reasonable and justified because the circumstances did not allow sufficient time for the contractor to conduct an adequate and independent investigation, see, e.g., *Eastern Tunneling, supra*, *Frederick Snare, supra*. See also *In re D. Federico Company*, 16 B.R. 282 (D. Mass. 1981) (holding that estimate disclaimer language may not be effective where the awarding authority possesses superior knowledge—i.e., relevant information about site conditions—which is not disclosed to bidders, including non-disclosure of a data report upon which the estimates were derived); *Ray D. Lowder, Inc. v. North Carolina State Highway Commission*, 217 S.E.2d 682 (N.C. Ct. App. 1975) (allowing contractor to obtain adjustment in unit price in circumstances in which the contractor experienced an over 2,000 percent overrun in estimated quantities and in which the awarding authority did not disclose to the contractor the actual data report on which estimates were derived), *cert. denied*, 218 S.E.2d 467 (N.C. 1975); *Alpert v. Commonwealth*, 258 N.E.2d 755 (Mass. 1970) (holding that contractor was entitled to equitable adjustment for additional quantities beyond those estimated in bid and contract documents in circumstances in which the Department of Public Works had advised bidders that they had been given "complete information," but department had not disclosed all information in its possession—including boring samples and analysis; and in which the contractor was not informed that borings were only taken in limited area of project site and provided an inadequate basis for department's quantity estimates). See generally M. Loulakis, *Construction Management Law and Practice*, §4.7, p. 97 & n. 37 (New York: John Wiley & Sons, Inc., 1995).

Clearly, in many jurisdictions, a substantial amount of confusion, uncertainty, and controversy surrounds the issue of enforceability of disclaimer-type provisions, especially when used in conjunction with differing site conditions clauses. One noted commentator has observed that if the contract is clear that the contractor is assuming the risk of accuracy/completeness of specifically disclaimed information, the law should enforce the disclaimer. J. Sweet, *Legal Aspects of Architecture, Engineering and the Construction Process*, 5th ed. (St. Paul, MN: West Publishing Co., 1994), p. 402. In any event, given the uncertainty regarding the enforceability of disclaimer-type provisions, it is probable that prudent bidders will necessarily include some contingency for the risk associated with the disclaimed information, thereby increasing the contract price regardless whether the risk materializes and even if the contractor is allowed an equitable adjustment under the differing site conditions clause due to the unenforceability of the disclaimer. In addition, the inclusion of a disclaimer provision determined to be effective to preclude recovery by the contractor, may lead to a higher incidence of contractor claims against the design and construction management professional due to the absence of any remedy under the owner-contractor agreement. See, e.g., *M. Miller Co. v. Dames*

& Moore, 18 Cal. Rptr. 13 (Cal. Ct. App. 1961) (allowing contractor to pursue negligence claim against soil engineers based on content of soil reports, despite the fact that owner's agreement with contractor disclaimed accuracy of reports made available to contractor, on the ground that disclaimer in owner-contractor agreement did not shield soil engineer from liability because soil engineer was not party to owner-contractor agreement). See also J. Sweet, *Legal Aspects of Architecture, Engineering and the Construction Process*, 5th ed. (St. Paul, MN: West Publishing Co., 1994), p. 552, and Section 10.2.4.

41. In addition to the cases referenced in Note 40, *supra*, dealing with estimate disclaimers, there are several cases which illustrate these principles. See, e.g., *J. F. White Contracting Co. v. M.B.T.A.*, 666 N.E.2d 518 (Mass. App. Ct. 1996) (rejecting contractor's claim for adjustment in compensation amount specified for allowance items due to substantial increase in actual quantities for allowance items beyond those estimated in bid and contract documents; in part, court denied recovery because contract documents clearly provided that actual costs for performance of allowance items could exceed the allowance amount specified for that work due to variations between estimated and actual quantities of the work and that, in such cases, the stipulated allowance amount would establish the basis of compensation; court also held that the inclusion of estimated quantities for allowance items and the allowance cost for those items did not create any implied warranties in view of disclaimer contained in contract documents), *review denied*, 670 N.E.2d 966 (Mass. 1996).

D. Federico Co., Inc. v. Commonwealth, 415 N.E.2d 855 (Mass. App. Ct. 1981) (contractor sought equitable adjustment due to "substantial underestimation by the Commonwealth's designer, a private engineering firm, of the amount of excavation and replacement fill that would be required to lay a proper base for [certain] roads . . ."; no test borings or soil analysis were done for the purpose of preparing the estimate; the engineering firm based its estimate purely on the assumption that because roads had been constructed previously with no difficulty in the same vicinity, it would not be necessary to excavate beyond an average depth of 12 inches; contractor sought an equitable adjustment in the unit price on the basis that the specified unit price did not adequately compensate it for the work it performed in excavating and filling borrow given the actual quantities encountered; the court rejected this claim stating that the contract made clear that the excavation unit price was to control regardless of actual quantities and that the estimates shown for unit-price items were for purposes of bid-comparison only and were not guaranteed; the court also rejected any theory of implied warranty on the ground that the contract provisions precluded such a warranty in stating that estimated quantities were furnished solely for bid-comparison purposes and were not guaranteed; court concluded that the objective of such provisions was to preclude any warranty as to the accuracy of estimates and to place on the bidder the risk that subsoil conditions might be more adverse than the parties had anticipated).

Brant Construction Co., Inc. v. Metropolitan Water Reclamation District of Chicago, 967 F.2d 244 (7th Cir. 1992) (even though the actual quantities exceeded estimated quantities by more than 7 percent, contractor could not recover beyond stipulated unit price; court rejected contractor's claim that the owner breached an implied warranty that the plans it furnished would be adequate to determine the quantity of material to be excavated; in substance, court held that despite owner's general implied warranty obligation, the contractor was not entitled to recovery beyond stipulated unit price because the contract documents stated that the plans were approximations and declared that the quantity of material to be excavated would be the ultimate quantity, not estimated quantities).

Brown Brothers, Inc. v. The Metropolitan Government of Nashville and Davidson County, 877 S.W.2d 745 (Tenn. Ct. App. 1993) (contractor claimed that it was entitled to renegotiate unit prices because government breached warranties in grossly underestimating quantities of rock to be excavated; court rejected the claim on the basis of disclaimers contained in the contract documents with respect to the accuracy of estimates and which warned bidders they should "rely exclusively upon their own estimates, investigation and other data which are necessary for full and complete information upon which the proposal may be based").

Candee Construction Company, Inc. v. South Dakota DOT, 447 N.W.2d 339 (S.D. 1989) (contractor sought equitable adjustment in specified unit prices due to substantial increase in estimated quantities; court rejected contractor's claim on the basis that the estimated quantities were "clearly and repeatedly identified as estimates only" and contractor was warned that the actual quantities may be higher than those estimated), *cert. denied*, 494 U.S. 1067 (1990).

In several cases involving alleged inaccurate quantity estimates, contractors have asserted claims against both the owner and the owner's independently retained consulting engineer. As in the cases involving claims against *only* the owner, the outcome of such cases has depended on whether the estimated quantities were represented as only approximations and/or whether the owner disclosed relevant information (for example, data reports) on which the estimates were based. See, e.g., *Foster Wheeler Enviresponse, Inc. v. Franklin County Convention Facilities Authority, et al.*, 678 N.E.2d. 519 (Ohio 1997) (affirming dismissal of environmental remediation contractor's negligent misrepresentation claim against owner's engineering consultant; claim was based upon allegations that engineer misrepresented (a) the estimates of tar-contaminated soil—140 cubic yards—set forth in bid and contract documents and (b) that the contractor would be compensated for quantities of contaminated soil excavated in excess of the 140 cubic yard estimate without the need for prior written authorization from the owner; in affirming dismissal of the claim, the court rejected both theories on the grounds that the construction contract stated the quantities set forth were "to be considered as approximate and are to be used for the comparison of bids only," thereby operating as a disclaimer, and rendering the contractor's reliance thereon unreasonable as a matter of law with respect to the anticipated quantity of contaminated soil; and the limitations upon the engineer's authority to approve extra work contained in the construction contract precluded, as a matter of law, the contractor from relying upon any alleged representation of the engineer regarding the lack of need for written orders from the owner; finally, the court also relied upon the economic loss doctrine, see Section 10.1.2.1, as justification for denying the contractor's claim against the engineer); *Bilotta Construction Corp. v. Village of Mamaroneck*, 199 A.D.2d 230 (N.Y. App. Div. 1993) (rejecting claim of contractor against owner and engineer on the basis of misrepresentation of information regarding elevations; in so holding, court held that the contract documents made clear that data concerning elevations was approximate only, with no guaranty of accuracy; that the existing grade elevations were approximations only and that the approximations were not to be relied on by contractor, and that contractor was obligated to conduct its own investigation prior to entering into contract; and, therefore, any reliance by contractor on such data was not reasonable and could not support misrepresentation claim against either owner or engineer since contractor did not establish justifiable reliance upon that data); *Zontelli & Sons, Inc. v. City of Nashwauk*, 373 N.W. 2d 744 (Minn. 1985) (claim against an owner and consulting engineer for negligence and misrepresentation in the preparation of estimates upon which contractor bid on a unit-price basis; bid and contract documents contained no information regarding subsurface soil conditions; contractor asserted that it encountered more concrete than estimated and that the amount of unsuitable material to be excavated was substantially understated and, moreover, was of a different quality than anticipated; contractor claimed that it had to pay more than four times the estimated unit price cost for performance of work and the city denied compensation for cost of work beyond unit prices bid; court refused to limit recovery to unit prices bid on theory that the facts presented an "extreme case" involving all "kinds of overruns and otherwise changed conditions, the extent and effect of which [were] so unusual as not to have been contemplated by the parties at the time of contracting;" the court also found the engineer negligent on the basis that the engineer had "drastically underestimated the quantity of concrete and unsuitable materials to be removed; had failed to indicate the unusual characteristics of those materials; and in other respects); *APAC-Carolina, Inc. v. Greensboro-High Point Airport Authority*, 431 S.E.2d 508 (N.C. Ct. App. 1993) (the contractor asserted claim against owner and engineer due to substantial overruns in undercutting required; court refused to allow claim against engineer on the basis that the contractor had failed to establish any justifiable reliance on information contained in the contract documents because contractor had failed to adequately inspect available information, and the contract clearly stated that any quantities mentioned in the contract were merely estimates), *rev. denied*, 438 S.E.2d 197 (N.C. 1993); *Raymond International, Inc. v. Baltimore County, Maryland*, 412 A.2d 1296 (Md. Ct. Spec. App. 1980) (claim by contractor against owner and engineer for fraud, negligent misrepresentation and breach of warranty; contractor claimed that estimated quantities were inaccurate and that they were grossly underestimated resulting in the unit price representing an inadequate measure of compensation; court allowed recovery against the engineer on the basis that the estimates were materially wrong and substantially inaccurate and engineer reasonably should have known information was inaccurate based upon its prior involvement at the site), *cert. denied*, 449 U.S. 1013 (1980); *L. Loyer Construction Co. v. City of Novi*, 446 N.W.2d 364 (Mich. Ct. App. 1989) (contractor asserted claim against owner and engineer; contractor claimed that the work required to complete the project varied from estimates set forth in contract documents upon which contractor based its bid; and that the quantity variations necessitated the use of unplanned, less efficient methods of excavation and use of a different off-site dump than originally contemplated; that the specified unit prices did not adequately compensate contractor for the additional quantities of work encountered. Court rejected contractor's claim on the grounds that owner did not guaranty the representation as to soil conditions; that the risk that soil and water conditions might be different than expected was expressly assumed by contractor; that the quantities listed in the proposal were not guaranteed to be actual quantities; that contractor had a duty to investigate the specific subsurface conditions, as well as the site generally; and that excavation and the disposal of the disputed work items were to be paid at the contract unit price, not the contractor's actual cost), *appeal denied*, 435 Mich. 871 (1990); *Costanza Construction Corp. v. City of Rochester*, 147 A.D.2d 929 (N.Y. App. Div. 1989) (contractor asserted breach of contract, negligence, breach of warranty and equitable adjustment claims against owner and engineer due to alleged gross inaccuracies in estimates of cubic yards of rock to be excavated; contractor sought compensation beyond unit prices stipulated in contract; court rejected contractor's claims because the contract documents clearly included a disclaimer by the city of any responsibility for the accuracy or completeness of information on the drawings concerning existing conditions, including rock, and obligated the contractor to satisfy itself as to site conditions; accordingly, as a matter of law, the contractor could not establish that it had justifiably relied upon information in the contract documents, thereby warranting dismissal of claims against owner and engineer), *appeal dismissed*, 541 N.E.2d 429 (N.Y. 1989).

42. The entire subject of what information regarding site conditions (actual or anticipated) to furnish or make available to bidders; whether to include that information within the definition of "contract documents"; whether to designate that information (or any portion thereof) as "reference" documents; whether to utilize disclaimers in connection with the specific portions of the information disclosed (see discussion in Note 40, *supra*); whether to furnish or make available to the contractor both data and interpretations; and whether to include all or some of the information in the definition of "contract documents," is the subject of much controversy and discussion in the underground design and construction community. See Opinion Papers on "Geotechnical Reports in Underground Construction," an unpublished collection of papers presented at the April 21, 1996, Underground Technology Research Counsel Forum in Washington, D.C.

Despite the lack of universal consensus on all aspects of the disclosure subject, it appears reasonably clear that owners (and their representatives) should not withhold or conceal from bidders relevant (and especially accurate or reliable) information regarding site conditions. See *Geotechnical "Differing Site Conditions,"* Geotechnical Engineering Notebook, GT Guideline No. 15, (U.S. DOT Federal Highway Administration, April 1996). If there are reservations about the reliability or accuracy of some or all of that information, selective and specific disclaimers and/or appropriate qualifications should be utilized. Nondisclosure, however, is rarely the appropriate response and, more often than not, will generate contractor claims based on fraud, misrepresentation, impossibility of performance, or superior knowledge of the owner regarding site conditions.

Finally, the effectiveness of a disclaimer-type provision in negating and/or precluding a contractor's ability to recover for a differing site condition may be undercut in circumstances where the evidence discloses that the owner and/or the professional consultant failed to disclose material information that was peculiarly in the "superior" knowledge or exclusive possession of the owner and/or the owner's professional consultant (i.e., not generally or reasonably available to bidders within the bid period). See, e.g., *Richardson Electrical Co., Inc. v. Peter Francese & Sons, Inc.*, 484 N.E.2d 108, 110–111 (Mass. App. Ct. 1985) (ruling that electrical subcontractor was entitled to payment for extra work required to dig trench to accommodate telephone cable where construction plans and specifications provided by owner, and relied on by subcontractor in making bid, failed to give notice that subcontractor would be required to dig trench to accommodate telephone cable);

Gulf Contracting v. Bibb County, 795 F.2d 980, 982 (11th Cir. 1986) (allowing general contractor to pursue action for negligent misrepresentation against project architects and engineers, who, although notified of existence of subsurface debris through a soil test report, failed to reveal the problem in the plans and specifications upon which general contractor relied in making its bid, where removal of the debris required that general contractor expend extra time and expense); *In re D. Federico Co.*, 16 B.R. 282, 287–89 (Bankr. D. Mass. 1981) (although instructions to bidders contained exculpatory clause advising prospective bidders to acquaint themselves with conditions at the site, where owner failed to adequately describe in bid documents massive ruins that lay below the mud line that contractor would be required to excavate before rebuilding bulkhead piers, owner must compensate general contractor for extra work "up to the amount that a reasonable bidder who had access to the [undisclosed information] prior to the submission of its bid would have found necessary for the performance of its work") *aff'd*, 723 F.2d 122 (1st Cir. 1983); *Frederick Snare Corp. v. Maine-New Hampshire I.B.A.*, 41 F. Supp. 638 (D.N.H. 1941) (allowing contractor to recover on differing site conditions claim despite contractor's failure to comply with contract provision requiring that each bidder fully inform himself of site conditions, because engineers who drafted the inaccurate specifications knew that proposed bidders would have no opportunity to check the information contained in the bidding plan during the bidding period); *Raymond International, Inc. v. Baltimore County, Maryland*, 412 A.2d 1296 (Md. Ct. Spec. App. 1980) (contractor who encountered conditions on site that differed from those represented in contract documents allowed to recover, despite existence of clause in contract requiring contractor to independently verify information upon which it based its bid because engineer knew or should have known that the representations in the contract documents regarding the conditions of the site were inaccurate in view of its prior involvement at the site), *cert. denied*, 449 U.S. 1013 (1980); *Ray D. Lowder, Inc. v. North Carolina State Highway Commission*, 217 S.E.2d 682 (N.C. Ct. App. 1975) (allowing contractor to obtain equitable adjustment, despite existence of disclaimer in contract documents, where contractor experienced an excessive overrun in estimated quantities and where awarding authority did not disclose to the contractor the actual data report upon which the estimates were derived), *cert. denied*, 218 S.E.2d 467 (N.C. 1975).

43. See, e.g., *W.F. Magann Corporation v. Diamond Manufacturing Company, Inc.*, 580 F. Supp. 1299, 1310-1313 (D.S.C. 1984) (allowing dredging subcontractor to pursue, and recover on, an action based on allegations that specifications provided by owner in bid documents were defective because they contained an incorrect classification of soil borings and did not adequately reflect amount of material to be dredged) *aff'd in part, rev'd in part*, 775 F.2d 1202 (4th Cir. 1985). See also Section 10.2.2.3.

As one commentator has noted, the scenario of the typical differing site conditions claim is as follows:

> A contractor encounters unanticipated additional problems in some phase of the execution of the work. He then attempts to recover the additional costs as a differing or concealed/changed condition under the contract. If the contract will not support his cause, he is likely to pursue it on the basis that the

problems are attributable to inadequate or incorrect information supplied by the owner.

S. Stein, *Construction Law*, ¶5.07[2] [d] [ii] (New York: Matthew Bender, 1997). For further discussion regarding combined differing site conditions and defective design claims, see Section 10.2.2.3.

44. See, e.g., *Granite Construction Co. v. United States*, 24 Cl. Ct. 735 (1991) (additional cost due to performance difficulties resulting from contractor's construction means/methods and/or equipment problems do not justify entitlement under differing site conditions provision); *Appeal of M.A. Mortenson Company*, 90-1 B.C.A. (CCH) ¶22,313 (Sept. 15, 1989) (denying contractor recovery on Type II differing site conditions claim on ground that the contractor's methods and means of construction, or equipment or other performance difficulties, as distinct from physical site conditions of an unknown or unusual character, caused the increased costs claimed); *Appeal of Heintz Construction Company*, 74-1 B.C.A. (CCH) ¶10,562 (Mar. 19, 1974) (owner responded to contractor's differing site conditions claim by arguing that contractor should not be allowed equitable adjustment because contractor's chosen means and methods of construction were unreasonable based on contract documents; court rejected owner's argument, finding no basis to conclude that contractor should have anticipated from contract drawings that required aggregate could not be produced by contractor's crushing operation). See also *Hydromar Corp. v. U.S.*, 25 Cl. Ct. 555 (1992) (ruling that where contractor was given discretion concerning procedures and equipment to be used on project, contractor's selection of construction method which turned out to be inefficient did not amount to a differing site condition), *aff'd*, 980 F.2d 744 (Fed. Cir. 1992); *Appeal of Hydromar Corp.*, 89-3 B.C.A. (CCH) ¶21,898 (May 16, 1989) (partially denying contractor's differing site conditions claim based on a finding that the contractor's use of inappropriate dredging tools contributed to the contractor's performance problems and its claimed increased costs).

45. Frequently, and unfortunately, disputes and claims arise on a construction project due to disappointed expectations of, and/or unknowing assumptions by the owner, design professional, and contractor as to the risk of defective design, unconstructability, or improper, inadequate, or ineffective construction means and methods or equipment. See generally D. Cohen, "Minimizing Conflicts Between Design Professionals and Contractors," *Construction Briefings*, 2nd Series, No. 95-11 (Washington, DC: Federal Publications, Inc., October 1995). Because of the close interaction on subsurface projects between the design of the completed project and the construction means and methods utilized in the construction process, the frequency and severity of these disputes intensifies on subsurface projects. For further discussion, see Section 10.2.2.2.

46. The term "defective design" encompasses any act, error, omission, or other deficiency in the design and, as such, is to be distinguished from the more restricted phrase and standard of "negligent" design. See Section 10.1.1.2. All "negligent" design is "defective," but not all "defective" design is "negligent." In order for "defective" design to be "negligent," there must be adequate proof that the "defective" design represents a departure from the "professional standard of care;" see Section 10.1.1.2.

47. For a general discussion of the implied warranty doctrine, see S. Stein, *Construction Law*, ¶5.07[2] [b] [i] (New York: Matthew Bender, 1997).
48. However, even if the design is defective, a contractor may not recover unless it can prove that it complied with the design; contractor recovery under the implied warranty will result only if the contractor complies with the design furnished by the owner. *Tyger Construction Co., Inc. v. United States*, 31 Fed. Cl. 177, 243 (1994).
49. If the contractor is presented with an obvious omission, inconsistency, or discrepancy in the design or contract documents, the contractor must, at a minimum, request clarification. *John F. Miller Co., Inc. v. George Fichera Construction Corp.*, 388 N.E.2d 1201 (Mass. App. Ct. 1979) (recognizing contractor's obligation to ask for clarification when presented with an obvious omission, but excusing plumbing subcontractor's failure in this regard based on state statute and business custom, which permitted plumbing contractor to confine its attention to plumbing section of specifications and not to look elsewhere for missing toilet room accessories requirements).
50. The term "design professional of record" (DPR) refers to a design professional who typically is engaged by a project owner to design and prepare construction documents (or other documents intended to be employed in the implementation or construction of the design) and to administer the construction (or other improvement objective) of the project that is to be performed by a contractor independently retained by the owner. A DPR is to be distinguished from a design professional who is engaged by a construction contractor and/or specialty trade contractor to design (or to provide other professional services relating to) specific construction means and methods, such as temporary shoring, temporary cofferdams, sheeting, bracing, temporary support systems, form work and associated support systems, and earth support systems. Compare the role of the engineer in *Levy v. Schnabel Foundation Co.*, 584 A.2d 1251 (D.C. App. 1991), who was hired by a contractor to design construction means and methods.
51. As one commentator has stated:

    > The work of construction is the expertise of the Contractor, who should be allowed latitude and leeway in the use of the construction means, methods, techniques, sequences, and procedures as are compatible with the project's requirements and will result in a project completed in accordance with the requirements of the Contract Documents. Control over the construction process, and the safety precautions of programs incident thereto, are within the Contractor's expertise and prerogative and are his sole responsibility.

    J. Clark, "Focus on Shop Drawings," EJCDC Doc. No. 1910-9-C, p. 4, (Alexandria, VA: EJCDC, 1985). See also S. Stein, *Construction Law*, ¶5.04[3][d] (New York: Matthew Bender, 1997).
52. Of course, a design professional may contractually undertake responsibility for the design and/or implementation of construction means and methods and, in those circumstances, would be responsible for negligence in the performance of that undertaking. See, e.g., *Wenzel v. Boyles Galvanizing Co.*, 920 F.2d 778, 781 (11th Cir. 1991) (design professional "specifically agreed to '[p]rovide, implement and

administer a site safety and health plan'"). Similarly, if a design professional, by actual conduct, becomes involved in the design or implementation of construction means and methods, he or she may be subject to liability based on alleged deficiencies and/or omissions in those construction means and methods. See, e.g., *Balagna v. Shawnee County*, 668 P.2d 157 (Kan. 1983) (holding that although architect-engineer had no contractual duty to control contractor's means and methods of construction and no contractual responsibility for safety of workers on job site, architect-engineer's knowledge, pursuant to inspection conducted by architect-engineer, that prescribed safety precautions were not followed at job site at time worker was fatally injured created duty to take some reasonable action to prevent injury to worker).

53. See, e.g., *Hanna v. Huer, Johns, Neel, Rivers & Webb*, 662 P.2d 243, 252 (Kan. 1983) (recognizing and following "the great weight of authority" supporting the rule that an architect does not by reason of its supervisory authority over construction assume responsibility for the day-to-day methods utilized by the contractor to complete the construction); *Transportation Ins. Co. v. Hunzinger Const. Co.*, 507 N.W.2d 136 (Wis. Ct. App. 1993) (negligence claims against architect properly dismissed on basis that suspension system for cement plaster soffit, which failed, constituted construction means/methods and that it was accepted and customary practice in construction industry that the details of the method of canopy suspension are left to general contractor and are not the responsibility of the architect); *Rodriguez v. Universal Fastenings Corp.*, 777 S.W.2d 513 (Tex. Ct. App. 1989) (affirming jury verdict in favor of engineer on claims that engineer was negligent in the design and inspection of a certain lift station under construction at the time of plaintiff's decedent's accident; plaintiff had contended that it was the responsibility of engineer to assure that adequate temporary support was installed prior to commencement of the concrete pour for lift station; pointing to standard contract provisions making contractor responsible for construction means and methods, the court rejected that contention); *Wells v. Stanley J. Thill & Assocs., Inc.*, 452 P.2d 1015, 1018 (Mont. 1969) (affirming summary judgment in favor of engineer on grounds that engineer was not responsible for failure of contractor to shore trench during excavation since the duty of the engineer "was to see that a certain end result was eventually accomplished, namely that the product, as finally constructed and turned over to the city, met the plans and specifications. . . . [t]he engineer's real interest was not in the actual construction but in a completed project in accordance with the plans and specifications"); *Dillard v. Shaughnessy, Fickel & Scott Architects*, 864 S.W.2d 368 (Mo. Ct. App. 1993) (affirming summary judgment in favor of architect and structural engineer on basis that the latter were not responsible for construction means and methods, including whether a wall was properly reinforced or braced during construction); *National Foundation Company v. Post, Buckley, Schuh & Jernigan, Inc.*, 465 S.E.2d 726 (Ga. Ct. App. 1995) (affirming summary judgment in favor of engineering firm on personal injury claim that alleged engineering firm was negligent in failing to design and place temporary handrails or barricades in a work site area; recognizing that the duty for workers' safety rests upon contractors who exercised control and supervisory responsibility on job site, and engineering firm contractually exercised no such control or supervisory responsibility); *Yocum v. City of Minden*, 649 So. 2d 129 (La. Ct. App. 1995) (affirming lower court's ruling in favor of engineer that engineer had no contrac-

tual duty to warn of dangerous condition of work site where contract expressly prohibited engineer from guiding general contractor in contractor's means and methods).

54. For a general discussion of responsibilities in the shop drawing process, see generally J. Clark, "Focus on Shop Drawings," EJCDC Doc. No. 1910-9-C (Alexandria, VA: EJCDC, 1985); R. Rubin and M. Ressler, "To Build a Better Mousetrap—The Search to Define Responsibility for Shop Drawing Review," *The Construction Lawyer* (ABA Press, April 1985); D. Ellickson, "Shop Drawings for Construction, Who Really is Responsible," 24 *Real Property, Probate and Trust Journal* 611 (Winter 1990).

55. The term "design concept" represents an important qualification on the DPR's shop drawing review role and is intended to restrict the focus on the DPR's review to the issue of whether the final, completed-in-place construction and/or installation of the components described or depicted in the shop drawings will function as a whole in the manner and in accordance with the standards and criteria contemplated by the contract documents. See *C. W. Regan, Inc. v. Parsons, Brinckerhoff, Quade & Douglas*, 411 F.2d 1379 (4th Cir. 1969) (engineer's review of temporary bulkhead submission was confined to structural soundness and contractor remained solely responsible for protection of work and all field construction details with respect to design and construction of the temporary bulkhead); *Reich v. Simpson, Gumpertz & Heger, Inc.*, OSHRC Docket No. 89-1300, 1992 O.S.H.D. (CCH) ¶29,828 (Aug. 28, 1992) (acknowledging uncontroverted expert testimony that engineer's review of shop drawings and notation regarding temporary shoring during review did not constitute involvement in construction means and methods, since focus of engineer's review and notation was limited to structural integrity of the completed project in place and, specifically, "the local condition of loss of continuity of the deck as it was a single span condition").

56. Standard Form of Agreement Between Owner and Architect—1987 Edition, AIA Document B141.

57. J. Sweet, *Legal Aspects of Architecture, Engineering and the Construction Process*, p. 221, 5th ed. (St. Paul, MN: West Publishing Co., 1994).

58. Similarly, John Clark has admonished: "[I]t is important to remember that the data contained in a submittal, which relates to the means, methods, etc., of construction and neither pertains to the design concept, nor indicates compliance with the requirements of the Contract Documents, is not passed upon by the Engineer and thus should not be treated as amending, modifying or supplementing any parts of the Contract Documents. Being information that relates solely to the construction process, it is totally within the Contractor's control and completely his own responsibility." J. Clark, "Focus on Shop Drawings," EJCDC Doc. No. 1910-9-C, pp. 6–7 (Alexandria, VA: EJCDC, 1985).

59. For example, it has been held that a contractor's participation in "panel discussions" during the design development process will not shift the risk of defective design to the contractor. *Haehn Management Co. v. United States*, 15 Cl. Ct. 50, 56 (1988), *aff'd*, 878 F.2d 1445 (Fed. Cir. 1989). Moreover, a contractor was determined not to have assumed the risk of defective design when the government amended a specification using a contractor's suggestion that had been submitted in a value engineering proposal, but withdrawn by the contractor before the govern-

ment acted upon it. *Appeal of Tranco Industries, Inc.*, 78-2 B.C.A. (CCH) ¶13,307 (June 20, 1978).

60. *Austin Co. v. United States*, 314 F.2d 518 (holding that where plaintiff-contractor persuades defendant who has drawn the original contract specifications to change them in accordance with plaintiff's ideas, plaintiff has clearly assumed the risk of impossibility of performance of the contract), *cert. denied*, 375 U.S. 830 (1963); *Hawaiian Bitumuls & Paving v. United States*, 26 Cl. Ct. 1234, 1240 (1992) ("... when defendant has provided design specifications and drawings, and plaintiff persuades defendant to change them in favor of plaintiff's preferred specification, plaintiff assumes the risk that performance under its proposed specifications may be impossible. In general, the party originating the design specifications is responsible for losses suffered by the other parties due to defects in the specifications").

61. For another, similar formulation of the distinction between "design" and "performance" specifications, see *Aircraft Gear Corp. v. Kaman Aerospace Corp.*, 856 F. Supp. 446, 452 (N.D. Ill. 1994).

62. See also *Concrete Placing Co., Inc. v. United States*, 25 Cl. Ct. 369 (1992) (recognizing that "a contractor can obtain an equitable adjustment only if it can demonstrate that the specifications are design rather than performance specifications"), *aff'd*, 985 F.2d 585 (Fed. Cir. 1992); *Aircraft Gear Corp. v. Kamen Aerospace Corp.*, 856 F. Supp. 446, 451-55 (N.D. Ill. 1994) (where defective specification provided by contractor to subcontractor was a "performance" specification, subcontractor was responsible for additional expenses incurred by it as a result of the alleged defect).

63. See *Aleutian Constructors v. United States*, 24 Cl. Ct. 372, 378-81 (1991) (although detailed specification for cartridge was provided to contractor, because it was left to the contractor to use its own "judgment, experience and know-how" in determining how to manufacture certain aspects of cartridge, court shifted to contractor the responsibility for the means and methods selected to achieve the end result). See also *Aircraft Gear*, 856 F. Supp. at 451-55 (Note 62, *supra*) (where subcontract placed on subcontractor duty of generating the actual manufacturing orders for project, court concluded that subcontractor was responsible for "exercis[ing] its ingenuity in achieving the standard of performance, in selecting the means and in assuming a corresponding responsibility for that selection," and therefore bore risk of defective design); *Norwood Manufacturing, Inc. v. United States*, 21 Cl. Ct. 300, 307-09 (1990) (rejecting contractor's argument that risk of its failure to perform should be shifted to owner because specifications were "design" specifications, based upon a finding contractor was given substantial latitude in shaping character and quality of end product, and therefore contractor assumed risk that its plans under discretionary portions of specifications might not comply fully with owner's intended use requirements); *Universal Contracting & Brick Painting Co. v. United States*, 19 Cl. Ct. 785, 793-95 (1990) (denying owner's motion for summary judgment on contractor's defective specifications claim, because an issue of fact existed regarding whether contract contained primarily "design" or "performance" specifications; specifically, court indicated that an inquiry into owner's conduct was relevant because owner may have allowed contractor greater discretion and greater deviation from specifications than plain reading of contract suggested); *Haehn Management Co. v. United States*, 15 Cl. Ct. at 56, (Note 59, *supra*) (refus-

ing to shift risk for defective design to contractor because requirements in specifications displaced contractor's discretion, and because the machinery and methodology for performing work were set out in precise detail).

In circumstances in which the design furnished by the owner is either incomplete or contains material omissions, and in which doubt exists as to whether responsibility for completion or development of the design rested with the owner or contractor, courts and administrative boards may consider whether the particular design issue falls within the expertise and qualification of the contractor (as distinct from the owner) as well as the "custom and the industry" with respect to the responsibility for the design of the particular matter. P. D'Aloisio, "The Design Responsibility and Liability of Government Contractors," 22 *Pub. Constr. L.J.* 515, 532-34 (1993). For example, because responsibility for the design of construction means and methods typically is assigned to the construction contractor, in *Appeal of CSH Contractors, Inc.*, 78-2 BCA, ¶13,250 (involving the collapse of a cantilever lintel) and *North American Construction Corp.*, 85-1 B.C.A. (CCH) ¶17,780 (Nov. 30, 1984) (involving the collapse of a wall), administrative boards held that the contractors are responsible for the collapses of the items being constructed because industry practices, as well as contract language, place design responsibility for temporary support and shoring of those items on the contractors. Similarly, in *Allied Contractors, Inc. v. United States*, 180 Ct. Cl. 1057 (1967), the Court of Claims held that retaining walls on a construction project collapsed during construction because the walls were not properly designed and that the failure to properly design the walls to support construction loads was an obvious error that the contractor either knew or should have known about and, as a result, the court ruled that the contractor was responsible for the design of the walls and, therefore, was liable for damages resulting from their collapse.

64. The *Mortenson* decision is discussed in detail by this author in the following article: D. J. Hatem, "Contractor's Right to Rely Upon 'Concept Drawings' in Estimating Project Cost," *The CA/T Professional Liability Reporter*, Vol. 1, No. 1, (Boston: Burns & Levinson LLP, Dec. 1995) p. 6–7.

65. Design responsibility may be reallocated to the contractor, and the owner's implied warranty obligation may be negated where the contractor assumes responsibility for the design, See *Wunderlich Contracting Co. v. United States*, 351 F.2d 956 (Ct. Cl. 1965) (holding that where (a) contractor was aware of possible discrepancies in the project's plans and specifications; (b) owner invited prospective bidders to offer comments with a view towards eliminating such possible discrepancies; and (c) owner warned prospective bidders that timely performance within the time specified in the contract documents could only be achieved under ideal conditions, contractor could not maintain action against owner for breach of implied warranty of completeness and accuracy, or where the contract documents contain a disclaimer); *Mooney's Inc. v. South Dakota Department of Transportation*, 482 N.W.2d 43 (S.D. 1992) (rejecting contractor's argument that owner created an implied warranty of accuracy by providing pit test data, where contract documents disclaimed accuracy of information and placed burden on contractor to determine nature and materials in pit). See also *Lewis v. Anchorage Asphalt Paving Co.*, 535 P.2d 1188 (Ala. 1975) (holding that in an "end result" contract where the owner provided no plans and specifications to the contractor, and the contractor agreed to complete the job in

a "workman-like manner," contractor had duty to warn owner of potential defects in project which contractor knew or should have known about).

The court's decision in *Willamette Crushing Co. v. State of Arizona*, 932 P.2d 1350 (Ariz. Ct. App. 1997), demonstrates how an owner's implied warranty obligation may be negated when a contractor assumes responsibility for a design. *Willamette* involved a state highway project. The contract documents contained a traffic control plan (TCP), which specified the stages of construction and was intended to permit maximum traffic during construction. After being awarded the contract, the contractor performed an independent analysis of the TCP contained in the contract documents, and concluded that the project could not be completed by the contract's due date if the specified TCP were utilized. The contractor asked for an extension of the contract's due date, but the state refused. The state did, however, approve the contractor's request to use an alternate TCP devised by the contractor. Using its own TCP, the contractor completed the project. Subsequently, the contractor asserted a breach of implied warranty claim against both the owner and the design engineer, alleging that the owner impliedly warranted accuracy of the TCP contained in the contract documents, and that the TCP contained in the contract documents was defective. The court denied the contractor's implied warranty claim, as a matter of law, because the contractor did not follow the TCP contained in the contractor documents, but rather completed the job using its own TCP.

66. The Massachusetts Appeals Court decision in *John F. Miller Co., Inc. v. George Fichera Construction Corp.*, 388 N.E.2d 1201 (Mass. App. Ct. 1979), illustrates this point. In *Miller*, the plans and specifications for a housing project called for a waste piping system that apparently did not comply with state plumbing code. The subcontractor on the project submitted its bid on the assumption that it could substitute a waste piping system different from that called for by the plans and specifications—one that had the advantage of complying with state codes. Noting that the subcontractor never notified the architect and the owner of any deviations in the specifications from applicable codes, the appellate court affirmed the lower court's denial of extra costs incurred by the subcontractor as a result of its installation of the substitute waste piping system.

67. See, e.g., *D. Federico Co., Inc. v. Commonwealth*, 415 N.E.2d 855, 857 (Mass. App. Ct. 1981) (holding that the contract documents' disclaimers specifically precluded warranty of, or contractor's reliance upon, estimates furnished by engineering firm); *Mooney's Inc. v. South Dakota Dept. of Transportation*, 482 N.W.2d 43 (S.D. 1992) (rejecting contractor's claim that, by providing pit test data in bid information, project owner implicitly warranted the accuracy of data on the ground that the contract documents contained disclaimer which provided that the owner "does not guarantee the quantity or quality of the [data]" in issue); *Daniel O'Connell's Sons, Inc. v. Commonwealth*, 212 N.E.2d 219 (Mass. 1965) (holding that contractor was not entitled to rely on geological data provided by owner because contract documents expressly disclaimed owner's responsibility for the accuracy of data, and required contractor to conduct an independent investigation regarding all such data). See Notes 40 and 41, *supra*.

68. In this regard, reference should be made to Paragraph 3.2.1 of AIA Document A201, General Conditions of the Contract for Construction (1987 edition), which provides that:

> The Contractor shall carefully study and compare the Contract Documents with each other and with information furnished by the Owner . . . and shall at once report to the Architect errors, inconsistencies or omissions discovered. The Contractor shall not be liable to the Owner or Architect for damage resulting from errors, inconsistencies or omissions in the Contract Documents unless the Contractor recognized such error, inconsistency or omission and knowingly failed to report it to the Architect. If the Contractor performs any construction activity knowing that it involves a recognized error, inconsistency or omission in the Contract Documents without such notice to the Architect, the Contractor shall assume appropriate responsibility for such performance and shall bear an appropriate amount of the attributable costs for correction.

Another related provision typically contained in bid documents provides that:

> Each bidder shall promptly notify the Owner of any ambiguity, inconsistency or error he may discover upon examination of the Contract Documents, the site or local conditions. The submission of a bid will constitute a representation by the bidder that he has complied with every requirement of this paragraph and that the Contract Documents are sufficient in scope and detail to indicate and convey understanding of all terms and conditions for performance of the work of this contract.

69. See, e.g., *Wenzel v. Boyles Galvanizing Co.*, 920 F.2d 778, 781 (11th Cir. 1991) (design professional "specifically agreed to '[p]rovide, implement and administer a site safety and health program'").

70. The nature, purpose, and extent of a design and/or construction management professional's review/approval role in connection with a contractor's proposed construction means and methods could also result in an enhanced liability exposure for the professional. An important decision that emphasizes the importance of defining the nature, scope and purpose of the engineer's role in reviewing contractor design is *Morse/Diesel, Inc. v. Trinity Industries, Inc.*, 67 F.3d 435 (2nd Cir. 1995). The *Morse/Diesel* decision addressed the standard governing a design professional's review of construction means and methods proposed by a contractor. The *Morse/Diesel* case arose out of the construction of a hotel project in Manhattan's Times Square. The general contractor for the project sued the structural steel subcontractor for damages due to late performance of the steel work. The steel subcontractor counterclaimed. After 10 years of litigation, and a six-and-a-half week trial, the jury awarded the general contractor $26 million, to which the court added another $27 million in interest. After an appeal to the Second Circuit, the verdict was reversed and remanded.

    The central issue between the general contractor and steel subcontractor in *Morse/Diesel* was one of responsibility for the delay in performance of the steel erection work. The general contractor blamed the steel subcontractor based on evidence of fabrication errors, shipping mishaps, crane failures, and so on. The general contractor sought damages due to its acceleration of the work to recapture time allegedly lost due to delay in the structural steel work.

The steel subcontractor contended that the general contractor was responsible for the delay, and attributed much of the delay to the project architect and its structural engineering consultant (collectively, the "design professional"). According to the steel subcontractor, the design professional made "unwarranted demands regarding the installation of certain special steel trusses, known as Vierendeel trusses, and the use of additional bracing during erection."

Although the Second Circuit reversed the jury award for certain legal reasons (dealing with the adequacy of the trial judge's instructions to the jury), there are certain aspects of the trial judge's rulings in *Morse/Diesel* that were affirmed by the Second Circuit. Of particular interest are those rulings related to the authority of the design professionals with regard to the steel subcontractor's proposed construction means and methods.

As noted above, the steel subcontractor defended the general contractor's claims by introducing evidence that its proposed methods of erecting the Vierendeel trusses, and for temporarily bracing the structure, were adequate and that the design professionals' "directives" to use alternative methods were unreasonable, unnecessary, and the cause of substantial delay.

Prior to trial, however, the district court had ruled that the design professional's decisions with regard to the erection of the trusses and temporary bracing "must be accepted as final and valid under the contracts unless the [steel subcontractor] can show such decisions to have been the product of bad faith, fraud, or palpable mistake equivalent to bad faith." Based on this pretrial ruling, the trial judge refused to allow the steel subcontractor to introduce evidence on this issue for the jury's consideration unless, in the judge's estimation, that the proposed evidence "would allow a reasonable jury to conclude that [the] claims of bad faith (or the like) were correct ..." Based on that ruling and the standard of admissibility of evidence (described above), the trial judge effectively excluded all testimony of the steel subcontractor's expert as to his opinion that there was no "engineering basis" for the design professional's refusal to permit the steel subcontractor to use its own selected and planned steel erection methods and that the steel subcontractor's methods were "totally adequate." The trial judge ruled that the steel subcontractor's expert opinion was inadmissible because saying that the steel subcontractor's planned and selected erection methods were "totally adequate" (while true) did not demonstrate or reasonably tend to prove that the design professional's rejection of them was "the product of bad faith, fraud, or palpable mistake equivalent to bad faith."

On appeal, the steel subcontractor challenged the trial judge's evidentiary ruling on the grounds that: (1) under the contract documents, the steel subcontractor has the exclusive right to select and implement construction means, methods, sequences, and procedures ("construction means/methods") to be employed in the performance of the work, absent a special direction or requirement in the contract documents that a particular construction means/methods be employed by the subcontractor; (2) that the design professional's direction that a particular erection method and temporary bracing design be used by the subcontractor constituted an unwarranted intrusion into the exclusive province of the subcontractor regarding selection, design and implementation of construction means/methods; and (3) that the provisions of the contract documents empowering the design professionals to interpret the contract documents and judge the acceptability of the work do not extend to issues or disputes relating to construction means/methods (as distinct from work specifically required by the contract documents to be incorporated into the complete project).

In rejecting these arguments, the Second Circuit relied heavily on the following provisions of the contract documents, which explicitly conferred upon the design professionals the authority to review and approve the subcontractors proposed construction means and methods:

1. "The Subcontractor shall undertake and complete the Work under the direction and supervision of the Contractor, and to the entire satisfaction of the Contractor, the Owner and the Architect."
2. "Subcontractor will furnish tower cranes for erection of structural steel, metal deck, precast plank, and hosting of steel stairs. All load calculations and support methods must be submitted for approval ... All reinforcing of the structure is required for the tower cranes is included by the Subcontractor. Details and location of reinforcing must be submitted for approval."
3. "After award [of contract]: ... (3) Erection procedure (for review and scope): Submit data to illustrate the erection procedure, including the erection, bolting and welding sequence, temporary staying and bracing required, and clarifying information and calculations requested by the Architect/Engineer."

In affirming the trial judge's evidentiary ruling, the Second Circuit summarily stated that "whether the [subcontractor's] proposed methods for erecting the Vierendeel trusses and temporary bracing were sound and workable, and whether the [design professionals] demands were unnecessary, are ultimately irrelevant. It was therefore appropriate at trial to preclude [the subcontractor's] construction expert from stating his opinions on such questions. To say the [design professionals have] discretionary authority, of course, is not to say [that they have] unbounded discretion ... [T]he district court ruled that the [design professionals'] decisions' must be accepted as final and valid under the contract, unless [the subcontractor] can show such decisions to have been the product of bad faith, fraud, or palpable mistake equivalent to bad faith'."

See also *D.C. McClain, Inc. v. Arlington County*, 452 S.E.2d 659 (Va. 1995) (engineer's review and approval of contractor's proposed construction means/methods relating to post-tensioning of a bridge did not relieve contractor of its responsibility for those construction means/methods). The *McClain* decision is discussed in detail in D. J. Hatem, "Engineer's Review and Approval of Contractor's Instruction Means/Methods Submission Regarding Post-Tensioning of Bridge does not Relieve Contractor of Responsibility for Construction Means/Methods," *The CA/T Professional Liability Reporter*, Vol. 1, No. 1, (Boston: Burns & Levinson LLP, December 1995), pp. 10–12.

71. P. Partridge and V. Noletto, Jr., "Construction Management: Evolving Roles and Exposure of Construction Managers and Architects/Engineers," in *Legal Handbook for Architects, Engineers and Contractors*, Vol. 9 (Deerfield, IL: Clark Boardman Gallagher, 1993), p. 169.

72. See *Secretary of Labor v. Kulka Construction Management Corp.*, OSHRC Docket No. 88-1167, 1992 O.S.H.D. (CCH) ¶29,829 (Aug. 28, 1992) (design professional who contractually agreed to "[p]rovide recommendations and information

... regarding the assignment of responsibilities for safety precautions and programs" substantially supervised the work site and was subject to the construction standards of 29 CFR §1926); *Secretary of Labor v. CH2M Hill Central, Inc.*, 1997 WL 197011 (O.S.H.R.C. April 21, 1997), *appeal docketed*, No. 97-2504 and No. 97-2503, (7th Cir. 1997). (ruling that occupational safety and health standards for construction applied to engineering firm, notwithstanding contract language expressly disclaiming safety responsibility, because engineering firm (1) possessed broad responsibilities in relation to construction activities, including both contractual and de facto authority relating directly to the work of the trade contractors, and (2) was directly and substantially engaged in activities integrally connected with safety issues).

73. Compare *Secretary of Labor v. Simpson, Gumpertz & Heger, Inc.*, 3 F.3d 1 (1st Cir. 1993) (affirming administrative law judge's refusal to fine or penalize engineer with respect to citation arising out of accident at construction site because engineer had limited involvement/presence on construction site, and therefore construction site was not a "place of employment" that engineer had a duty under OSHA to protect) and *Secretary of Labor v. Foit-Albert Associates, Architects & Engineers, P.C.*, 1997 WL 197016 (O.S.H.R.C. April 21, 1997) (affirming administrative law judge's refusal to fine or penalize engineering firm for OSHA citation because engineering firm had no contractual authority to direct the activities of the trade contractors, and had no contractual responsibility over safety means and measures; although engineering firm performed inspection activities that may have unavoidably addressed safety issues, it did not manifest the ability to control or direct matters of safety, and therefore did not exercise "substantial supervision" over the construction work that would have brought it under the OSHA construction work safety regulations and would have required it to protect the project site) with *Secretary of Labor v. CH2M Hill Central, Inc.*, 1997 WL 197011 (O.S.H.R.C. April 21, 1997), *appeal docketed*, No. 97-2504 and No. 97-2503, (7th Cir. 1997). (reversing administrative law judge's refusal to fine engineering firm hired as consultant on tunneling project with respect to citation arising out of explosion at construction site; finding that occupational safety and health standards for construction applied to engineering firm, notwithstanding contract language expressly disclaiming safety responsibility, because engineering firm (1) possessed broad responsibilities in relation to construction activities, including both contractual and *de facto* authority relating directly to the work of the trade contractors, and (2) was directly and substantially engaged in activities integrally connected with safety issues) and *Secretary of Labor v. Kulka Construction Management Corp.*, OSHRC Docket No. 88-1167, 1992 O.S.H.D. (CCH) ¶29,829 (Aug. 28, 1992) (penalizing construction manager for OSHA violation because construction manager exercised "substantial supervisory authority" over the construction site).

See generally A. Sapper, "OSHA and The Design Professional," *Civil Engineering*, (April 1994), p. 60.

74. In addition, the degree of involvement and control of an envionmental professional over construction means and methods and/or safety issues in connection with environmental remediation projects similarly may subject the environmental professional to strict liability exposure and/or fines or penalty assessments under federal and state environmental laws. Compare *Kaiser Aluminum & Chemical Corp. v. Catellus Development Corp.*, 976 F.2d 1338 (9th Cir. 1992) (ruling that excavator

on housing development project who spread contaminated soil over uncontaminated areas of property exercised sufficient control over development phase of the project to be held liable under environmental law which penalizes those who "own" or "operate" a facility at the time hazardous waste is disposed at the facility) with *City of North Miami, Florida v. Berger*, 828 F.Supp. 401 (E.D. Va. 1993) (ruling that engineering firm that did not have actual control or authority to control landfill operation could not be held liable as an "operator" of a facility under environmental statute penalizing those who owned and operated facilities at time hazardous waste was disposed at facility; noting that engineering firm did not exercise control over any waste disposed in landfill, and did not possess authority to control, supervise, or direct actual, day-to-day operations of project). See generally D. J. Hatem, "Potential Liability of Environmental Professionals: Risk Management and Contractual Responses," *Avoiding Environmental Malpractice*, MCLE (April 30, 1996).

75. See generally R. F. Cushman and D. Tortorello, eds. *Differing Site Condition Claims* (New York: John Wiley & Sons, Inc., 1992), pp. 68–70; C. Allen Foster, et al., eds., *Construction and Design Law*, Chapter 12, Section 12.2b.3 (Charlottesville, VA: The Mitchie Company, 1989).

76. See, e.g., *Richardson Electrical Co., Inc. v. Peter Francese & Son, Inc.*, 484 N.E.2d 108 (Mass. App. Ct. 1985) (owner of water pumping station held liable for extra costs incurred by subcontractor for digging a trench to accommodate telephone cable where specifications failed to give subcontractor notice that it would be required to dig such a trench; contract documents contained no disclaimer of accuracy or reliance on information provided by owner); *John F. Miller Co., Inc. v. George Fichera Construction Corp.*, 388 N.E.2d 1201 (Mass. App. Ct. 1979) (holding that subcontractor on public works contract was entitled to recover extra compensation for costs incurred in performing work not specifically identified in section of contract documents dealing with his trade); *Alpert v. Commonwealth*, 258 N.E.2d 755 (Mass. 1970) (owner of public project held liable for breach of warranty for failing to disclose boring information withheld where "Notice to Contractors" expressly stated that "complete information" was available, and where prior to submitting bid contractor requested all information and was informed that "all information" was contained in the plans and specifications furnished to contractor); *Montrose Contracting Co., Inc. v. County of Westchester*, 80 F.2d 841 (2d Cir. 1936) (holding that owner impliedly warranted that tunnel was substantially a free-air job by providing contractor with specifications which would be adequate only if tunnel was to be built in free-air, and that contractor could recover damages sustained in relying on such warranty), *cert. denied*, 298 U.S. 662 (1936); *Blue Bell Inc. v. Cassidy*, 200 F. Supp. 443 (N.D. Miss. 1961) (holding that contractor was not liable to owner for building collapse caused by faulty design of footings on which columns rested, and/or bad soil conditions, where contractor followed plans and specifications furnished by owner without showing of negligence). See also *Midwest Dredging Co. v. McAninch Corp.*, 424 N.W.2d 216 (Iowa 1988) (allowing contractor to recover on breach of implied warranty claim based on allegations that owner implicitly warranted that plans and specifications were accurate and that borrow was hydraulically dredgeable); *Appeal of R. L. Coleman & Co.*, 1963 B.C.A. ¶3748, ASBCA No. 7334 (April 30, 1963) (rejecting contractor's differing site conditions claim based on allegations that water table encountered was mate-

rially different from that shown on plans or indicated in contract documents, but holding owner liable for breach of warranty for faulty design). See also *Appeal of Engineered Systems, Inc.*, 80-2 B.C.A. (CCH) ¶14,458 (Nov. 30, 1978) (allowing contractor to recover on Type I differing site conditions claims based on allegations that the owner (1) implicitly warranted that subsurface conditions were such that a particular work method (jacking) could be successfully employed; and (2) implicitly warranted the accuracy of contract drawings pertaining to the location of underground cables). See generally J. C. McBride and T. J. Touhie, *Government Contracts: Cyclopedic Guide to Law, Administration, Procedure*, ¶29.70[4] (New York: Matthew Bender & Co., 1996).

77. Several court decisions have acknowledged the interrelationship between claims based on differing site conditions and defective design in the context of subsurface projects. In many combined differing site conditions and defective design claims, it is common for the same evidence to be offered by the contractor in support of both the differing site conditions and defective design components.

In *Iacobelli Construction, Inc. v. County of Monroe*, 32 F.3d 19 (2nd Cir. 1994), a general contractor asserted a combined differing site conditions and breach of warranty claim against a public owner. In substance, the contractor contended that it incurred more subsurface water inflows during construction of a tunnel than it had anticipated. The contractor filed suit, asserting breach of contract and breach of warranty claims against the owner, and a negligence claim against the consulting engineers who were involved in the design and construction administration phases of the project.

With regard to the contractor's differing site conditions claim, the court summarized the requirements of proof for a so-called Type I differing site condition as follows:

> To prevail on a Type I differing site conditions claim, a plaintiff must show: (1) the contract documents affirmatively indicate subsurface conditions; (2) she acted as a reasonably prudent contractor in interpreting the contract documents; (3) she reasonably relied on the indications of subsurface conditions in the contract; (4) the subsurface conditions actually encountered differed materially from those indicated in the contract; (5) the actual subsurface conditions were not reasonably foreseeable; and (6) her damage was attributable to the materially different subsurface conditions.

Iacobelli at 24.

In declining to dismiss the claim, the court held that the contractor had presented sufficient evidence in support of its differing site conditions claim to demonstrate that there were genuine issues of material fact that required a trial for final resolution. With respect to the contractor's claim based on breach of warranty, the court considered the same evidence in light of the contractor's allegations that the owner "warranted the accuracy of its plans and specifications and the fitness of its plans and specifications for the purposes intended." The court noted the close link between the differing site conditions claim and the breach of warranty claim and held that there were sufficient factual issues to preclude (prior to trial) dismissal of the contractor's breach of warranty claims.

The contractor's companion negligence claim against the consulting engineers (which was based on allegations of negligence in the design, construction inspection and claims administration phases of the project) was dismissed on the basis of the statute of limitations.

See also *Neal & Co. v. City of Dillingham et al.*, 923 P.2d 89 (Alaska 1996) (contractor submitted differing site conditions claim and defective specifications claim against owner/city, alleging that it encountered difficulties in excavation during the project because of unexpected soil conditions; court affirmed dismissal of both claims based on contractor's failure to satisfy notice requirements; of note, court found no need to distinguish between differing site conditions claim and defective specifications claim, as both claims arose out of allegations that the composition of the soil encountered differed from what the data summary predicted); *Appeal of Clearwater Constructors, Inc.*, 96-2 BCA (CCH) ¶28,495 (Aug. 8, 1996) (acknowledging interrelatedness and alternative nature of differing site conditions and defective design claims, the board held that both claims failed due to lack of probative evidence that contractor relied on information contained in the contract documents in bidding the project).

78. See D. J. Hatem, "Errors/Omissions Cost Recovery Claims," *The CA/T Professional Liability Reporter*, Vol. 1, No. 4, (Boston: Burns & Levinson LLP, June 1996), pp. 1–10. In addition, should the owner settle or otherwise pay a combined differing site conditions/defective design claim solely on the basis of the defective design portion, (or on terms that leave ambiguous or unclear whether the settlement or payment was due to allegations of the defective design), grant or other funding reimbursement for the settlement or payment may be jeopardized or totally precluded. This is because, typically, grants or other funding programs will not reimburse an owner for settlements or payments based on design or other professional negligence, as distinct from purely differing site conditions. See J. P. Gould, "Geotechnology in Dispute Resolution," *Journal of Geotechnical Engineering*, (July 1995), pp. 523–24.

79. The focus of the discussion in this section of the chapter is limited to *professional liability* risk associated with the design professional's involvement in the design-build delivery method and does not address other types of risks related to the design-build process, particularly those related to the performance of construction work and the furnishing of equipment and materials. In addition to professional liability risk, the design professional involved in the design-build delivery method may (depending on its contractual status) have certain other liability and contractual risk—such as payment obligations to vendors and subcontractors; strict liability; product liability; liquidated damages exposure; increased exposure for statutory environmental and OSHA penalties and liabilities; or performance obligations in the event of construction contractor default or termination, of a non-professional nature, the discussion of which is beyond the scope of this chapter. H. G. Block, "As the Walls Came Tumbling Down: Architects' Expanded Liability Under Design-Build/Construction Contracting," 17 *J. Marshall L. Rev.* 11 (1984).

80. See discussion in Section 10.2.2.2(A) regarding circumstances in which an owner may assume some liability when it prescribes or otherwise becomes involved in design.

81. See, e.g., EJCDC "Standard General Conditions of the Construction Contract," ¶3.5.1 (Alexandria, VA: EJCDC, 1996).

82. See R. F. Cushman, C. M. Jacobsen, and P. J. Trimble, *Proving and Pricing Construction Claims*, 2nd ed., §12.5 (New York: John Wiley & Sons, Inc., 1996).

83. In addition, surety companies for construction contractors have been extremely reluctant to provide performance bonds for design services obligations. See T. H. Asselin and L. B. Stout, "Legal Exposure of the Design/Build Participants: The View of the General Contractor," in *The Construction Lawyer*, C. A. Meeker, ed., (ABA Press, August 1995).

84. By limiting the design-builder's design adequacy risk to the negligence-based professional standard of care, this approach also has the advantage of aligning that risk assumption with the coverage typically afforded under most professional liability insurance policies. See Note 3, *supra*.

85. For example, in *Appeal of M. A. Mortenson Co.,* 93-3 B.C.A. (CCH) ¶26,189 (June 30, 1993), the board found the government liable for estimated quantities of material indicated in conceptual design drawings, even though the contract awarded was a design-build contract under which the contractor had responsibility for completing the design and preparing the construction drawings. The board reasoned that the contractor had no choice but to rely on conceptual drawings in preparing a firm price for the work. In these types of circumstances, the extent to which design adequacy risk remains with the owner or is transferred to the design-builder depends, to some degree, on the distinction between design and performance specifications. This distinction and the risk allocation consequences, particularly in the context of subsurface construction projects, are discussed in detail in Section 10.2.2.1(A).

86. See M. L. Loulakis and O. J. Shean, "Risk Transference in Design-Build Contracting," *Construction Briefings*, 2nd Series (Washington DC: Federal Publications, Inc., April 1996).

87. See, e.g., R. F. Cushman and G. C. Hedemann, *Architect and Engineer Liability: Claims Against Design Professionals*, 2nd ed. (New York: John Wiley & Sons, Inc., 1995).

88. See generally H. G. Block, "As the Walls Came Tumbling Down: Architects Expanded Liability Under Design-Build/Construction Contracting," 217 *J. Marshall L. Rev.* 11 (1984).

89. See Notes 72 through 74, *supra*, and accompanying text.

90. See T. H. Asselin and L. B. Stout, "Legal Exposure of the Design/Build Participants: The View of the General Contractor," *The Construction Lawyer*, C. A. Meeker, ed., (ABA Press, August 1995).

91. In addition to demonstrating the importance of a properly drafted scope of services provision, the *Maddox* decision also emphasizes the legal distinction between liability under an implied warranty obligation and negligence liability. (See Section 10.2.3.1) Although this distinction is not explicitly discussed in the *Maddox* decision, in reviewing the jury's award on the issue of errors contained in the bidding information provided by Benham to Maddox, the Eighth Circuit imposed liability on Benham on the basis of implied warranty, as distinct from negligence.

On appeal, Benham argued Maddox had no contractual right to recover for errors contained in bidding information provided by Benham to Maddox, because Benham never warranted the accuracy of the bidding information that it supplied to

Maddox. The Eighth Circuit, however, rejected this argument, noting that Benham repeatedly assured Maddox that it was qualified to prepare the bidding estimates, and that it had the manpower and expertise to perform its work. Relying on Missouri law, which provides that when a company represents itself as being able to do work of a particular character, a warranty is implied that the work will be performed properly, the court concluded that Benham's assurances created an implied warranty, allowing liability for bidding errors. Thus, liability was predicated on breach of warranty, and not on any negligence attributable to the engineers.

92. D. B. Desai, "Design Build—Back to the Future," *AUA News*, Vol. 11, No. 3, (September 1996), pp. 19, 24.

93. See R. J. Essex, ed., "Geotechnical Baseline Reports for Underground Construction—Guidelines and Practices," prepared by the Technical Committee on Geotechnical Reports of the Underground Technology Research Council (New York: ASCE, 1997). There are potential professional liability concerns associated with the use of the baseline approach. For further discussion, see D. J. Hatem, "Geotechnical Baselines: Professional Liability Implications," *The CA/T Professional Liability Reporter*, Vol. 3, No. 1 (Boston: Burns & Levinson LLP, October 1997); and D. J. Hatem, "Geotechnical Baselines and Geotechnical Reality," *The CA/T Professional Liability Reporter*, Vol. 2, No. 2, (Boston: Burns & Levinson LLP, December 1996) pp. 1–8. For further discussion regarding the use of baseline geotechnical assumptions in the design-build contracting process, see M. L. Loulakis and O. J. Shean, "Risk Transference in Design-Build Contracting," *Construction Briefings*, 2nd Series (Washington, DC: Federal Publications, Inc., 1996); and J. D. West, R. A. Schechter, R. D. Fennema and S. Tell, *Design-Build Construction* (Washington, DC: Federal Publications, Inc., 1997).

94. As one commentator has observed:

> Initially, the contractor feels the impact of unforeseen subsurface site conditions. Contract clauses may or may not clearly provide both the substance and procedure for determining responsibility ... The subsurface condition may impact the entire design creating potential liability for the designers—the architect, engineer and geotechnical engineers—since major design changes may be required to overcome the condition.

G. Jones, "The U.S. Perspective on Procedures for Subsurface Ground Conditions Claims," *The Int'l Construction L. Rev.*, 155, 170 (Lloyds of London Press, Ltd., 1990).

Similarly, another commentator has noted:

> [T]here is a clear negative aspect [to changed conditions claims] that puts the geotechnical engineer of a project at risk for failing to meet the the standard of care in their site characterization. With only a modest amount of imagination, an attorney can convert what is truly a [differing site conditions] claim into a suit against the project owner and design professionals. The owner's geotechnical engineers should recognize that it is greatly in their own interest to produce a pre-bid investigation that will avoid a claim. If a claim eventuates, then the engineer should strive to resolve that claim

> equitably and promptly and not encourage it to escalate to the next level. Note that federal backers will often contribute to settlement of a [differing site conditions] claim, but they certainly will not choose to share the cost if the finding is of negligence and defective design.

J. P. Gould, "Geotechnology in Dispute Resolution," *Journal of Geotechnical Engineering* (July 1995), pp. 523–24 (Reprinted by permission of the publisher; copyright 1995 ASCE).

95. For detailed discussions of the topic of fair allocation of risk in the specific context of subsurface conditions, see R. Smith, "Risk Identification and Allocation Saving Money by Improving Contracts and Contracting Practices," *The Int'l L. R.* 40 (1995); T. Neff, "Contract Management," *World Tunneling*, (November 1993); G. Brierley and B. Cavan, "The Risks Associated with Tunneling Projects," *Tunneling Technology*, (June 1987); *Owner's Guide to Saving Money by Risk Allocation*, Joint Publication of Associated General Contractors of America and American Consulting Engineers Council (June 1990). D. J. Hatem, "Differing Site Conditions: Liability Precautions for Design and Construction Management Professionals," *Defense Couns. J.* (October 1994), p. 555 discusses how fair allocation of risk for subsurface conditions will also, from a legal standpoint, minimize the risk of design and construction management professional liability. See generally M. Loulakis, *Construction Management Law and Practice*, Section 4.7, (New York: John Wiley & Sons, Inc., 1995) pp. 95–106. For a discussion of the role of legal counsel in the risk allocation process, see D. J. Hatem, "Risk Allocation and Dispute Resolution On Construction Projects: Roles and Challenges For Legal Counsel," *The CA/T Professional Liability Reporter*, Vol. 2, No. 4 (Boston: Burns & Levinson LLP, June 1997).

96. See Sections 10.1.2.1 and 10.1.2.2. See also Moorhous and Maillet, "Identifying Causes of Failure, *Journal of Management in Engineering* (ASCE, May 1994); *Foster Wheeler Enviresponse, Inc. v. Franklin County Convention Facilities Authority, et al.*, 678 N.E.2d. 519 (Ohio 1997) (environmental remediation contractor pursued negligent misrepresentation claim against owner's engineering consultant in circumstances in which construction contract provisions precluded contractor's recovery against owner, but was denied recovery). *M. Miller Co. v. Dames & Moore*, 18 Cal. Rptr. 13 (Cal. Ct. App. 1961) (allowing contractor to pursue negligence claim against soil engineers based on content of soil reports, despite the fact that owner's agreement with contractor disclaimed accuracy of reports made available to the contractor, on the ground that the disclaimer in owner-contractor agreement did not shield soil engineer from liability because soil engineer was not a party to owner-contractor agreement; court also rejected soil engineers' argument that contractor could not pursue negligence claim against soil engineer because there was no privity between soil engineers and contractor); *Davidson and Jones, Inc. v. County of New Hanover*, 255 S.E.2d 580 (N.C. Ct. App. 1979) (holding that disclaimer in contractor's agreement with owner, which provided that contractor was to "rely entirely on its own judgment in submitting its bid" and was to "conduct a site inspection to have a complete understanding of all existing conditions relating to the work," did not bar contractor's negligent misrepresentation claim against project engineer), *rev. denied*, 259 S.E.2d 911 (N.C. 1979). See generally J. Monsees et al., "The Perspective of Engineering Firms: GDSRs Are Not

Working as Envisioned" in *Opinion Papers on "Geotechnical Reports in Underground Construction,"* an unpublished collection of papers presented at the April 21, 1996 Underground Technology Research Counsel Forum in Washington, D.C. J. Sweet, *Legal Aspects of Architecture, Engineering and the Construction Process*, 5th ed. (St. Paul, MN: West Publishing Co., 1994), p. 552; J. P. Gould, "Geotechnology in Dispute Resolution," *Journal of Geotechnical Engineering*, (July 1995) pp. 525–26.

97. As discussed in Section 10.1.2.1, these types of situations in which risk has been allocated (even if unfairly) between contracting parties—i.e., the owner and contractor—with respect to a particular risk (here, subsurface conditions) represent excellent candidates for the imposition of the barrier to third-party claims against the professional interposed by the economic loss doctrine.

As one commentator has stated:

> Undoubtedly, arguments can be made for a disclaimer system or a system that uses a [differing site conditions] clause. Courts should effectuate *whichever* choice has been made. If it appears clear that the contractor has, happily or not, accepted the risk of unforeseen subsurface conditions, the court should effectuate that risk allocation and not seek to destroy it by tortured interpretation generated by a belief that it is unconscionable for the owner to place these risks on the contractor. Only if it is clear that the contractor was not made aware of the risk or if the imposition of the risk violates the obligation of good faith and fair dealing should the court entertain a result that does violence to the loss distribution selected and expressed in the contract.
>
> Similarly, courts should *not* allow a contractor to maintain a tort action against a geotechnical engineer [or another design and construction management professional, for that matter] if it is clear that the contractor was expected to assume this risk. Allowing the contractor to transfer this risk to the geotechnical engineer [or, for that matter, any other design or construction management professional] through a tort action frustrates the efficiency of the system selected.

J. Sweet, *Legal Aspects of Architecture, Engineering and the Construction Process*, 5th ed., Chapter 25, (St. Paul, MN: West Publishing Co., 1994), p. 552, (Reprinted by permission; copyright 1994 by West Publishing Company. All rights reserved.)

98. C. Allen Foster, et al., eds., *Construction and Design Law*, Chapter 12, Section 12.2c (Charlottesville, VA: The Mitchie Company, 1989).

99. See, e.g., *Davidson and Jones, Inc. v. County of New Hanover*, 255 S.E.2d 580 (N.C. Ct. App. 1979) (architect held liable to contractor in negligence for failing to make reasonable and proper examinations and inspections of soil conditions of the project site and of the foundations of adjoining structures), *cert. denied*, 259 S.E.2d 911 (N.C. 1979); *Zontelli & Sons, Inc. v. City of Nashwauk*, 373 N.W.2d 744 (Minn. 1985) (holding engineer is obligated to indemnify city for contract claims due to engineer's negligence in failing to conduct a proper investigation that resulted in drastic underestimates in the quantity of concrete and unsuitable material to be excavated and removed by contractor, and a misrepresentation that

the highway in question was a municipal as opposed to a state highway, thereby misleading bidders with respect to the scope of the work required to be performed).

100. See generally *Guardian Constr. Co. v. Tetra Tech Richardson, Inc.*, 583 A.2d 1378 (Del. Super. Ct. 1990) (holding, by adopting the parameters of §552 of the Restatement (Second) of Torts (1977) (see Note 27, *supra*), that contractor and subcontractor were entitled to pursue claim against design engineer for foreseeable economic losses, regardless of whether or not privity existed, when design engineer negligently miscalculated tidal heights and project benchmarks, which were used to develop project plans and specifications and which were known and intended to be for the guidance of those contractors); *Gulf Contracting v. Bibb County*, 795 F.2d 980 (11th Cir. 1986) (holding that contractor was entitled to pursue claim against architect and engineer for cost of removing alleged unanticipated debris on theory that architect and engineers were negligent in failing to disclose existence of subsurface debris in question in bid and contract documents, and that the absence of a contractual or privity relationship between contractor and architect/engineer would not, as a matter of law, preclude recovery against architect/engineer). See also *Appeal of R.L. Coleman & Co.*, 1963 B.C.A. 3748, ASBCA No. 7334 (April 30, 1963) (rejecting contractor's differing site conditions claim against government based on allegations that water table depicted on plans differed materially from condition actually encountered, but holding government liable for breach of warranty for faulty design); *Appeal of Engineered Systems, Inc.*, 80-2 B.C.A. (CCH) ¶14,458 (Nov. 30, 1978) (contractor allowed to recover from government/owner on differing site conditions claim where drawings furnished by government/owner's engineer failed to accurately depict the location of cables on the project site).

101. See *Fondedile, S.A. v. C.E. Maguire*, 610 A.2d 87 (R.I. 1992) (contractor asserted negligence claim against design professional alleging that the design professional negligently failed to disclose movement of a sea wall in contract documents; court rejected contractor's claim finding that even if design professional withheld such information, contractor could not establish a causal link between its alleged damages and design professional's allegedly negligent conduct). See generally, J. C. McBride and T. J. Touhie, *Government Contracts: Cyclopedic Guide to Law, Administration, Procedure*, ¶29.70[4] (New York: Matthew Bender, 1991). See also *Richardson Electrical Co., Inc. v. Peter Francese & Son, Inc.*, 484 N.E.2d 108 (Mass. App. Ct. 1985). Although the *Richardson* decision did not directly involve a claim against a design professional (the owner's claim against the engineer was held in abeyance pending the outcome of the contractor's claim against the owner), that decision does recognize the right of a subcontractor to sue for differing site conditions under circumstances where (a) the owner and its engineer knew of the alleged differing site conditions prior to the construction of the project, but did not disclose the information to bidders; and (b) the subcontractor was found to have exercised reasonable diligence in preparing its bid. Significantly, the court in *Richardson* observed that the relevant contract documents did not contain any disclaimer as to site information provided in the contract documents. See also *Frederick Snare Corp. v. Maine-New Hampshire Interstate Bridge Authority*, 41 F. Supp. 638 (D.N.H. 1941) (contractor allowed to recover on differing site conditions claim asserted against owner of project, despite the existence of contract provision requiring contractor to "investigate" and base his bid on his own opinion of the conditions, where engineers knew that no accurate soundings could be made to

determine subsurface conditions in the four-day bidding period, but withheld that information).

102. See, e.g., *Ohio Valley Contractors, Inc. v. Board of Education*, 391 S.E.2d 891 (W. Va. 1990) (holding that contractor stated valid claim against owner due to (a) ambiguities in the contract documents prepared by owner's architect concerning the authority of the architect who authorized additional excavation beyond that included in the contractor's bid; and (b) the existence of factual issues as to whether the architect was authorized on behalf of the owner to direct the contractor to perform additional excavation).

103. See, e.g., *Bilotta Construction Corp. v. Village of Mamaroneck*, 199 A.D.2d 230 (N.Y. App. Div. 1993) (contractor asserted claim against owner and engineer on the basis of misrepresentation of information regarding estimates of elevations; court rejected contractor's claim, holding that, based on disclaimers in contract documents, and provision requiring contractor to conduct its own investigation prior to entering into the contract, any reliance by the contractor on such data was not justifiable, and therefore could not support a misrepresentation claim against either the owner or the engineer); *Zontelli & Sons, Inc. v. City of Nashwauk*, 373 N.W.2d 744 (Minn. 1985) (contractor asserted claim against owner and consulting engineer for negligence and misrepresentation in preparation of estimates on which contractor bid, alleging that it encountered more concrete than estimated and that the amount of unsuitable material to be excavated was substantially understated and was of a differing quality than anticipated; court found engineer negligent because, among other things, engineer had "drastically underestimated" the quantity of concrete and unsuitable materials to be removed and had failed to indicate the unusual characteristics of those materials); *APAC-Carolina, Inc. v. Greensboro-High Point Airport Authority*, 431 S.E.2d 508 (N.C. Ct. App. 1993) (contractor asserted claim against owner and engineer due to substantial overruns in undercutting required; court refused to allow claim against engineer on the basis that the contractor had failed to adequately inspect available information, the contract clearly stated that any quantities mentioned in the contract were merely estimates, and, therefore, the contractor had failed to establish any justifiable reliance on information contained in the contract documents), *rev. denied*, 438 S.E.2d 197 (N.C. 1993); *Raymond International, Inc. v. Baltimore County, Maryland*, 412 A.2d 1296 (Md. Ct. Spec. App. 1980) (claim by contractor against owner and engineer arising out of allegations that quantities estimated in contract documents were inaccurate and grossly underestimated, resulting in the unit price representing an inadequate measure of compensation; court allowed recovery against engineer on the basis that the estimates were materially wrong and substantially inaccurate and engineer reasonably should have known information was inaccurate based upon engineer's prior involvement at the site), *cert. denied*, 449 U.S. 1013 (1980); *L. Loyer Construction Company v. City of Novi*, 446 N.W. 2d 364 (Mich. Ct. App. Ct. 1989) (contractor asserted claim against owner and engineer claiming that the work required to complete the project varied from estimates set forth in the contract documents on which the contractor based its bid, and that the quantity variations necessitated the use of unplanned, less efficient methods of excavation and use of a different off-site dump than originally contemplated. Court rejected contractor's claim holding that owner did not guaranty the representation as to soil conditions; that the risk that soil and water conditions might be different than expected was expressly assumed

by contractor; that the quantities listed in the proposal were not guaranteed to be actual quantities; and that contractor had a duty to investigate the specific subsurface conditions, as well as the site generally), *appeal denied*, 435 Mich. 871 (1990); *Costanza Construction Corp. v. City of Rochester*, 147 A.D.2d 929 (N.Y. App. Div. 1989) (contractor asserted breach of contract, negligence, breach of warranty and equitable adjustment claims against owner and engineer alleging gross inaccuracies in estimates of cubic yards of rock to be excavated; court rejected contractor's claim because the contract documents clearly included a disclaimer by owner of any responsibility for accuracy or completeness of information on the drawings concerning existing conditions, including rock, and obligated contractor to satisfy itself as to site conditions; accordingly, as a matter of law, the contractor could not establish that it had justifiably relied on information in the contract documents, thereby warranting dismissal of claims against owner and engineer), *appeal dismissed*, 541 N.E.2d 429 (N.Y. 1989).

104. See, e.g., *Northrup Contracting, Inc. v. Village of Bergen*, 527 N.Y.S.2d 670 (N.Y. Sup. Ct. 1986); *W. H. Lyman Constr. Co. v. Village of Gurnee*, 403 N.E.2d 1325 (Ill. App. Ct. 1980). In *Northrup*, a contractor sued an engineering firm and project owner, alleging that the engineering firm was negligent in (a) the preparation of test pits that failed to disclose high groundwater within the area of the contractor's work; and (b) the administration of the construction contract—specifically, the engineer's delay in addressing and resolving the contractor's differing site conditions claim based on the alleged unanticipated high groundwater, and in recommending termination of the contractor due to nonperformance (as a result of the unresolved nature of the contractor's claim). The court held that the contractor stated a valid legal claim against the engineering firm not only with respect to the claim of negligence in the preparation of contract documents, but also with respect to the claim that the engineer was negligent in the manner in which it administered the construction contract and addressed (or, more precisely, failed to timely address) the contractor's differing site conditions claim. Similarly, in the *Lyman* case, the court, while rejecting the contractor's differing site conditions claim on the rationale that the contractor had failed to undertake an adequate site investigation, did permit the contractor to proceed to trial against an engineering firm based on the contractor's allegations that the engineering firm had negligently administered the performance of the construction contract by withholding approval for many months of the contractor's method of sealing the manhole basis, and otherwise failing to adequately address the contractor's differing site conditions claim.

In *Brinderson Corp. v. Hampton Roads Sanitation Dist.*, 825 F.2d 41 (4th Cir. 1987), a contractor was allowed to pursue a differing site conditions claim against an owner despite the absence of the timely submission of the written notice required by the differing site conditions clause in light of the fact that "the owner's engineers were fully informed of all of the contractor's [information as it developed concerning the differing site condition claim]." The court held that the engineer's awareness of the contractor's differing site conditions claim would be imputed to the owner (given that the engineer was the owner's representative and agent during the construction phase of the project); and, accordingly, the contractor's failure to provide timely written notice to the owner as required by the contract documents did not defeat the contractor's claim. Although not clear from the court's decision in *Brinderson*, it appears that the engineer did not notify the owner of the engineer's awareness of the contractor's differing site conditions claims and, under such circumstances, the engineer could be found liable to the owner for the contractor's claim on the theory of indemnification. See *Zontelli & Sons v. City of Nashwauk*, 353 N.W. 2d 600 (Minn. Ct. App. 1984) (ruling that an owner may obtain indemnification from its project engineer for owner's payment of contractor's differing site condition claim, based on finding that engineering firm breached its duty to owner to provide satisfactory plans and specifications).

105. See, e.g., *Victor M. Solis Underground Utility & Paving Co. v. City of Laredo*, 751 S.W.2d 532 (Tex. Ct. App. 1988), *writ denied* (June 28, 1989). *Solis* involved the design and construction of a storm sewer system. The contractor in *Solis* brought an action for tortious interference with a contractual relationship against an engineering firm (and its project engineer) for recommending the contractor's termination. Under the terms of its contract with the city, the engineering firm was obligated to continuously inspect the contractor's work and report any failures in conforming to the project specifications. The contractor knew of this duty to inspect through his contract with the city. Evidence was presented at trial that the contractor was uncooperative and did not meet the plans and specifications. The court enumerated the elements of tortious contractual interference as: (1) the defendant maliciously interfered with a contractual relationship; and (2) the defendant did so without legal justification or excuse. In applying these elements to the facts of the case, the court held that, since the contractor's work had not met the specifications, there was no evidence of maliciousness and the project engineer was justified in exercising his superior legitimate interest by recommending termination. Thus, neither element of the cause of action had been satisfied. In addition, the court discussed interference actions, and distinguished between interference caused by third-party strangers and interference caused by a servant or agent. Concluding that the engineering firm was clearly the city's agent, the court stated that "[j]ust as an instructed verdict in favor of the city on the interference action would have been proper, because the cause of action must fail, so it was in the case of its agents." However, there was a strong dissent to this portion of the opinion. The dissent argued that the engineering firm never claimed to be anything but an independent contractor; therefore, the "servant rule" did not apply. The dissent further argued that since the engineering firm was to serve as referee (with binding decisions) in all questions arising under the contract between the contractor and the city, it could not be the city's agent because it would be utterly inconsistent for a servant to act as an impartial referee for his master. See also *Forte Brothers, Inc. v. National Amusement, Inc.*, 525 A.2d 1301 (R.I. 1987) (holding that contractor can maintain cause of action for negligence against supervising architect/site engineer for failing to accurately measure removal of mass boulders by plaintiff, to report removal to the owner, and to authorize payment to plaintiff, when contractor reasonably and directly relied on architect/engineer professionally rendering its contractual duty to do so); *Fitzpatrick Constr. Corp. v. County of Suffolk*, 138 A.D.2d 446 (N.Y. App. Div. 1988) (in an action by contractor against engineering firm alleging intentional interference with contractual relations, court held that contractor's vague reference to intentional course of conduct was insufficient to substantiate intentional interference claim against engineering firm), *appeal denied in part*, 534 N.E.2d 315 (1988); *Costanza Constr. Corp. v. City of Rochester*, 135 A.D.2d 1111 (N.Y. App. Div. 1987) (holding that intrusion that is negligent or incidental to a lawful purpose, but not intentional, is insufficient

to maintain a claim against architect/engineering firm for tortious interference with a contractual relationship).

106. See *Fontaine Bros., Inc. v. Springfield*, 617 N.E.2d 1002 (Mass. App. Ct. 1993) (upholding determination by architect that contractor was not entitled to additional compensation for removal of below grade debris where contract documents gave architect final review authority), *review denied*, 621 N.E.2d 685 (Mass. 1993).

107. Unreasonable delay in addressing and/or resolving such claims may subject the design professional to liability. See Note 104, *supra*.

108. For an excellent collection of case histories and associated risk management lessons and recommendations in the area of subsurface investigation, design, and construction, see *Case Histories of Professional Liability Losses* (ASFE, 1996).

109. See "Important Information About Your Geotechnical Engineering Proposal" (ASFE, 1992) and "Important Information About Your Geotechnical Engineering Report" (ASFE, 1992).

110. Attorneys who represent owners have advised them to bridge the gap of liability exposure between the owner's implied warranty obligation to the contractor and the design professional's more restricted "professional standard of care" liability exposure, by requiring that design professionals accept certain contractual provisions, such as that the design professional will perform his or her services in accordance with the highest professional standards and quality; that the design professional has the specialized skill to recognize all specific dangers that may arise to third parties (including construction workers); that the design professional "will supply error free, and that it will assure that suitable materials are used in the project, that it will assume sole responsibility for any damages, loss, claim or expense caused by the use of unsuitable materials approved by it, [and] that the project will be built in a workmanlike and professional manner"; and that the design professional will be liable for all errors and omissions relative to its services. See T. A. Colvig and S. L. Cali, "The Liability Gap Between Owner and Architect" Chapter 10 in *1997 Wiley Construction Law Update*, N. J. Sweeney, ed., (New York: John Wiley & Sons, Inc. 1997).

111. Guidelines for the preparation of geotechnical reports and studies are addressed in Chapter 3 of this book. Additional guideline references include: J. P. Gould, "Geotechnology in Dispute Resolution," *Journal of Geotechnical Engineering*, (July 1995), pp. 525–26; Terry Mellors, "Referencing Ground Conditions," *WT Review*, (November 1995), pp. 367–370; R. J. Essex, *Means of Avoiding and Resolving Disputes During Construction*, (Woodward-Clyde Consultants, October 1994); R. J. Essex, ed., "Geotechnical Baseline Reports for Underground Construction Guidelines and Practices," prepared by the Technical Committee on Geotechnical Reports of the Underground Technology Research Council (New York: ASCE, 1997). Many of these guidelines recommend a constructability review of such reports and studies by a contractor qualified and experienced in the bidding and construction processes. See *Geotechnical "Differing Site Conditions,"* Geotechnical Engineering Notebook, GT Guideline No. 15, (USDOT Federal Highway Administration, April 1996).

11

INSURANCE FOR SUBSURFACE PROJECTS

Timothy F. Gilmartin, Esq.
Stone & Webster Engineering Corporation

11.0 INTRODUCTION

The purpose of this chapter is to examine the various types of insurance policies available to a design professional, construction manager, or contractor involved in subsurface projects and how those policies can mitigate and assist in managing the various risk exposures involved with those projects. Consultation with qualified and experienced insurance professionals and legal counsel is recommended for all interested parties, since this chapter is not intended as a substitute for that consultation. In addition, this chapter contains generalizations about insurance coverage; in any specific context, the actual insurance policy terms should always be consulted.

11.1 RELATIONSHIP OF CONTRACTS TO INSURANCE

In order to determine the risk exposure and what insurance may be necessary, the contract for the intended services must be reviewed in detail, and the following questions answered:

Subsurface Conditions: Risk Management for Design and Construction Management Professionals, edited by David J. Hatem
ISBN 0471-15607-8

- *What types of services are being performed—professional, construction,* or *both*?

  The types of services that are to be performed dictate what insurance coverage should be procured. Generally, where professional services are to be performed, professional liability insurance should be obtained, along with workers' compensation insurance and general and automobile liability insurance. Where construction work is to be performed, workers' compensation insurance, general and automobile liability insurance, and builder's all-risk insurance would be appropriate.

- *For whom are the services being directly performed—the owner, contractor, designer,* or *another party*?

  When contracting for services, the major contractual and legal risks are usually undertaken by the owner for whom the services are being performed. If the design professional's agreement is directly with the ultimate owner, then contractual limitations of those risks can be mutually agreed upon. If, however, the design professional is contracting with the construction contractor, then there is no contractual relationship between the design professional and the owner. This lack of contractual agreement with the owner presents a problem in that the only contractual limitations of liability that may apply to the design professional's services are those passed down to the design professional through the contractor's contract with the owner. Insurance and legal protections may be the only protection to the design professional against a claim by the owner.

- *What are the contractual and legal risks to the client and to third parties*?

  Figure 11.1 briefly describes the various contractual and legal risks faced by entities performing professional and construction services.[1]

  Certain of these risks are business risks only, and are usually uninsurable. For example, completing the work for a fixed price and within a certain budget is usually a business risk. Standard of care and indemnifi-

To Client:	*To Third Parties*:
• Price	• Legal Liability Common Law Statutes
• Schedule	
• Scope	
• Warranties	
• Guarantees	
• Indemnities	
• Standard of Care	
• Legal Liability Common Law Statutes	

Figure 11.1 Contractual and legal risks.

To Client :	*To Third Parties* :
• Negotiated Limiations of Liability — Remedies — Dollar Limits — Time Limits • Limitations Provided by Law — Negligence Standard — Statute of Repose — Statute of Limitations	• Limitations Provided by Law — Negligence Standard — Statute of Repose — Statute of Limitation

Figure 11.2 Limitations on contractual and legal risks.

cation obligations, which are based on negligence or contractual requirements obligating that services conform to accepted standards of care, can be insured. Legal liability arising out of the negligent performance of services is generally insurable. If professional negligence causes injury to the client or a third party, the insurer should respond to protect the professional.

- *What limitations of liability apply to the contract and legal risks to the client and to third parties?*

  Various limitations of liability may provide protection to a design professional or contractor against claims arising from their work. Summarized in Figure 11.2 are various limitations that may be available.

  Parties to an agreement may mutually agree on acceptable limitations of liability in their contract. Professionals, for instance, may attempt to limit their warranty obligations to reperformance or redesign of any deficient services, or to limit recovery against them to a specific dollar amount or particular type of damage. Contractors may assume responsibility for defective material or faulty workmanship up to a certain dollar limit based on a percentage of contract price (and for a limited period of time). Indemnity obligations for claims of third parties against an owner may be limited to a specific time period and dollar amount (including the amount of available insurance proceeds). Mutual waivers of consequential damages, as well as overall limitations of liability whether arising under the contract or at law, may be agreed upon between contracting entities (see Chapter 2, Section 2.3, and Chapter 10, Section 10.1.1.1).

  Through statutory enactment, states have imposed certain time limits for the assertion of various types of claims, including contract claims or negligence-based claims. *Statutes of limitations* limit the time in which a claim can be made. The clock usually starts running upon the occurrence of the act or omission upon which the claim is predicated, or the discovery of the alleged injury or damage. In contrast, *statutes of repose* set specific and absolute time limitations for the assertion of claims with a starting date of some distinctively and objectively defined point, such as substantial completion of a project. The time periods for both types of statutes vary from state to state, with periods generally in the three- to six-year range for limitations periods and in the six- to ten-year range for repose periods.

Contractual and legal limitations of liability can protect a professional or contractor from liability claims only to a limited extent, and beyond the protections afforded by these limitations, typically do not apply to claims by third parties. The purchase of insurance in various forms and limits is essential for any business to transfer risk and protect its assets and the personal assets of its officers, directors, employees, and shareholders from liabilities related to its professional and/or business operations.

11.2 TYPES OF INSURANCE COVERAGE

The purchase of insurance is one method to transfer to an insurer some of the risk associated with engineering and construction projects. Insurance is an important component of an effective risk management program, as are many of the other components previously discussed in other chapters of this book, such as prudent client selection and the development and use of appropriate contract terms. While an important component of a risk management program, insurance is neither intended, nor should be considered, a *substitute* for such a program. Traditional types of insurance coverage addressing risk exposures are shown in Figure 11.3.

Each of these will be discussed below.

Coverage	Exposure
Professional Liability	Damage and injuries resulting from negligent design or other negligent professional services
Commercial General Liability	Third-party bodily injury and property damage, faulty construction workmanship
Workers' Compensation and Employer's Liability	Injury to workers employed by the design professional or contractor
Builder's Risk	Physical damage to the work during construction, whether property of owner, contractor, or supplier
Construction Equipment	Damage to contractor's construction equipment used in project execution
Environmental/Contractor's Pollution Liability	Third-party environmental bodily injury and property damage and cleanup costs resulting from release of pollutants

Figure 11.3 Types of insurance coverage.

11.3 PROFESSIONAL LIABILITY INSURANCE

Professional liability insurance protects the design and construction management professional against claims arising from negligent errors, acts, or omissions in the performance of his or her professional services. This type of insurance is intended to provide coverage for defense costs, as well as indemnification for damages paid to others for claims covered under the insurance. A negligent act, error, or omission claim against the professional typically involves an allegation that the professional did not meet the professional standard of care in the performance of professional services. The professional standard of care for design and construction management professionals is that which is consistent with the degree of care and skill ordinarily exercised by others in the same profession currently practicing under similar circumstances. (For a detailed discussion of the standard of care, see Chapter 10, Section 10.1.1.2.) This standard provides the test for determining whether a professional has a legal liability for damages resulting from his or her services. If the design professional has performed the services in accordance with the professional standard of care, he or she should not be legally liable for the damages (unless the professional has agreed to perform in accordance with a higher standard). If he or she is determined to have breached that standard, any resulting liability typically will be covered under the professional liability policy. However, if the professional is determined to be liable on the basis of a contractual (or warranty) standard that exceeds the negligence-based professional standard of care, there is probably no coverage since professional liability insurance will respond only to the extent of the professional's actual or alleged negligence.

To be covered by a professional liability policy, a claim against a professional must arise out of the performance of professional services. To some degree, what constitutes "professional services" depends on the particular practice of the professional, what that professional discloses to the insurer in the application for insurance, and/or how the policy defines that term. The policy's declaration page typically will reference the professional services being covered, e.g., "engineering, design, and construction management." It may also reference the application for the insurance completed by the professional. The application should state broad categories of services performed by the professional, which will help to alleviate any subsequent disputes with underwriters as to whether the "professional service" from which the claim arises falls within the scope of coverage.

Key concepts of professional liability policies are as follows:

Limits of liability—The limits of the professional liability insurance are stated "per claim and aggregate per policy period," e.g., "Limit: $1,000,000 per claim and aggregate per year." Deductibles are stated "per claim" and are not generally capped by an aggregate, e.g., "Deductible: $100,000 per claim." Some policies may have a self-insured retention (SIR) instead of a deductible. Under an SIR, the insured is responsible for paying defense costs and damages directly until such time as the limit of SIR is reached; the insurer's risk does not attach until the insured has satisfied its SIR obligation.

The limit of liability includes indemnification for "damages" (compensatory only) and "claims expenses" (attorney fees and other costs and expenses in handling the claim). Thus, both indemnification and claims expense payments serve to reduce the available limit of liability.

Claims-made coverage—These policies are written on a "claims made and circumstances reported" basis. In general, these policies cover only those claims reported during the policy period or any extended (or discovery) period stated in the policy. Also, if, during the policy period, the insured gives notice to the insurer of a "circumstance likely to give rise to a claim," this usually preserves coverage (assuming available coverage limits) under the policy in effect on the date of that notice even if the related claim is not made until after the policy expires. A claim is generally defined as a demand against the insured for money or services. To be covered, a claim must relate to an actual or alleged negligent act, error, or omission committed subsequent to the policy "retroactive" or "prior acts" date. This date can be the date on which the insured first obtained this type of insurance, the date on which the insured's practice commenced, or some other date agreed to by the insurer and insured.

Generally, in order to maintain coverage for prior acts, the professional must continue to renew the policy annually by submission of a renewal application. Typically, the renewal application will ask the policyholder to list all claims against the insured made within the last ten years and, after inquiry of principals, partners, officers, employees, or directors, any circumstances likely to be the basis of a claim under the policy. If claims or circumstances are known and not reported prior to the policy renewal, then there will be no coverage under the renewal policy for these past claims or circumstances. It is, therefore, important to identify in the policy who is deemed to have "knowledge" of a claim or circumstance on behalf of the insured. It should only be the person responsible for purchasing the insurance, e.g., the risk manager and/or the general counsel or senior level officers of the company. A lower-level employee should not be allowed to jeopardize the insurance for the entire company by inadvertently failing to disclose or withholding information about a known claim or circumstance likely to give rise to a claim.

Legal liability—Professional liability policies "pay on behalf" of the insured all sums for which the insured is legally obligated to pay as damages arising out of alleged or actual negligent acts, errors, or omissions of the insured in the performance of professional services. To be liable for damages, the insured must have performed in a manner that was below the "standard of care" reasonably required of other professionals performing similar professional services.

Claims settlement—Under a professional liability policy, an insured has the right to consent to a settlement by the insurance company. If an insured does not agree to the settlement, then the insurer may reserve its rights not to pay any additional damages or claims expenses over the proposed settlement amount.

11.3.1 Major Exclusions

Liability of others assumed by the insured under a contract unless such liability would have existed even in the absence of the contract.

Comment: The professional's indemnity and standard-of-care contractual obligations should be limited to the negligence-based professional standard of care (i.e., legal liability). There is no coverage for assuming liability for another's negligence, whether that of the owner or some third party; or for assuming liability under an indemnification or other contract provision that is based on a standard that exceeds the negligence-based professional standard of care (see Chapter 10, Section 10.1.1.2).

Dishonest acts committed by or at the direction of the insured.

Comment: Should not apply to insured who had no knowledge of dishonest act.

Punitive damages, fines, and penalties or client's refusal to pay fees.

Comment: The absence of coverage in this regard should be carefully considered in negotiating contractual provisions dealing with liability and/or fines and penalties, especially in the environmental context.

Express warranties or guarantees, e.g., guaranteeing a specific amount of production from a plant.

Comment: Should not apply if liability would have applied in absence of the exclusion (i.e., liability based upon a departure from the negligence-based professional standard of care would be covered). The professional should not assume responsibility for meeting specific performance levels (especially those that exceed the professional standard of care), as such responsibilities will not be covered unless the professional would be legally liable for them.

Personal injury sustained by the insured's employees.

Comment: Applies only to employees of insured, since this liability risk would be covered by workers' compensation insurance.

Insured vs. Insured.

Comment: Provides good argument not to name client as insured on the practice policy. (See Hatem, 1996.)

Professional services insured under a project policy.

Comment: At a minimum, the professional's practice policy should apply if the project policy is canceled, exhausted, or the insurer goes bankrupt. Try to negotiate a practice policy that is in excess of and difference in conditions to (and if possible, difference in limits to) the protection under the project policy, in the event that the project policy is exhausted, or claims are not covered due to exclusions.

Pollution liability.

Comment: Exclusions of coverage with respect to liability arising out of pollution claims are quickly disappearing, although some limitations still apply. Remaining exclusions generally apply only to wrongful acts prior to 1986, pollution from any property owned or rented to the insured, or the actual generation, transportation, storage, disposal, or arranging for disposal of pollutants.

Requiring, advising, or maintaining insurance policies and surety bonds.

Comment: The professional should be careful, especially if providing construction or program management services; the failure of the construction or program manager to advise the owner regarding insurance or surety matters or to require or ensure that a construction contractor is providing insurance or bonds could lead to a professional claim against the construction or program manager, which would not be insured.

11.3.2 How Much Professional Liability Insurance to Buy?

Deciding on how much professional liability insurance a firm should buy is subject to many factors. What do clients require of the professional? Clients generally require only minimum limits, although some may ask for a higher aggregate or a specific project policy if they feel the professional's policy limit of liability is inadequate or that the policy could be exhausted by claims from other projects. How large are projects on which your firm is providing services at any given point in time? Generally, the larger the related construction value, the higher the limits needed. What professional services are provided, and on what types of projects? Consulting or engineering services related to commercial projects would not require the same limits as engineering services related to chemical, petroleum, or power projects. Do the services involve especially sensitive third-party risks? Projects involving large interaction with the general public would require higher limits. Finally, a professional firm needs to determine how much insurance it can afford in relation to its revenue. Although consultation with a knowledgeable insurance broker is helpful in this regard, to a large degree, the determination of the amount of policy limits is substantially influenced by the firm's tolerance for risk.

11.3.3 Market Ratings and Capacity

An insurer's market rating, capacity, and participation in the professional liability marketplace can vary from year to year. If a contractor or designer is interested in pursuing project professional liability insurance, he or she should check with an insurance broker to verify the current rating, capacity, and participation of an insurer in the professional liability marketplace.

11.3.4 Project Professional Liability Insurance

A project professional liability policy is a dedicated policy for one particular project. Typically, project policies have fixed limits over a large deductible or self-insured retention and apply to all claims made during the design and construction phase, plus a predetermined "discovery" or extended reporting period after the project completion for reporting claims (up to five years).

There are many advantages to a project professional liability policy, including:

- Dedicated policy limits for the life of the project, which are not subject to reduction or exhaustion due to claims from other projects.
- The premium has a guaranteed rate, making it easier to pass the cost along to the owner as a direct expense.
- Limits in excess of $50 million are available.
- Consistent coverage and limits for all design firms on the project, including those uninsured or underinsured.
- Under most circumstances, the project policy is non-cancelable.
- Single point of claims responsibility, including joint defense agreements among insureds.
- Claims can be reported in an extended period up to five years (if selected) after completion of construction.

Figure 11.4 illustrates a project policy.

Typical project professional liability limits—Where an owner requires separate limits of coverage for its, or where the professional determines that its practice policy does not provide sufficient coverage, then the guidelines shown in Figure 11.5 may be used to determine the project specific limit to purchase.

Cost—Depending on the limits and deductible purchased, premium can range from 0.25 percent to 1.00 percent of construction values. Also affecting premium are the contract terms, design team experience, project complexity, type, and legal jurisdiction.

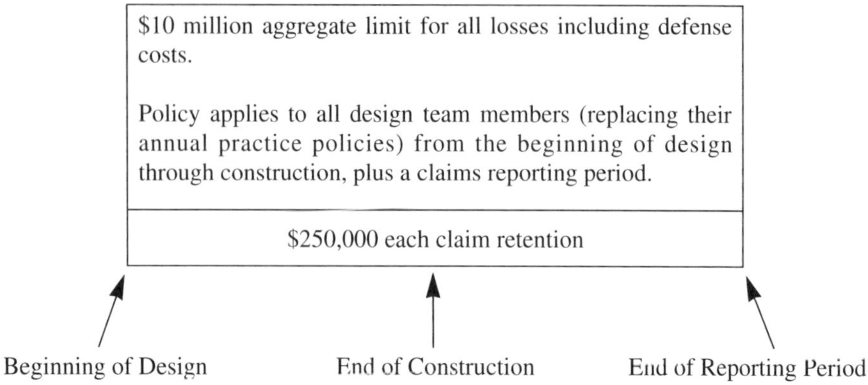

Figure 11.4 Project policy.

11.4 COMMERCIAL GENERAL LIABILITY INSURANCE

Commercial general liability insurance is the latest version of many variations of general liability insurance. Two policy forms are currently in use: an occurrence-based policy and a claims-made policy. The occurrence-based policy is generally used and preferred by most insureds.

Under the Coverage A (Bodily Injury and Property Damage) insuring agreement, the insurer agrees to pay those sums the insured becomes legally obligated to pay as damages as a result of bodily injury or property damage to which the insurance applies. The bodily injury or property damage must be caused by an occurrence that takes place in the coverage territory.

The obligation of the insurer to "pay those sums" does not mean the insured needs to pay first, only that the insured has an obligation to pay. Such legal obligation arises from civil liability of the insured for either unintentional (negligent) or intentional torts under common law, statute, or contract. Damages comprise those sums of money that the law imposes as compensation resulting from bodily injury or compensation for damage to property, including loss of use.

Bodily injury includes sickness or disease sustained by a person, including death, resulting from any of these. Property damage means physical injury to

Construction Values	Typical Limits
Under $25 million	$3 million to $5 million
$25 million to $100 million	$5 million to $10 million
$100 million to $1 billion	$10 million to $25 million
Over $1 billion	$25 million or more

Figure 11.5 Typical project professional liability limits.

tangible property, including resulting loss of use of that property, and loss of use of tangible property that is not physically injured.

The bodily injury or property damage must be caused by an "occurrence." An "occurrence" means an accident, including continuous or repeated exposure to substantially the same generally harmful conditions. Whether an event is sudden and definite in time and place, or one that is continuous and difficult to pinpoint in time or place, does not matter. What matters is whether the bodily injury or property damage results without the insured's foresight or anticipation.

The occurrence must also take place in the "coverage territory," which is generally limited to the United States (including territories and possessions), Puerto Rico, and Canada. For goods or products made or sold in the coverage territory, the injury or damage may occur in all parts of the world.

Major exclusions to this type of insurance include:

Expected or Intended Injury

This insurance does not apply to bodily injury or property damage expected or intended from the standpoint of the insured.

Comment: It is the intentional *injury*, rather than the intentional *act*, that is excluded.

Contractual Liability and Professional Liability

The insurance does not apply to bodily injury or property damage for which the insured is obligated to pay damages by reason of the assumption of liability in a contract or agreement.

There are two exceptions to the exclusion: (1) liability assumed under an "insured contract"; or (2) liability the insured would have in the absence of the contract or agreement.

Comment: The exceptions to the exclusion provide broader coverage than a professional liability policy that provides coverage only if the insured was liable in the absence of a contract or agreement. An "insured contract" includes that part of any other contract agreement pertaining to a business under which the professional assumes the tort liability of another party to pay for bodily injury or property damage to a third person. Tort liability is a liability that would be imposed by law in absence of any contract or agreement. An insured contract *does not* include indemnity of an architect, engineer, or surveyor for injury or damage arising out of their professional services. If the insured is an architect, engineer, or surveyor, an insured contract *will not* include any contract or agreement to assume liability for injury or damage arising out of the insured's rendering or failure to render professional services. This should be qualified so as not to apply to those contracts where the insured also provides construction services. This professional liability exclusion in the definition of an insured contract applies to "injury or damage," whereas the coverage section of the policy covers only "bodily injury or property damage." This exclusion should be looked at very carefully especially where an insured is involved in construction activity, either directly or with subcontractors or in a joint venture.

Damage to Property

A current version of this exclusion reads as follows:

"Property damage" to:

(1) Property you own, rent, or occupy;

(2) Premises you sell, give away, or abandon, if the "property damage" arises out of any part of those premises;

(3) Property loaned to you;

(4) Personal property in the care, custody, or control of the insured;

(5) That particular part of real property on which you or any contractors or subcontractors working directly or indirectly on your behalf are performing operations, if the "property damage" arises out of those operations; or

(6) That particular part of any property that must be restored, repaired, or replaced because "your work" was incorrectly performed on it.

Comment: Paragraph (1) excludes damage to both real and personal property that the insured owns, rents, or occupies. The purpose of the exclusion is to avoid coverage for an exposure that can be insured by some form of property insurance, such as builder's risk.

Paragraph (2) excludes coverage for damage to property that has been sold, given away, or abandoned. However, this exclusion does not apply to "your work" (work or operations performed by or on behalf of the named insured).

Paragraph (3) excludes both real and personal property loaned to the professional.

Paragraph (4) excludes only personal property and not real property in the insurer's care, custody, and control. An insured needs to arrange some other form of insurance, such as property insurance, to cover the personal property of others in his or her care, custody, or control.

Paragraph (5) excludes only damage to a particular part of any "real" property. It does not exclude damage to personal property. For example, if a subcontractor drops a beam and damages the professional's other work, the professional would not have coverage for damage to the beam, but he or she would have coverage for any resulting damage to other property. Builder's or installation risk insurance should be provided to cover this exclusion and insulate the professional liability insurance from what is really property damage exposure.

In paragraph (6), the chief applicability of the faulty workmanship exclusion is for property away from the insured's premises, and is applicable only to ongoing operations (i.e., not to completed operations).

Damage to "Your Product"

The exclusion of damage to the named insured's products applies to:

"Property damage" to "your product" arising out of it or any part of it.

The definition of "your product" determines the applicability of the exclusions.

"Your product" means:

Any goods or products, other than real property, manufactured, sold, handled, distributed, or disposed of by:

(1) You;

(2) Others trading under your name; or

(3) A person organization whose business or assets you have acquired; and

"Your product" also includes:

a. Warranties or representations made at any time with respect to the fitness, quality, durability, performance, or use of "your product"; and

b. The providing of or failure to provide warnings or instructions.

Comment: Work done by contractors on buildings or other structures would not be excluded, since section (a) states that the exclusion applies to "other than real property."

The warranty part of the definition of "your product" does not imply warranty coverage; only bodily injury or property damage resulting from such warranty is covered, except to the product itself.

Providing or failing to provide warnings or instruction is also included in the definition of "your product." The intent of this definition is to clarify that product suits based on a failure to warn are subject to the aggregate limit that applies to product/completed operations.

Damage to "Your Work"

"Property damage" to "your work" arising out of it or any part of it and included in the "products completed operations hazard."

This exclusion does not apply if the damaged work or the work out of which the damages arises was performed on your behalf by a subcontractor.

The term "your work" is defined as:

a. Work or operations performed by you or on your behalf; and

b. Materials, parts, or equipment furnished in connection with such work or operations.

Also included in the definition is the same reference to warranties and failure to work included in the damage to "your product" exclusion.

Comment: The most important feature of the current "damage to your work" exclusion is that it *does not* apply if work, or the work out of which the damage arises, was performed by a subcontractor. An example of this is where a general contractor erected a building using the services of many subcontractors, including one for all of the electrical work. Subsequent to the building being occupied, faulty wiring causes a fire, which destroys the building. If the general contractor is sued, he or she will be protected by the policy, since the work out of which the damage arises was performed by a subcontractor on behalf of the general contractor and, as such, is not excluded. If, instead, the loss had originated in work performed by the general contractor, there would not be any coverage for any work performed by the general contractor. The general contractor could face an uninsured loss if the property insurer on the building subrogates against the general contractor. This is one reason to try to negogiate a waiver from the owner for any liability for damage to the project once the coverage under the builder's risk insurance is completed.

Damage to Impaired Property or Property Not Physically Injured

This exclusion is comparable to the failure to perform exclusion of previous general liability policies.

The current exclusion applies to:

"Property damage" to "impaired property" or property that has not been physically injured, arising out of:

(1) A defect, deficiency, inadequacy or dangerous condition in "your product" or in "your work"; or

(2) A delay or failure by you or anyone acting on your behalf to perform a contract or agreement in accordance with its terms.

This exclusion does not apply to the loss of use of other property arising out of sudden and accidental physical injury to "your product" or "your work" after it has been put to its intended use.

"Impaired property" means tangible property, other than "your product" or "your work," that cannot be used or is less useful.

Comment: If a product made by an insured is incorporated into another product made by another manufacturer, and if the insured's product is defective or deficient and this results in the other product being less useful or totally unusable, the first part of the definition of "impaired property" is met. If the other product can be restored to use by repairing or replacing the insured's product, both parts of the definition are satisfied and the exclusion applies.

The second part of the exclusion refers to cases in which an insured fails to fulfill a contract in accordance with its terms. If an insured contracts with another party to deliver a part to be incorporated into a final product and fails to do so, the insured will not have coverage for any resulting claim for loss of use of the final product, because the final product can be restored to use by fulfilling the terms of the contract.

The exception to the exclusion states that the exclusion does not apply to loss of use of other property arising out of sudden and accidental physical injury to the insured's product or work after it has been put to its intended use.

11.4.1 Occurrence versus Claims-Made Triggers

The trigger of the occurrence policy is bodily injury or property damage that *occurs during the policy period.* If a party is injured by an insured's product today, the occurrence policy in effect today will apply to the loss, whether a claim is made against the insured this year or some later year.

The trigger of the claims-made policy, on the other hand, is the first making of the claim against the insured during the policy period. Thus, if someone is injured today, but does not make a claim against the insured during the current policy period, there is no coverage under the current policy if the claim is made in a subsequent policy year unless the policy allowed the "reporting of circumstances likely to give rise to claim" and the insured did, in fact, provide notice of the injury to the current policy.

In general, insureds prefer occurrence-type coverage over claims-made. As long as the insured keeps occurrence-type coverage in effect for all times, bodily injury or property damage that occurs during those times should fall within the coverage under one or more of the insured's policies.

11.4.2 Policy Limits

Current forms of general liability insurance are subject to a "general aggregate" limit, which limits the total amount payable during the policy period for all coverages, except for products-completed operations claims, which are subject to a separate aggregate limit.

Limits of coverage are expressed as in the following example:

General aggregate	$2,000,000
Products-completed operations aggregate	$2,000,000
Personal and advertising liability	$1,000,000
Each occurrence	$1,000,000
Fire damage (any one fire)	$1,000,000
Medical expense (any one person)	$5,000

Reduction of the general aggregate limit does not affect the products-completed operations aggregate limit. Each aggregate limit represents a separate amount of insurance. Thus, an insured with a general aggregate limit of $2 million, as stated above, and a products-competed operations aggregate limit of $2 million could possibly collect up to a total of $4 million under the policy.

11.5 EXCESS/UMBRELLA LIABILITY INSURANCE

Excess/umbrella liability insurance allows firms to buy additional limits of insurance where their operations require it. In general, umbrella policies are excess over commercial general liability, automobile liability, and employer's liability insurance. Limits are per occurrence and aggregate. Although occurrence-type excess liability can be purchased for high limits of coverage, availability of coverage is limited to a "reported occurrence" type of policy. Under the reported occurrence policy, notice must be provided to the insurer of the occurrence during the policy period. If notice is provided, this preserves coverage under that policy. If no notice is provided and the policy expires, coverage ceases under that policy. The professional can get coverage under a subsequent policy if the occurrence happened after any applicable retroactive date in the subsequent policy.

11.6 BUILDER'S RISK INSURANCE

Builder's risk insurance covers physical damage to or loss of the project being constructed, except as excluded. Property covered by the policy includes materials, equipment, supplies, and machinery that goes into a project and generally covers such property while in transit, during on-site and off-site storage, during and after installation, testing, and until the project is accepted by the owner.

The limit of liability for builder's risk insurance should be the estimated replacement cost of the project, including direct material and labor, engineering, and supervisory costs, and continuing and expediting expenses. Sublimits generally apply to floods and earthquakes, and higher deductibles apply to testing and, in some coastal areas, to wind damage. On foreign projects, the builder's risk insurance is usually combined with a third-party liability policy and is called a "construction all risks" (CAR) policy. The limits of third-party coverage vary with project size and project location, with typical minimum limits of $5 million and maximum limits of $50 million or more. CAR policies also provide coverage during the warranty (maintenance) period, which is usually one or two years. This warranty coverage is provided for losses due to an error during the construction period or for losses caused during the provision of warranty services.

Major exclusions include:

- Faulty workmanship, defects in materials, or error in design; however, there is coverage for any resultant physical damage.

Comment: Design professionals and contractors should attempt to be included as additional insureds, including waivers of subrogation, on a builder's risk policy provided by the owner or general contractor. This gives the design professional and contractor coverage if there is physical damage caused by a design error, faulty workmanship, or a defect in material, and prevents the insurer from subrogating against them for a covered loss. Some policies provide that even if a design professional is included as an additional insured on the policy, the insurer retains the right of subrogation against the design professional. This provision must be deleted from the policy.

- Consequential loss such as loss of use or occupancy.

  Comment: Coverage can be included in builder's risk insurance for delaying project startup due to physical damage to the work. Such coverage is bought in an amount that would cover an owner's lost profit, continuing expenses, and debt service for a specific time period, e.g., six months. Such coverage is rated separately from the property coverage of the builder's risk and is subject to a minimum deductible period of 30 days. To the extent an owner recovers under such delay in startup coverage, this may mitigate any liquidated damage payable by a contractor under the contract with the owner.

11.7 CONSTRUCTION EQUIPMENT INSURANCE

Coverage for loss or damage to construction equipment is usually on an "all risks" policy and is provided by the contractor who owns or leases the equipment. If the design professional has some supervisory role in the construction, he or she should try to be included in any waiver (including waiver of subrogation) provided by the contractor to the owner. This is particularly important where underground equipment, such as tunneling equipment, is involved, due to its high cost.

11.8 WORKERS' COMPENSATION AND EMPLOYER'S LIABILITY

Workers' compensation insurance provides benefits to employees as provided by the applicable state workers' compensation laws. The applicable state law is usually the law at the location of the project, although it may be the law at the place of hire of the employee. Coverage under workers' compensation insurance applies to bodily injury by accident and bodily injury by disease (including resulting death). Claims involving mental or emotional conditions are usually compensable if related to some physical injury; where the mental injury is unrelated to physical injury, the courts are divided on whether such

Workers' Compensation Class	Workers' Compensation Description	Standard Rate per $100.00 of Payroll	EMF	Net Rate per $100.00 of Payroll
6251	Tunneling	$25.00	.80	$20.00

Figure 11.6 Example of premium calculation.

claims are compensable. In addition to benefits paid to employees, workers' compensation also provides a duty to defend the insured in any proceeding for benefits payable under the policy. The cost of workers' compensation insurance varies from state to state, by the type of work being done, and in the amount of expended losses for that type of work. In addition, a contractor's individual experience affects the ultimate cost through the application of an experience modification factor (EMF) against the standard premium for the type of work performed. If a contractor has good experience, the EMF will be less than 1.0. An example of how premium is calculated is shown in Figure 11.6. Having a low EMF helps contractors in competitive bidding situations and provides owners with an indication of an insured's good safety record.

The "employer's liability coverage" part of the workers' compensation policy protects the insured against liability imposed by law for injury to employees in the course of employment that is not compensable under the workers' compensation section. The coverage corresponds to liability coverage and specific limits of coverage are purchased, such as $1 million per accident and $1 million aggregate for disease. The most common forms of claims against an insured that are defended by employer's liability are the third-party over-actions against the insured. This happens when an injured employee of the insured sues a third party for his injuries and that third party names the insured (employee) in the lawsuit, claiming the insured was negligent in causing the injury.

If an insured is subject to federal workers' compensation benefit laws, this coverage must be added to the workers' compensation policy. Examples of this would be the Longshore and Harbor Workers' Compensation Act, the Defense Base Act, and the Jones Act.

11.9 CONTRACTORS' OPERATIONS AND PROFESSIONAL SERVICES ENVIRONMENTAL INSURANCE

Two forms of coverage have developed in response to remediation and evaluation services in connection with cleaning up hazardous waste sites.

Contractors' pollution liability coverage provides insurance for contractors performing remediation work and pays losses due to bodily injury, property damage, or cleanup costs. The insured becomes legally obligated to pay, provided the pollution condition commences after the policy retroactive date and the loss is unexpected and unintended from the standpoint of the insured.

Pollution errors and omissions liability coverage provides insurance for professionals performing professional services in connection with remediation and pays losses due to personal injury, property damage, or cleanup costs that the insured becomes legally obligated to pay, which are caused by any negligent act, error, or omission in the performance of insured's professional services.

Both forms of coverage are usually on a "claims-made" basis with an extended reporting period, although occurrence-type coverage is becoming available.

Major exclusions include:

- Liability under workers' compensation or personal injury to employee of the insured
- Punitive, exemplary, or multiple damages; civil or criminal fines
- Contractual liability, unless liability would have attached in the absence of such contract or agreement
- Nuclear hazards

11.10 DESIGN-BUILD RISK AND INSURANCE ISSUES[2]

There are several variations in the structure and participants of design-build projects (see Figure 11.7). Responsibility for design-build projects may be assumed by:

- Joint venture between contractor and designer
- One entity providing both design and construction
- Contractor hiring design firm as subcontractor
- Design firm hiring contractor as subcontractor
- One entity subcontracting both design and construction

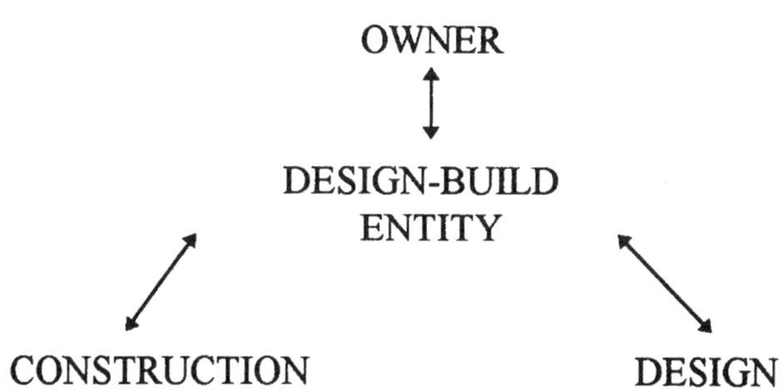

Figure 11.7 Design-build interrelationships.

Coverage	Contractor	Design Firm
General Liabiity	Evidence of coverage to owner and add owner as additional insured.	Evidence of coverage to contractor and add contractor as additional insured.
Professional Liability	Design firm can give only limited professional contractual coverage to contractor. In addition, many design firms carry low limits that are spread over all of their projects and coverage varies significantly from firm to firm. To be properly protected, contractors may need their own professional liability policy. Coverage is currently available (see note below).	Evidence coverage to owner. Limited contractual coverage for contractor. Will not add anyone as additional insured.
Surety	If project is bonded, contractor can issue bonds as required. Bond rates are higher, however, for design-build work. Surety should agree to use equivalent of class "B" rates.	Not applicable.
Builder's Risk	Coverage might be placed by owner or contractor. May or may not include the design firm as insured and include "soft cost" coverage.	Insurer may sue if it pays a loss caused by design firm. Design Professional should try to be included as an insured with a waiver of subrogation.

Contractor's Professional Liability—Few contractors buy professional liability insurance, but many face professional exposures by providing services such as construction management and performing limited design work. As a design-build firm, a contractor increases its professional exposure by becoming the contracting entity between the owner and design firm and "overseeing' design services. Coverage is available from several underwriters including CNA, St. Paul, Zurich and Lloyd's of London, to insure both the contractor's direct liability and contingent liability in the event the design firm's coverage is insufficient.

Figure 11.8 Contractor as design-builder.

Figure 11.8 illustrates insurance coverage available to the contractor as design-builder.

Figure 11.9 discusses coverage for the design firm as design-builder.

Figure 11.10 examines the contractor/design firm joint venture as design-builder.

11.11 SPECIMEN INSURANCE ARTICLE—PRIME CONTRACT

The following is a sample of an insurance article from a contract involving both design and construction. It is comprehensive in its requirements and represents what the designer/contractor may be required to provide for various projects.

Coverage	Contractor	Design Firm
General Liabiity	Evidence of coverage to design firm and add design firm as additional insured.	Evidence of coverage to owner and add owner as additional insured.
Professional Liability	Design-build scenario does not increase contractor's liability when acting as a subcontractor.	Evidence coverage to owner. Will not add anyone as additional insured. Contractual liability is limited to professional negligence arising from design firm's acts and the acts of its subconsultants.
Surety	If job is bonded, contractor can issue bonds directly to the owner or with the design firm as dual obligee. Bond rates are higher, however, for design-build work. Surety would agree to use equivalent of class "B" rates.	If job is bonded, design firm can work with the contractor to issue bonds as dual obligee. Bond rates are higher, however, for design-build work. Surety should agree to use equivalent of class "B" rates.
Builder's Risk	Owner, contractor or design firm may arrange coverage. Additional coverage enhancements to protect design firm should not affect contractor.	Owner, contractor or design firm may arrange coverage. Design firm should request that it be named as an insured, subrogation is waived, "soft" costs are covered and consequential damages arising from a "professional" loss are covered.

Design Firm's Additional "Construction" Liability—Normally, design firms have no "construction" risks and attempt to limit contractual liability to a negligence standard. As a design-builder, all of this changes. Some additional risks are insurable, such as contractual liability, injuries to subcontractor employees and completed operations (general liability). Others, however, are not. For example, risk of delay, performance risks, and warranty risks.

Figure 11.9 Design firm as design-builder.

In addition, specific language is added with regard to the owner's insurance requirements, which can be used to protect the designer/contractor from the risks of property damage to existing or completed facilities. Italicized portions represent key features of the different coverages.

11.11.1 Insurance and Bonds

A. *General.* Contractor shall, at all times during the period in which this Agreement is in force, provide and maintain the types and limits of insurance and bonds as set forth in this Article. Contractor shall procure at its own expense and maintain in full force and effect during the life of this Agreement with responsible insurance companies authorized to do business in the State of ________________, the types and limits of insurance as set forth in this Article. Such companies shall have a Best's Insurance Reports rating of B++ and class VIII or better

Coverage	Contractor	Design Firm
General Liabiity	Normally, contractor's coverage will be evidenced to owner and add owner as additional insured. For joint loss, will only cover contractor's interest in design-build entity. Design-build entity could opt to obtain its own coverage for the project.	Coverage will insure design firm's interest in design-build entity.
Professional Liability	Design firm can give only limited contractual coverage to contractor. Contractor may need its own professional coverage to be protected. As mentioned above, coverage is currently available. Also, design build entity could obtain its own coverage for the project.	Professional liability insurance can protect design firm and its interest in the joint venture and serve as evidence of coverage to the owner. Some underwriters may add design-build entity as an additional insured. Design-build entity could obtain its own coverage.
Surety	If job is bonded, contractor may still issue bonds directly to the owner or the design-build entity may become bonded. Surety underwriter will normally require joint and several indemnification be signed by contractor and design firm. Can be a concern because liability could exceed percentage ownership in design-build entity. Also, bond rates are higher for design-build work. Surety should agree to use equivalent of class "B" rates.	If job is bonded, contractor may still issue bonds directly to the owner or the design-build entity may become bonded. Surety underwriter will normally require joint and several indemnification be signed by contractor and design firm. Can be a concern because liability could exceed percentage ownership in design-build entity. Also, bond rates are higher for design-build work. Surety should agree to use equivalent of class "B" rates.
Builder's Risk	Owner, contractor or design firm may arrange coverage. Additional coverage enhancements to protect design firm should not affect contractor.	Owner, contractor or design firm may arrange coverage. Design firm should request that it be named as an insured, subrogation is waived, "soft" costs are covered and consequential damages arising from a "professional" loss.

Figure 11.10 Contractor/Design firm joint venture as design-builder.

or shall be of recognized responsibility satisfactory to Owner, and Financing Parties, if any. Capitalized terms used in this article and not otherwise defined in this Agreement shall have the meanings generally ascribed to them in the commercial insurance industry in the United States. Contractor, at its own cost, may purchase any additional insurance it believes necessary to protect its interests.

B. *Contractor's Insurance*

1. **Performance Bond.** Contractor shall furnish a Performance Bond (i) in the amount of the Firm Fixed Price, (ii) in the favor of the Owner, (iii) approved by the Owner, (iv) executed by the Contractor and a corporate surety authorized and admitted to do business in the State of ______________ and licensed by the State of ______________ to execute bonds as surety, and approved by Owner, and (v) conditioned on the faithful performance of the Work in accordance with this Agreement. This bond shall remain in effect until the expiration of the Primary Warranty Period or the Extended Warranty Period, whichever is later if the surety on any bond furnished by Contractor is declared a bankrupt or becomes insolvent or its right to do business is terminated in any state where any part of the Project is located or it ceases to meet the requirements of this section. Contractor shall within ten (10) days thereafter substitute another bond and surety, both of which must be acceptable to Owner.

2. **Workers' Compensation Insurance.** Contractor shall maintain Workers' Compensation insurance to cover statutory limits of the Workers' Compensation laws of the State of ______________ (and any other state in which the Work is to be performed) including USL&H coverage, where applicable, and Employer's Liability (including Occupational Disease) coverage with limits of not less than $1,000,000, which shall cover all of Contractor's employees engaged in the Work.

3. **Commercial General Liability Insurance.** Contractor shall maintain Commercial General Liability insurance written on an occurrence basis and with a combined single limit of not less that $2,000,000 *per occurrence and aggregate*. Such insurance shall include coverage for Products-Completed Operations, Broad Form Contractual Liability, Broad Form Property Damage and Personal Injury Liability, Premises/Operations, Independent Contractor Liability, and Explosion, Collapse and Underground Hazards Coverage and Hostile Fire Liability. Such Commercial General Liability Insurance shall include an endorsement to provide for a standard severability of interests and cross liability clause (including Completed Operations). Coverages for completed operations shall remain in effect for a period of a least two (2) years following Final Acceptance.

4. **Automobile Liability Insurance.** Automobile liability insurance (including coverage for owned, non-owned, and hired automobiles) used in connection with the Work shall be obtained by Contractor in an amount not less than $1,000,000 per accident for combined bodily injury, property damage, or death.

5. **Aircraft Insurance.** If performance of the Work requires use of an aircraft that is owned, leased, or chartered, Contractor and Subcontractors shall carry aircraft insurance or require the aircraft owner to carry insurance in an amount not less that $5,000,000 Combined Single Limit per occurrence including passenger and crew.

6. **Umbrella or Excess Liability Insurance.** Contractor shall maintain Umbrella/Excess Insurance on an occurrence basis covering claims in excess of the underlying insurance described in Sections 2, 3, 4, and 5 with a $25,000,000 minimum limit *per occurrence and aggregate*.

7. **Construction Equipment Insurance.** Contractor shall maintain insurance against loss or damage to any construction tools or equipment used in the performance of the work whether owned, borrowed or hired by Contractor.

8. **Professional Liability Insurance.** Contractor shall maintain Professional Liability insurance with a limit of $5,000,000 *each claim and aggregate*, subject to a *deductible* not to exceed $100,000 *per claim*. Such coverage shall be maintained until two (2) years after Final Acceptance of the Facility.

9. **Builder's All-Risk Insurance.** On or prior to thirty (30) days after the Effective Date of this Agreement, Contractor shall obtain and thereafter at all times during performance of the Work maintain *Builder's All-Risk Insurance*. Coverage shall remain in effect until Final Acceptance of the Work. Such Builder's All-Risk shall insure as named insured Contractor, its affiliates and Contractor's Subcontractors of any tier, including Engineer, Owner, and affiliates of Owner, and the Financing Parties as their interest may appear, and shall cover all property in the course of construction, including the Work, materials and equipment, miscellaneous equipment and furnishings (other than equipment property covered under Contractor's equipment floater) from physical loss or damage caused by perils covered by a Builder's All-Risk form or equivalent coverage. Such insurance shall include mechanical and electrical breakdown coverage during start-up and testing and other operation of the Facility prior to Final Acceptance of the Facility. *The limit of liability shall be the full replacement cost of the Work* then at risk, including primary cost of equipment plus freight. The required deductible for all such insurance shall not exceed $25,000. Any required payment of the deductible shall be the responsibility of Contractor. *The Builder's All-Risk coverage shall not contain an exclusion for resultant damage caused by faulty workmanship, design or materials. Such insurance shall provide for a waiver of the underwriters' right to subrogation against Owner, Contractor and their affiliates, Subcontractors, including Engineer, and Financing Parties.*

10. **Delay in Opening Insurance.** On or prior to thirty (30) days after the Effective Date of this Agreement, Contractor shall obtain and thereafter at all times during the performance of the Work, maintain Delay in Opening Insurance. Such Delay in Opening Insurance shall insure Owner for the actual loss sustained due to any delay in start-up or delay in operations. The policy limits shall be not less than $20,000,000 with the period of coverage continuing for no less than twelve (12) months after the Guaranteed Completion Dates; *provided that, such delayed opening results from perils to the Facility otherwise covered by the Builder's All-Risk insurance*. The Delay in Opening Insurance shall be of sufficient limits and duration to ensure that Owner will be reimbursed for its actual loss sustained, including additional interest expense and certain fixed operating expenses to be agreed to by Owner and Financing Parties, if any. The coverage shall be subject to a waiting period of no more than thirty (30) days. *Proceeds of the Delay in Opening Insurance actually received by Owner shall be used to defray or offset Owner's losses or costs in lieu of a portion of Contractor's obligation to pay Delay Liquidated Damages, if any*, if Contractor fails to achieve Substantial Completion on or before the Substantial Completion Date.

C. *Endorsements.* All policies of liability insurance to be maintained by Contractor shall be written or endorsed to include the following:

1. With respect to Worker's Compensation/Employer's Liability Insurance, to provide that the insurer (I) *shall waive for the benefit of Owner (1) all rights of subrogation against Owner*, its affiliates, or their directors, officers, employees or agents for payment under such policies.

2. With respect to General Liability, Automobile Liability and Excess/Umbrella Insurance, to provide that the insurance *shall waive (I) any right of recovery which the insurer may have or acquire against Owner*, its affiliates, or their directors, officers, employees or agents for payment under such policies.

3. To provide a severability of interest and Cross Liability Clause.

4. That the insurance shall be primary and not excess to or contributing with any insurance or self-insurance maintained by Owner.

5. With the exception of the insurance required under Section B.2 to *name Owner, its affiliates, and their directors, officers, employees and agents as additional insureds* with respect to any injury or damage arising from any work or services performed on Owner's premises.

D. *Loss Payable.* Losses, if any, covered by All-Risk, or Delay in Opening Insurance shall be payable to Owner, Financing Parties, Contractor and Subcontractors as additional named insurers, as their respective interests may appear. The policies shall be endorsed to state that, notwithstanding anything to the contrary, the interest of the Financing Parties shall not be invalidated by any act or failure to act on the part of Owner. Contractor or any other Person, and the Financing Parties shall be insured regardless of any breach or violation by Owner, Contractor or any other Person, of any warranties, declarations or conditions contained in such insurance policies. The Parties agree that in the event of physical loss of the Project, any insurance proceeds under this section shall be reserved in a segregated account until such time that it has been determined whether or not to repair, replace or rebuild the damaged portion of the Project.

E. *Subcontractor Insurance.* Contractor shall cause all Subcontractors to procure at their own expense and maintain in full force and effect during the time which they are involved in performance of the Work adequate coverage in accordance with Contractor's normal practice but not less than as provided herein. Prior to the commencement of work by any Subcontractor, or as requested from time to time, Contractor shall provide Owner acceptable evidence of such insurance.

F. *Contractor's Waiver.* Contractor releases, assigns and waives any and all rights of recovery against Owner, the Financing Parties, and their respective affiliates, subsidiaries, employees, successors' permitted assigns, insurers and underwriters, because of the existence of deductible clauses or inadequacy of limits of any policies of insurance maintained or required to be maintained pursuant to this Agreement.

G. *Certificates.* On or prior to thirty (30) days after the Effective Date of this Agreement, Contractor shall furnish to the Owner certificates of insurance from each insurance carrier showing that the above required insurance is in full force and effect, the amount of the carrier's liability thereunder, and further providing that the insurance will not be canceled, changed, or not renewed until the expiration of a least thirty (30) days (or ten (10) days in the case of cancellation due to non-payment of premiums) after written Notice of such cancellation, change or nonrenewal has been received by Owner. Certificates of insurance submitted under this section shall be in form and content acceptable to Owner.

H. *Cost of Premium.* It is expressly agreed and understood that the cost of premiums and deductibles for insurance set forth in this Article shall be borne by Contractor and all insurance provided by Contractor shall be endorsed to provide that Owner shall have no liability for the payment of any premium.

I. *Owner's Rights.* If Contractor fails to provide or maintain any insurance required hereunder, Owner shall have the right, but not the obligation, to provide or maintain any such insurance, and to deduct the cost thereof from any amounts due and payable to Contractor (including, without limitation, the Firm Fixed Price), or, in the event there are no such amounts due and payable, Contractor shall reimburse Owner for such costs on demand. Should any of the policies required to be maintained become cancelled, Contractor shall immediately procure replacement coverage. The failure of Contractor to procure such replacement coverage which is within the reasonable control of Contractor (so as to provide continuous coverage) shall constitute a default hereunder.

J. *No Limitation of Liability.* The required coverages referred to and set forth in this Article shall in no way affect, nor are they intended as a limitation of Contractor's liability with respect to its performance of the Work.

K. *Owner's Insurance. Owner agrees to maintain operating property insurance on its existing property* at the Project Site during construction and thereafter and *Owner waives all rights of recovery including insurer's rights of subrogation against Contractor and any of its Subcontractors, including Engineer, for loss or damage covered by such operating property insurance.*

After Acceptance of the Project, Owner agrees to maintain operating property insurance on the completed Project and Owner hereby waives all rights of recovery including insurer's rights of subrogation against a Contractor and any of its Subcontractors, including Engineer, for loss or damage covered by such operating property insurance.

11.12 WRAP-UP INSURANCE PROGRAMS

A wrap-up insurance program provided either by the owner or by the contractor is a centralized and controlled insurance and loss control program applicable to a specific construction project or a series of related projects. The program generally covers all participating contractors and subcontractors of any tier and, for large projects, can generate significant premium savings over individual contractor provided insurance.

Insurance coverages provided under a wrap-up insurance program can include:

- Workers' compensation
- Employers' liability
- Commercial general liability (including completed operations)

- Excess liability
- Builder's risk insurance
- Professional liability
- Pollution liability

The benefits of a wrap-up insurance program are significant, and include:

- Broad insurance protection with uniform limits and coverages for all insureds
- Significant project savings through a loss-sensitive or similar program
- Reduced subrogation and cross-liability actions
- Coordinated and improved safety program to minimize losses and achieve a more effective safety record, resulting in ultimate premium savings
- Centralized claims reporting and administration
- Cash flow advantages resulting from reduced premium payments
- Extended protection for claims occurring after completion of construction

To enhance the potential savings under a wrap-up insurance program, a loss-sensitive program can be used for workers' compensation and general liability insurance. The final premium costs for these coverages are determined by the payment of reduced fixed costs of the insurers, plus actual losses and loss handling charges. Catastrophic losses would be limited in the loss-sensitive program. Excess liability, builder's risk, professional liability, and pollution liability can achieve cost savings by leveraging economies of scale and showing a controlled exposure to underwriters. Over the past few years, the insurance industry has come to view wrap-ups as extremely attractive business, and underwriters compete very aggressively for this business. Under these conditions, it is possible to negotiate a program that both minimizes costs, if loss results are favorable, and eliminates premium penalties, if adverse loss results are realized.

11.13 CONCLUSION—KEY INSURANCE/LIABILITY ISSUES ON PROJECTS

1. Limit contractual indemnity obligations:
   a. To your negligence or those for whom you are legally liable. Professional liability insurance may not cover broader forms of indemnity.
   b. To claims of third parties against owner (indemnitee). Do not let indemnity become additional warranty.
   c. To a specific time period. Do not let indefinite indemnity preclude the use of applicable statute of repose.

2. Property insurance:
   a. Builder's risk should name the engineer and the contractor as additional insureds, with a specific waiver of subrogation. There should be no right to subrogate against the engineer for losses due to design error or against the contractor for faulty workmanship.
   b. The policy should include coverage for resultant damage due to error in design, faulty workmanship, or defect in material.
   c. Upon project completion, get a waiver from the owner (including waiver of subrogation) for any loss or damage covered by permanent property insurance on the completed project. Completed operations coverage in commercial general liability insurance may not provide full coverage for this exposure, and property insurers are eager to subrogate.
3. If you are providing engineering services to the owner, try to be included as an indemnitee on the contractor's indemnity and also as an additional insured on the contractor's liability insurance. This will help protect against claims of the contractor's employees. Also, obtain a waiver of subrogation on workers' compensation policies of contractors.
4. Buy adequate limits and types of insurance for adequate durations:
   a. Professional liability generally covers negligence claims arising out of professional acts and is written on a "claims-made" basis.
   b. Commercial general liability generally covers physical activities, and is written on an "occurrence" basis.

NOTES

1. For a discussion of legal aspects of these risks, see generally Chapter 10.
2. For a discussion of other design-build issues, see Chapter 10, Section 10.2.3.

REFERENCES

D. J. Hatem, "Owner as 'Additional Insured' Under the Design Professional Liability Policy: Can It Be Done and, If So, How?", *The CA/T Professional Liability Reporter*, Vol. 2, No. 1 (Boston: Burns & Levinson LLP, June 1996), pp. 1–8.

O. Shean and D. Patkin, *Construction Insurance: Coverage and Disputes* (Charlottesville, VA: The Mitchie Company, 1994).

APPENDIX TO CHAPTER 10

THE BASELINE APPROACH AND CONSTRUCTION MEANS AND METHODS

Careful consideration should be given to risk management and professional liability implications associated with the use of the geotechnical or contractual baseline approach (the "baseline approach") to address construction means and methods and related contractor performance matters. Generally speaking, the baseline approach is utilized on subsurface projects "to establish a contractual statement of the geotechnical conditions anticipated to be encountered during underground and subsurface construction."[1] Similarly, it has been stated that "[t]he basic purpose of . . . baselines is to define concisely the geotechnical 'indications of the contract' for administration of the [differing site conditions] clause. These baselines establish a contractual basis for the allocation of geotechnical risk; they are not necessarily intended to define the 'single correct' interpretation of geotechnical conditions."[2] In substance, the purpose of the baseline approach is to assist in the allocation of risk for site conditions by facilitating the definition—qualitatively and/or quantitatively—of those site conditions that should be anticipated and included within the contractual obligations and contract price of the construction contractor.

Although the baseline approach is relatively new,[3] it is generally acknowledged that the use of baselines should be restricted to the definition (qualitative or quantitative) of *physical site conditions* anticipated to be encountered, as distinct from the definition and/or prescription of either: (a) the construction means and methods to be utilized by the contractor; and/or (b) the performance expectations, requirements, standards, or other related matters that may directly or indirectly implicate the contractor's construction means and methods, equipment, and/or safety precautions and programs. For example, it has been observed that:

> In general, a baseline report that predicts performance is asking for trouble, as this is only partly determined by geologic conditions, and is also partly determined

Subsurface Conditions: Risk Management for Design and Construction Management Professionals, edited by David J. Hatem
ISBN 0471-15607-8

> by the contractor's means and methods, which the engineer can not predict. Items such as TBM production rate, utilization, penetration rate or cutter wear; overbreak; or a precise description of what will be encountered at a specific location or a level of detail beyond what is needed should not be given. There may be exceptions to this, but the owner needs to understand the risks involved in the detailed monitoring required to verify means and methods.[4]

While the establishment of a baseline may be predicated upon assumptions about the construction means and methods to be selected by the contractor,[5] or may be established in a manner intended to influence the contractor's selection of construction means and methods,[6] it is generally recommended by those with substantial experience in subsurface projects that the baseline not address construction means and methods, contractor equipment, or related contractor performance considerations.[7]

Despite the general admonition that baselines should relate solely to the definition of physical site conditions as distinct from contractor performance expectations, requirements, or standards, inevitably there will be circumstances in which owners request (or, as a matter of professional judgment, engineers conclude) that baselines be developed for certain contractor performance expectations, requirements, or standards. Should the engineer decide to utilize the baseline concept for such a purpose, the issue will become whether that can be accomplished in a manner that limits (in the sense of defining, restricting, qualifying, and managing) the owner's and professional's risk exposure for construction means and methods and safety-related issues.

What are the relevant risk management/professional liability considerations from the standpoint of a consulting engineer providing ("traditional") engineering services on behalf of the project owner in these circumstances? Clearly, the owner typically would need the services of the consulting engineer to advise and assist in the development and implementation (through document provisions) of the baseline approach and would request (if not require) that the consulting engineer provide the necessary services in those respects. Rejecting the owner's request for such services may not be a viable or realistic option from a "client relations" or "business" perspective.

However, to a significant degree, providing the requested services may: (a) constitute (or be deemed to constitute) an involvement of the consulting engineer in construction means and methods and safety precautions and programs; (b) require the consulting engineer to perform services or become involved in areas beyond those specifically required in its agreement with the owner; (c) result in "dislocation" or imbalancing of the contractor's otherwise exclusive responsibility for construction means and methods and safety-related issues due to the consulting engineer's involvement in the latter areas; and (d) expose the consulting engineer to the risk of civil liability and/or OSHA exposure.[8]

Given the owner objectives sought to be achieved by the baseline approach and the risk management and professional liability considerations and concerns from the standpoint of the consulting engineer, how can the baseline approach be developed and implemented in these circumstances in a manner that may reasonably accomplish the owner's objectives, provide reasonable risk management controls for the consulting engineer's professional liability concerns, and potentially mitigate the consulting engineer's risk of OSHA exposure?[9]

Documentation Between Owner and Consulting Engineer

The owner's request to the consulting engineer for specific involvement in the development and implementation of the baseline approach should be documented, especially if compliance with this request is not within the scope of the basic services required of the consulting engineer. While such a request may have financial implications in terms of the consulting engineer's entitlement to additional compensation, there are other equally, if not more, important interests of a risk management/professional liability nature presented by such a request. Specifically, as noted above, such a request may result in (or may later be alleged to have resulted in) the consulting engineer's involvement in construction means and methods and/or safety issues—thereby subjecting the consulting engineer to increased risk of both civil liability and OSHA exposure. The limited nature of any such involvement should be documented so as to establish and memorialize the defined scope and extent of that involvement. In addition, the *purpose* of that involvement should be confirmed in writing. Specifically, in relation to the utilization of the baseline approach in the differing site condition context, the purpose would be to advise and assist the owner in developing, defining, and contractually implementing a basis for acknowledging a differing site condition, and to establish the basis for compensating the contractor for certain *minimum*, or *baseline*, performance standards required to address that condition. In other words, the purpose is directed to the definition and implementation of a contractual baseline due to a differing site condition, and the establishment of a basis for equitable adjustment under the contract due to that condition—not safety-related issues *per se*.

This documentation between the owner and consulting engineer may be effectuated by an amendment to the agreement between them or, at a minimum, through correspondence between them. The documentation should state clearly that the consulting engineer's specific, defined, and limited involvement in this particular issue does not create or imply any broader obligations of the consulting engineer with respect to any other construction means and methods and/or safety-related matters; nor does the consulting engineer's involvement in the particular construction means and methods issue negate, diminish, or otherwise affect or alter the contractor's otherwise exclusive and complete responsibility under the contract documents for construction means and methods and safety-related issues.

Contract Document Provisions Relating to Baseline

Statement of Purpose of the Contract Document Baseline Provisions: The pertinent baseline provisions of the contract documents should contain an explicit statement as to the purpose of those provisions—that is, to acknowledge the differing site condition and to define certain *minimum* or *baseline* performance standards that the contractor is required to implement for the purpose of addressing that condition and establishing the basis of an equitable adjustment. It may also be helpful to state that no differing site condition claim predicated upon the condition described in the baseline will be granted unless the contractor has fully complied with and implemented the minimum contractor performance requirements or standards set forth in the contract documents.

It may be preferable that the relevant contract document provisions: (a) have a title or description that reflects the pertinent site conditions, rather than one that reflects safety-related requirements; and (b) be stand-alone in nature—that is, not characterized as a revision to safety or related specifications or similar portions of the contract documents.

Text of Contract Document Baseline Provision: The text of the baseline provision should make clear that *minimum* or *baseline* performance requirements or standards are stated in order to establish a clearly defined basis for an equitable adjustment or compensation for certain (i.e., not the universe of) measures or standards required to be undertaken by the contractor. In this regard: (a) the more detailed or prescriptive the description of the minimum requirements or standards, the greater the risk that the consulting engineer may later be determined to have assumed responsibility for the adequacy (in a total or aggregate sense) of the measures required to address the differing site condition; and (b) the less discretion given to the contractor to develop supplemental and/or other necessary measures, the greater risk of professional liability and potential OSHA exposure for the consulting engineer. As such, some may consider it preferable to express those *minimum* requirements or standards in a "performance" manner (leaving the development of complete details to the contractor's discretion within broad guidelines), even though that approach may result in a less precisely defined baseline for compensation purposes.

Contract Document Provisions Maintaining and Reinforcing Contractual Risk Allocation for Construction Means and Methods and Safety Precautions and Programs

The contract document provisions pertaining to the baseline should clearly state and emphasize that the contractor performance requirements and standards set forth therein to address the site condition that is the subject of the baseline are *minimum* (or baseline) in nature, and that the contractor and/or its subcontractors remain solely and completely responsible for the design and implementation of all appropriate construction means and methods and safety precautions and programs, including without limitation compliance with all applicable federal, state and local, and industry rules and regulations regarding the safe conduct of the work and protection of persons and property. In this regard, it may be useful to explicitly reference and thereby reinforce the pertinent contract document provisions that allocate those responsibilities to the contractor and its subcontractors.

In addition, the contract document baseline provisions should also state that neither the owner's issuance of the baseline, nor any of the actions, performance of services and/or recommendations of the consulting engineer in the development, or negotiation and/or preparation of the terms of the baseline shall relieve the contractor of its exclusive contractual responsibility regarding construction means and methods and safety precautions and programs.

NOTES

1. R. J. Essex, ed., "Geotechnical Baseline Reports for Underground Construction—Guidelines and Practices," prepared by the Technical Committee on Geotechnical Reports of the Underground Technology Research Council (New York: ASCE, 1997), p. 1.
2. S. H. Bartholomew, et al., "Position Paper by the DRB Panel," in *Opinion Papers on "Geotechnical Reports in Underground Construction,"* an unpublished collection of papers presented at the April 21, 1996 Underground Technology Research Counsel Forum in Washington, D.C. For a discussion of the professional liability implications associated with the distinction between geotechnical reality and the baseline definition of geotechnical conditions, see D. J. Hatem, "Geotechnical Baselines and Geotechnical Reality," *The CA/T Professional Liability Reporter*, Vol. 2, No. 2 (Boston: Burns & Levinson LLP, December 1996), pp. 1–8.
3. See generally R. J. Essex (Note 1, *supra*).
4. Gregg Korbin, et al., "Consultants' Panel Report" in *Opinion Papers on "Geotechnical Reports in Underground Construction,"* an unpublished collection of papers presented at the April 21, 1996 Underground Technology Research Counsel Forum in Washington D.C.
5. For example, R. J. Essex has noted that "[b]aseline statements regarding anticipated ground behavior should be presented in context with the means and methods selected by the contractor. The baseline statements should be clear that the ground can be expected to behave differently with alternate tools, methods, sequences and equipment." R. J. Essex (Note 1, *supra*), p. 25.
6. See S. H. Bartholomew, (Note 2, *supra*), p. 4.
7. Specifically, it has been recommended that the baseline definition exclude "[d]iscussion of the details of construction methods and equipment, rates of production, etc., that lie within the contractor's field of expertise, except those which are contractual requirements or are absolutely necessary for establishing the baselines and ensuring that the baselines are understandable." S. H. Bartholomew (Note 2, *supra*), p. 5. For additional discussion of the risk allocation implications associated with prescribing construction equipment on subsurface projects, see J. Reilly, "Owner Responsibilities in the Selection of Tunnel Boring Machines with Reference to Contractual Requirements and Construction Conditions," unpublished paper presented at the World Tunnel Conference/International Tunneling Association 23rd General Assembly, in Vienna, April, 1997.

8. For a general discussion of the interrelationship between civil liability and OSHA exposure for design professionals, see D. J. Hatem, "Administrative, Regulatory, and Registration Proceedings Involving Design Professionals," in *1991 Wiley Construction Law Update*, S. M. Goldblatt, ed., (New York: John Wiley & Sons, Inc., 1991).
9. The question posed in the text is explored in detail in D. J. Hatem, "Geotechnical Baselines: Professional Liability Implications," *The CA/T Professional Liability Reporter*, Vol. 3, No. 1 (Boston: Burns & Levinson LLP, October 1997). In that article, this author focuses the discussion on the specific context of the methane gas differing site condition involved in *Secretary of Labor v. CH2M Hill Central, Inc.*, 1997 WL 197011 (O.S.H.R.C. April 21, 1997), *appeal docketed*, No. 97-2504 and 97-2503 (7th Cir. 1997), (the "*CH2M Hill* case"), and the baseline approach utilized by the owner and consulting engineer and incorporated in a construction contract modification intended to address that condition. Although the baseline approach was not specifically designated as such in the project at issue in *CH2M Hill*, a close review of the relevant facts, and the purpose and text of the contract modification demonstrate the essential characteristics of the baseline approach.

INDEX